Undergraduate Topics in Computer Science

Series editor

Ian Mackie

Advisory Board

Samson Abramsky, University of Oxford, Oxford, UK
Chris Hankin, Imperial College London, London, UK
Mike Hinchey, University of Limerick, Limerick, Ireland
Dexter C. Kozen, Cornell University, Ithaca, USA
Andrew Pitts, University of Cambridge, Cambridge, UK
Hanne Riis Nielson, Technical University of Denmark, Kongens Lyngby, Denmark
Steven S. Skiena, Stony Brook University, Stony Brook, USA
Iain Stewart, University of Durham, Durham, UK

Undergraduate Topics in Computer Science (UTiCS) delivers high-quality instructional content for undergraduates studying in all areas of computing and information science. From core foundational and theoretical material to final-year topics and applications, UTiCS books take a fresh, concise, and modern approach and are ideal for self-study or for a one- or two-semester course. The texts are all authored by established experts in their fields, reviewed by an international advisory board, and contain numerous examples and problems. Many include fully worked solutions.

More information about this series at http://www.springer.com/series/7592

Michael Oberguggenberger
Alexander Ostermann

Analysis for Computer Scientists

Foundations, Methods, and Algorithms

Second Edition

Translated in collaboration with Elisabeth Bradley

 Springer

Michael Oberguggenberger🆔
University of Innsbruck
Innsbruck
Austria

Alexander Ostermann🆔
University of Innsbruck
Innsbruck
Austria

ISSN 1863-7310   ISSN 2197-1781   (electronic)
Undergraduate Topics in Computer Science
ISBN 978-3-319-91154-0   ISBN 978-3-319-91155-7   (eBook)
https://doi.org/10.1007/978-3-319-91155-7

Library of Congress Control Number: 2018941530

This Springer imprint is published by the registered company Springer Nature Switzerland AG
The registered company address is: Gewerbestrasse 11, 6330 Cham, Switzerland

Preface to the Second Edition

We are happy that Springer Verlag asked us to prepare the second edition of our textbook *Analysis for Computer Scientists*. We are still convinced that the algorithmic approach developed in the first edition is an appropriate concept for presenting the subject of analysis. Accordingly, there was no need to make larger changes.

However, we took the opportunity to add and update some material. In particular, we added hyperbolic functions and gave some more details on curves and surfaces in space. Two new sections have been added: One on second-order differential equations and one on the pendulum equation. Moreover, the exercise sections have been extended considerably. Statistical data have been updated where appropriate.

Due to the essential importance of the MATLAB programs for our concept, we have decided to provide these programs additionally in Python for the users' convenience.

We thank the editors of Springer, especially Simon Rees and Wayne Wheeler, for their support during the preparation of the second edition.

Innsbruck, Austria
March 2018

Michael Oberguggenberger
Alexander Ostermann

Preface to the First Edition

Mathematics and mathematical modelling are of central importance in computer science. For this reason the teaching concepts of mathematics in computer science have to be constantly reconsidered, and the choice of material and the motivation have to be adapted. This applies in particular to mathematical analysis, whose significance has to be conveyed in an environment where thinking in discrete structures is predominant. On the one hand, an analysis course in computer science has to cover the essential basic knowledge. On the other hand, it has to convey the importance of mathematical analysis in applications, especially those which will be encountered by computer scientists in their professional life.

We see a need to renew the didactic principles of mathematics teaching in computer science, and to restructure the teaching according to contemporary requirements. We try to give an answer with this textbook which we have developed based on the following concepts:

1. algorithmic approach;
2. concise presentation;
3. integrating mathematical software as an important component;
4. emphasis on modelling and applications of analysis.

The book is positioned in the triangle between mathematics, computer science and applications. In this field, algorithmic thinking is of high importance. The algorithmic approach chosen by us encompasses:

a. development of concepts of analysis from an algorithmic point of view;
b. illustrations and explanations using MATLAB and maple programs as well as Java applets;
c. computer experiments and programming exercises as motivation for actively acquiring the subject matter;
d. mathematical theory combined with basic concepts and methods of *numerical analysis*.

Concise presentation means for us that we have deliberately reduced the subject matter to the essential ideas. For example, we do not discuss the general convergence theory of power series; however, we do outline Taylor expansion with an estimate of the remainder term. (Taylor expansion is included in the book as it is an indispensable tool for modelling and numerical analysis.) For the sake of readability, proofs are only detailed in the main text if they introduce essential ideas and contribute to the understanding of the concepts. To continue with the example above, the integral representation of the remainder term of the Taylor expansion is derived by integration by parts. In contrast, Lagrange's form of the remainder term, which requires the mean value theorem of integration, is only mentioned. Nevertheless we have put effort into ensuring a self-contained presentation. We assign a high value to *geometric intuition*, which is reflected in a large number of illustrations.

Due to the terse presentation it was possible to cover the whole spectrum from foundations to interesting *applications of analysis* (again selected from the viewpoint of computer science), such as fractals, L-systems, curves and surfaces, linear regression, differential equations and dynamical systems. These topics give sufficient opportunity to enter various *aspects of mathematical modelling*.

The present book is a translation of the original German version that appeared in 2005 (with the second edition in 2009). We have kept the structure of the German text, but took the opportunity to improve the presentation at various places.

The contents of the book are as follows. Chapters 1–8, 10–12 and 14–17 are devoted to the basic concepts of analysis, and Chapters 9, 13 and 18–21 are dedicated to important applications and more advanced topics. The Appendices A and B collect some tools from vector and matrix algebra, and Appendix C supplies further details which were deliberately omitted in the main text. The employed software, which is an integral part of our concept, is summarised in Appendix D. Each chapter is preceded by a brief introduction for orientation. The text is enriched by computer experiments which should encourage the reader to actively acquire the subject matter. Finally, every chapter has exercises, half of which are to be solved with the help of computer programs. The book can be used from the first semester on as the main textbook for a course, as a complementary text or for self-study.

We thank Elisabeth Bradley for her help in the translation of the text. Further, we thank the editors of Springer, especially Simon Rees and Wayne Wheeler, for their support and advice during the preparation of the English text.

Innsbruck, Austria   Michael Oberguggenberger
January 2011   Alexander Ostermann

Contents

Numbers

1

The commonly known rational numbers (fractions) are not sufficient for a rigorous foundation of mathematical analysis. The historical development shows that for issues concerning analysis, the rational numbers have to be extended to the real numbers. For clarity we introduce the real numbers as decimal numbers with an infinite number of decimal places. We illustrate exemplarily how the rules of calculation and the order relation extend from the rational to the real numbers in a natural way.

A further section is dedicated to floating point numbers, which are implemented in most programming languages as approximations to the real numbers. In particular, we will discuss optimal rounding and in connection with this the relative machine accuracy.

1.1 The Real Numbers

In this book we assume the following number systems as known:

$$\mathbb{N} = \{1, 2, 3, 4, \ldots\} \quad \text{the set of natural numbers;}$$
$$\mathbb{N}_0 = \mathbb{N} \cup \{0\} \quad \text{the set of natural numbers including zero;}$$
$$\mathbb{Z} = \{\ldots, -3, -2, -1, 0, 1, 2, 3, \ldots\} \quad \text{the set of integers;}$$
$$\mathbb{Q} = \left\{ \tfrac{k}{n} \; ; \; k \in \mathbb{Z} \text{ and } n \in \mathbb{N} \right\} \quad \text{the set of rational numbers.}$$

Two rational numbers $\frac{k}{n}$ and $\frac{\ell}{m}$ are equal if and only if $km = \ell n$. Further an integer $k \in \mathbb{Z}$ can be identified with the fraction $\frac{k}{1} \in \mathbb{Q}$. Consequently, the inclusions $\mathbb{N} \subset \mathbb{Z} \subset \mathbb{Q}$ are true.

© Springer Nature Switzerland AG 2018
M. Oberguggenberger and A. Ostermann, *Analysis for Computer Scientists*,
Undergraduate Topics in Computer Science,
https://doi.org/10.1007/978-3-319-91155-7_1

Let M and N be arbitrary sets. A *mapping* from M to N is a rule which assigns to each element in M exactly one element in N.[1] A mapping is called *bijective*, if for *each* element $n \in N$ there exists *exactly one* element in M which is assigned to n.

Definition 1.1 Two sets M and N have *the same cardinality* if there exists a bijective mapping between these sets. A set M is called *countably infinite* if it has the same cardinality as $\mathbb{N}$.

The sets $\mathbb{N}$, $\mathbb{Z}$ and $\mathbb{Q}$ have the same cardinality and in this sense are *equally large*. All three sets have an infinite number of elements which can be enumerated. Each enumeration represents a bijective mapping to $\mathbb{N}$. The countability of $\mathbb{Z}$ can be seen from the representation $\mathbb{Z} = \{0, 1, -1, 2, -2, 3, -3, \ldots\}$. To prove the countability of $\mathbb{Q}$, Cantor's[2] diagonal method is being used:

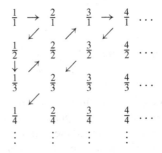

The enumeration is carried out in direction of the arrows, where each rational number is only counted at its *first* appearance. In this way the countability of all positive rational number (and therefore all rational numbers) is proven.

To visualise the rational numbers we use a line, which can be pictured as an infinitely long ruler, on which an arbitrary point is labelled as *zero*. The integers are marked equidistantly starting from zero. Likewise each rational number is allocated a specific place on the real line according to its size, see Fig. 1.1.

However, the real line also contains points which do not correspond to rational numbers. (We say that $\mathbb{Q}$ is *not complete*.) For instance, the length of the diagonal d in the unit square (see Fig. 1.2) can be measured with a ruler. Yet, the Pythagoreans already knew that $d^2 = 2$, but that $d = \sqrt{2}$ is not a rational number.

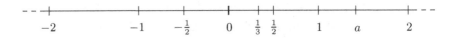

Fig. 1.1 The real line

---

[1] We will rarely use the term mapping in such generality. The special case of *real-valued functions*, which is important for us, will be discussed thoroughly in Chap. 2.
[2] G. Cantor, 1845–1918.

Fig. 1.2 Diagonal in the unit square

Proposition 1.2 $\sqrt{2} \notin \mathbb{Q}$.

Proof This statement is proven indirectly. Assume that $\sqrt{2}$ were rational. Then $\sqrt{2}$ can be represented as a reduced fraction $\sqrt{2} = \frac{k}{n} \in \mathbb{Q}$. Squaring this equation gives $k^2 = 2n^2$ and thus k^2 would be an even number. This is only possible if k itself is an even number, so $k = 2l$. If we substitute this into the above we obtain $4l^2 = 2n^2$ which simplifies to $2l^2 = n^2$. Consequently n would also be even which is in contradiction to the initial assumption that the fraction $\frac{k}{n}$ was reduced. □

As it is generally known, $\sqrt{2}$ is the unique positive root of the polynomial $x^2 - 2$. The naive supposition that all non-rational numbers are roots of polynomials with integer coefficients turns out to be incorrect. There are other non-rational numbers (so-called transcendental numbers) which *cannot* be represented in this way. For example, the ratio of a circle's circumference to its diameter

$$\pi = 3.141592653589793... \notin \mathbb{Q}$$

is transcendental, but can be represented on the real line as half the circumference of the circle with radius 1 (e.g. through unwinding).

In the following we will take up a pragmatic point of view and construct the missing numbers as decimals.

Definition 1.3 A finite decimal number x with l decimal places has the form

$$x = \pm d_0.d_1 d_2 d_3 \ldots d_l$$

with $d_0 \in \mathbb{N}_0$ and the single digits $d_i \in \{0, 1, \ldots, 9\}$, $1 \leq i \leq l$, with $d_l \neq 0$.

Proposition 1.4 (Representing rational numbers as decimals) *Each rational number can be written as a finite or periodic decimal.*

Proof Let $q \in \mathbb{Q}$ and consequently $q = \frac{k}{n}$ with $k \in \mathbb{Z}$ and $n \in \mathbb{N}$. One obtains the representation of q as a decimal by successive division with remainder. Since the remainder $r \in \mathbb{N}$ always fulfils the condition $0 \leq r < n$, the remainder will be zero or periodic after a maximum of n iterations. □

Example 1.5 Let us take $q = -\frac{5}{7} \in \mathbb{Q}$ as an example. Successive division with remainder shows that $q = -0.71428571428571...$ with remainders $5, 1, 3, 2, 6, 4, 5,$ $1, 3, 2, 6, 4, 5, 1, 3, \ldots$ The period of this decimal is six.

Each nonzero decimal with a finite number of decimal places can be written as a periodic decimal (with an infinite number of decimal places). To this end one diminishes the last nonzero digit by one and then fills the remaining infinitely many decimal places with the digit 9. For example, the fraction $-\frac{17}{50} = -0.34 = -0.3399999...$ becomes periodic after the third decimal place. In this way $\mathbb{Q}$ can be considered as the set of all decimals which turn periodic from a certain number of decimal places onwards.

Definition 1.6 The set of *real numbers* $\mathbb{R}$ consists of all decimals of the form

$$\pm d_0.d_1 d_2 d_3...$$

with $d_0 \in \mathbb{N}_0$ and digits $d_i \in \{0, ..., 9\}$, i.e. decimals with an infinite number of decimal places. The set $\mathbb{R} \setminus \mathbb{Q}$ is called the set of *irrational* numbers.

Obviously $\mathbb{Q} \subset \mathbb{R}$. According to what was mentioned so far the numbers

$$0.1010010001000010... \quad \text{and} \quad \sqrt{2}$$

are irrational. There are much more irrational than rational numbers, as is shown by the following proposition.

Proposition 1.7 *The set* $\mathbb{R}$ *is not countable and has therefore higher cardinality than* $\mathbb{Q}$.

Proof This statement is proven indirectly. Assume the real numbers between 0 and 1 to be countable and tabulate them:

$$1 \quad 0.d_{11} d_{12} d_{13} d_{14}...$$
$$2 \quad 0.d_{21} d_{22} d_{23} d_{24}...$$
$$3 \quad 0.d_{31} d_{32} d_{33} d_{34}...$$
$$4 \quad 0.d_{41} d_{42} d_{43} d_{44}...$$
$$. \quad ...$$
$$. \quad ...$$

With the help of this list, we define

$$d_i = \begin{cases} 1 & \text{if } d_{ii} = 2, \\ 2 & \text{else.} \end{cases}$$

Then $x = 0.d_1 d_2 d_3 d_4...$ is not included in the above list which is a contradiction to the initial assumption of countability. $\square$

Fig. 1.3 Babylonian cuneiform inscription YBC 7289 (Yale Babylonian Collection, with authorisation) from 1900 before our time with a translation of the inscription according to [1]. It represents a square with side length 30 and diagonals 42; 25, 35. The ratio is $\sqrt{2} \approx 1; 24, 51, 10$

However, although $\mathbb{R}$ contains considerably more numbers than $\mathbb{Q}$, every real number can be approximated by rational numbers to any degree of accuracy, e.g. π to nine digits

$$\pi \approx \frac{314159265}{100000000} \in \mathbb{Q}.$$

Good approximations to the real numbers are sufficient for practical applications. For $\sqrt{2}$, already the Babylonians were aware of such approximations:

$$\sqrt{2} \approx 1; 24, 51, 10 = 1 + \frac{24}{60} + \frac{51}{60^2} + \frac{10}{60^3} = 1.41421296..., $$

see Fig. 1.3. The somewhat unfamiliar notation is due to the fact that the Babylonians worked in the sexagesimal system with base 60.

1.2 Order Relation and Arithmetic on $\mathbb{R}$

In the following we write real numbers (uniquely) as decimals with an infinite number of decimal places, for example, we write 0.2999... instead of 0.3.

Definition 1.8 (Order relation) Let $a = a_0.a_1 a_2...$ and $b = b_0.b_1 b_2...$ be non-negative real numbers in decimal form, i.e. $a_0, b_0 \in \mathbb{N}_0$.

(a) One says that a is *less than or equal to* b (and writes $a \leq b$), if $a = b$ or if there is an index $j \in \mathbb{N}_0$ such that $a_j < b_j$ and $a_i = b_i$ for $i = 0, \ldots, j - 1$.

(b) Furthermore one stipulates that always $-a \leq b$ and sets $-a \leq -b$ whenever $b \leq a$.

This definition extends the known orders of $\mathbb{N}$ and $\mathbb{Q}$ to $\mathbb{R}$. The interpretation of the order relation $\leq$ on the real line is as follows: $a \leq b$ holds true, if a is to the left of b on the real line, or $a = b$.

The relation ≤ obviously has the following properties. For all $a, b, c \in \mathbb{R}$ it holds that

$$a \leq a \quad \text{(reflexivity)},$$
$$a \leq b \quad \text{and} \quad b \leq c \quad \Rightarrow \quad a \leq c \quad \text{(transitivity)},$$
$$a \leq b \quad \text{and} \quad b \leq a \quad \Rightarrow \quad a = b \quad \text{(antisymmetry)}.$$

In case of $a \leq b$ and $a \neq b$ one writes $a < b$ and calls *a less than b*. Furthermore one defines $a \geq b$, if $b \leq a$ (in words: *a greater than or equal to b*), and $a > b$, if $b < a$ (in words: *a greater than b*).

Addition and multiplication can be carried over from $\mathbb{Q}$ to $\mathbb{R}$ in a similar way. Graphically one uses the fact that each real number corresponds to a segment on the real line. One thus defines the addition of real numbers as the addition of the respective segments.

A rigorous and at the same time *algorithmic* definition of the addition starts from the observation that real numbers can be approximated by rational numbers to any degree of accuracy. Let $a = a_0.a_1a_2...$ and $b = b_0.b_1b_2...$ be two non-negative real numbers. By cutting them off after k decimal places we obtain two rational approximations $a^{(k)} = a_0.a_1a_2...a_k \approx a$ and $b^{(k)} = b_0.b_1b_2...b_k \approx b$. Then $a^{(k)} + b^{(k)}$ is a monotonically increasing sequence of approximations to the yet to be defined number $a + b$. This allows one to *define* $a + b$ as *supremum* of these approximations. To justify this approach rigorously we refer to Chap. 5. The multiplication of real numbers is defined in the same way. It turns out that the real numbers with addition and multiplication $(\mathbb{R}, +, \cdot)$ are a *field*. Therefore the usual rules of calculation apply, e.g., the distributive law

$$(a + b)c = ac + bc.$$

The following proposition recapitulates some of the important rules for ≤. The statements can easily be verified with the help of the real line.

Proposition 1.9 *For all $a, b, c \in \mathbb{R}$ the following holds:*

$$a \leq b \quad \Rightarrow \quad a + c \leq b + c,$$
$$a \leq b \quad \text{and} \quad c \geq 0 \quad \Rightarrow \quad ac \leq bc,$$
$$a \leq b \quad \text{and} \quad c \leq 0 \quad \Rightarrow \quad ac \geq bc.$$

Note that $a < b$ does *not* imply $a^2 < b^2$. For example $-2 < 1$, but nonetheless $4 > 1$. However, for $a, b \geq 0$ it always holds that $a < b \Leftrightarrow a^2 < b^2$.

Definition 1.10 (Intervals) The following subsets of $\mathbb{R}$ are called intervals:

$$[a, b] = \{x \in \mathbb{R} \; ; \; a \leq x \leq b\} \quad \text{closed interval};$$
$$(a, b] = \{x \in \mathbb{R} \; ; \; a < x \leq b\} \quad \text{left half-open interval};$$
$$[a, b) = \{x \in \mathbb{R} \; ; \; a \leq x < b\} \quad \text{right half-open interval};$$
$$(a, b) = \{x \in \mathbb{R} \; ; \; a < x < b\} \quad \text{open interval}.$$

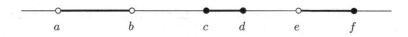

Fig. 1.4 The intervals (a, b), $[c, d]$ and $(e, f]$ on the real line

Intervals can be visualised on the real line, as illustrated in Fig. 1.4.

It proves to be useful to introduce the symbols $-\infty$ (minus infinity) and ∞ (infinity), by means of the property

$$\forall a \in \mathbb{R} : -\infty < a < \infty.$$

One may then define, e.g., the *improper* intervals

$$[a, \infty) = \{x \in \mathbb{R} \; ; \; x \geq a\}$$
$$(-\infty, b) = \{x \in \mathbb{R} \; ; \; x < b\}$$

and furthermore $(-\infty, \infty) = \mathbb{R}$. Note that $-\infty$ and ∞ are only *symbols* and *not* real numbers.

Definition 1.11 The *absolute value* of a real number a is defined as

$$|a| = \begin{cases} a, & \text{if } a \geq 0, \\ -a, & \text{if } a < 0. \end{cases}$$

As an application of the properties of the order relation given in Proposition 1.9 we exemplarily solve some inequalities.

Example 1.12 Find all $x \in \mathbb{R}$ satisfying $-3x - 2 \leq 5 < -3x + 4$. In this example we have the following two inequalities

$$-3x - 2 \leq 5 \quad \text{and} \quad 5 < -3x + 4.$$

The first inequality can be rearranged to

$$-3x \leq 7 \quad \Leftrightarrow \quad x \geq -\frac{7}{3}.$$

This is the first constraint for x. The second inequality states

$$3x < -1 \quad \Leftrightarrow \quad x < -\frac{1}{3}$$

and poses a second constraint for x. The solution to the original problem must fulfil both constraints. Therefore the solution set is

$$S = \left\{x \in \mathbb{R}; \; -\frac{7}{3} \leq x < -\frac{1}{3}\right\} = \left[-\frac{7}{3}, -\frac{1}{3}\right).$$

Example 1.13 Find all $x \in \mathbb{R}$ satisfying $x^2 - 2x \geq 3$. By completing the square the inequality is rewritten as

$$(x - 1)^2 = x^2 - 2x + 1 \geq 4.$$

Taking the square root we obtain two possibilities

$$x - 1 \geq 2 \quad \text{or} \quad x - 1 \leq -2.$$

The combination of those gives the solution set

$$S = \{x \in \mathbb{R} \; ; \; x \geq 3 \text{ or } x \leq -1\} = (-\infty, -1] \cup [3, \infty).$$

1.3 Machine Numbers

The real numbers can be realised only partially on a computer. In exact arithmetic, like for example in maple, real numbers are treated as symbolic expressions, e.g. $\sqrt{2} = \text{RootOf}(\_Z^2-2)$. With the help of the command evalf they can be evaluated, exact to many decimal places.

The floating point numbers that are usually employed in programming languages as substitutes for the real numbers have a fixed relative accuracy, e.g. *double precision* with 52 bit mantissa. The arithmetic rules of $\mathbb{R}$ are *not* valid for these machine numbers, e.g.

$$1 + 10^{-20} = 1$$

in double precision. Floating point numbers are standardised by the *Institute of Electrical and Electronics Engineers* IEEE 754-1985 and by the *International Electrotechnical Commission* IEC 559:1989. In the following we give a short outline of these machine numbers. Further information can be found in [20].

One distinguishes between single and double format. The single format (*single precision*) requires 32-bit storage space

V	e	M
1	8	23

The double format (*double precision*) requires 64-bit storage space

V	e	M
1	11	52

Here, $V \in \{0, 1\}$ denotes the sign, $e_{min} \leq e \leq e_{max}$ is the exponent (a signed integer) and M is the mantissa of length p

$$M = d_1 2^{-1} + d_2 2^{-2} + \ldots + d_p 2^{-p} \cong d_1 d_2 \ldots d_p, \quad d_j \in \{0, 1\}.$$

$$0 \qquad 2^{e_{\min}-1} \qquad 2^{e_{\min}} \qquad\qquad\qquad 2^{e_{\min}+1}$$

Fig. 1.5 Floating point numbers on the real line

This representation corresponds to the following number x:

$$x = (-1)^V 2^e \sum_{j=1}^{p} d_j 2^{-j}.$$

Normalised floating point numbers in base 2 always have $d_1 = 1$. Therefore, one does not need to store d_1 and obtains for the mantissa

$$
\begin{aligned}
\text{single precision} &\quad p = 24;\\
\text{double precision} &\quad p = 53.
\end{aligned}
$$

To simplify matters we will only describe the key features of floating point numbers. For the subtleties of the IEEE-IEC standard, we refer to [20].

In our representation the following range applies for the exponents:

	$e_{\min}$	$e_{\max}$
single precision	-125	128
double precision	-1021	1024

With $M = M_{\max}$ and $e = e_{\max}$ one obtains the largest floating point number

$$x_{\max} = \left(1 - 2^{-p}\right) 2^{e_{\max}},$$

whereas $M = M_{\min}$ and $e = e_{\min}$ gives the smallest positive (normalised) floating point number

$$x_{\min} = 2^{e_{\min}-1}.$$

The floating point numbers are *not* evenly distributed on the real line, but their *relative* density is nearly constant, see Fig. 1.5.

In the IEEE standard the following approximate values apply:

	$x_{\min}$	$x_{\max}$
single precision	$1.18 \cdot 10^{-38}$	$3.40 \cdot 10^{38}$
double precision	$2.23 \cdot 10^{-308}$	$1.80 \cdot 10^{308}$

Furthermore, there are special *symbols* like

$$
\begin{aligned}
&\pm\text{INF} \quad \ldots \quad \pm\infty\\
&\text{NaN} \quad\ \ldots \quad \text{not a number, e.g. for } \textit{zero divided by zero}.
\end{aligned}
$$

In general, one can continue calculating with these symbols without program termination.

1.4 Rounding

Let $x = a \cdot 2^e \in \mathbb{R}$ with $1/2 \leq a < 1$ and $x_{\min} \leq x \leq x_{\max}$. Furthermore, let u, v be two adjacent machine numbers with $u \leq x \leq v$. Then

$$u = \boxed{0 \mid e \mid b_1 \ldots b_p}$$

and

$$v = u + \boxed{0 \mid e \mid 00 \ldots 01} = u + \boxed{0 \mid e - (p-1) \mid 10 \ldots 00}$$

Thus $v - u = 2^{e-p}$ and the inequality

$$|\mathrm{rd}(x) - x| \leq \frac{1}{2}(v - u) = 2^{e-p-1}$$

holds for the optimal *rounding* $\mathrm{rd}(x)$ of x. With this estimate one can determine the *relative error* of the rounding. Due to $\frac{1}{a} \leq 2$ it holds that

$$\frac{|\mathrm{rd}(x) - x|}{x} \leq \frac{2^{e-p-1}}{a \cdot 2^e} \leq 2 \cdot 2^{-p-1} = 2^{-p}.$$

The same calculation is valid for negative x (by using the absolute value).

Definition 1.14 The number $\mathrm{eps} = 2^{-p}$ is called *relative machine accuracy*.

The following proposition is an important application of this concept.

Proposition 1.15 *Let $x \in \mathbb{R}$ with $x_{\min} \leq |x| \leq x_{\max}$. Then there exists $\varepsilon \in \mathbb{R}$ with*

$$rd(x) = x(1 + \varepsilon) \quad and \quad |\varepsilon| \leq \mathrm{eps}.$$

Proof We define

$$\varepsilon = \frac{\mathrm{rd}(x) - x}{x}.$$

According to the calculation above, we have $|\varepsilon| \leq \mathrm{eps}$. □

Experiment 1.16 (Experimental determination of eps) Let z be the smallest positive machine number for which $1 + z > 1$.

$$1 = \boxed{0 \mid 1 \mid 100 \ldots 00}, \quad z = \boxed{0 \mid 1 \mid 000 \ldots 01} = 2 \cdot 2^{-p}.$$

Thus $z = 2\,\mathrm{eps}$. The number z can be determined experimentally and therefore eps as well. (Note that the number z is called eps in MATLAB.)

In IEC/IEEE standard the following applies:

$$\text{single precision:} \quad \text{eps} = 2^{-24} \approx 5.96 \cdot 10^{-8},$$
$$\text{double precision:} \quad \text{eps} = 2^{-53} \approx 1.11 \cdot 10^{-16}.$$

In double precision arithmetic an accuracy of approximately 16 places is available.

1.5 Exercises

1. Show that $\sqrt{3}$ is irrational.
2. Prove the triangle inequality

$$|a + b| \leq |a| + |b|$$

for all $a, b \in \mathbb{R}$.

Hint. Distinguish the cases where a and b have either the same or different signs.
3. Sketch the following subsets of the real line:

$$A = \{x : |x| \leq 1\}, \quad B = \{x : |x - 1| \leq 2\}, \quad C = \{x : |x| \geq 3\}.$$

More generally, sketch the set $U_r(a) = \{x : |x - a| < r\}$ (for $a \in \mathbb{R}$, $r > 0$). Convince yourself that $U_r(a)$ is the set of points of distance less than r to the point a.
4. Solve the following inequalities by hand as well as with maple (using `solve`). State the solution set in interval notation.

(a) $4x^2 \leq 8x + 1$,

(b) $\dfrac{1}{3 - x} > 3 + x$,

(c) $|2 - x^2| \geq x^2$,

(d) $\dfrac{1 + x}{1 - x} > 1$,

(e) $x^2 < 6 + x$,

(f) $||x| - x| \geq 1$,

(g) $|1 - x^2| \leq 2x + 2$,

(h) $4x^2 - 13x + 4 < 1$.

5. Determine the solution set of the inequality

$$8(x - 2) \geq \frac{20}{x + 1} + 3(x - 7).$$

6. Sketch the regions in the (x, y)-plane which are given by

(a) $x = y$;   (b) $y < x$;   (c) $y > x$;   (d) $y > |x|$;   (e) $|y| > |x|$.

Hint. Consult Sects. A.1 and A.6 for basic plane geometry.

7. Compute the binary representation of the floating point number $x = 0.1$ in single precision IEEE arithmetic.

8. Experimentally determine the relative machine accuracy eps.
   Hint. Write a computer program in your programming language of choice which calculates the smallest machine number z such that $1 + z > 1$.

Real-Valued Functions

2

The notion of a function is the mathematical way of formalising the idea that one or more *independent quantities* are assigned to one or more *dependent quantities*. Functions in general and their investigation are at the core of analysis. They help to model dependencies of variable quantities, from simple planar graphs, curves and surfaces in space to solutions of differential equations or the algorithmic construction of fractals. One the one hand, this chapter serves to introduce the basic concepts. On the other hand, the most important examples of real-valued, elementary functions are discussed in an informal way. These include the power functions, the exponential functions and their inverses. Trigonometric functions will be discussed in Chap. 3, complex-valued functions in Chap. 4.

2.1 Basic Notions

The simplest case of a real-valued function is a double-row list of numbers, consisting of values from an *independent* quantity x and corresponding values of a *dependent* quantity y.

Experiment 2.1 Study the mapping $y = x^2$ with the help of MATLAB. First choose the region D in which the x-values should vary, for instance $D = \{x \in \mathbb{R} : -1 \le x \le 1\}$. The command

$$x = -1 : 0.01 : 1;$$

produces a list of x-values, the row vector

$$x = [x_1, x_2, \ldots, x_n] = [-1.00, -0.99, -0.98, \ldots, 0.99, 1.00].$$

© Springer Nature Switzerland AG 2018
M. Oberguggenberger and A. Ostermann, *Analysis for Computer Scientists*,
Undergraduate Topics in Computer Science,
https://doi.org/10.1007/978-3-319-91155-7_2

13

Using

$$y = x.^2;$$

a row vector of the same length of corresponding y-values is generated. Finally
$\texttt{plot(x,y)}$ plots the points $(x_1, y_1), \ldots, (x_n, y_n)$ in the coordinate plane and connects them with line segments. The result can be seen in Fig. 2.1.

In the general mathematical framework we do not just want to assign finite lists
of values. In many areas of mathematics functions defined on arbitrary sets are
needed. For the general set-theoretic notion of a function we refer to the literature,
e.g. [3, Chap. 0.2]. This section is dedicated to *real-valued functions*, which are
central in analysis.

Definition 2.2 A real-valued function f with *domain* D and *range* $\mathbb{R}$ is a rule which
assigns to every $x \in D$ a real number $y \in \mathbb{R}$.

In general, D is an arbitrary set. In this section,
however, it will be a subset of $\mathbb{R}$. For the expression *function* we also use the word *mapping* synonymously. A function is denoted by

$$f : D \to \mathbb{R} : x \mapsto y = f(x).$$

The *graph of the function* f is the set

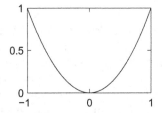

Fig. 2.1 A function

$$\Gamma(f) = \{(x, y) \in D \times \mathbb{R};\ y = f(x)\}.$$

In the case of $D \subset \mathbb{R}$ the graph can also be represented as a subset of the coordinate
plane. The set of the actually assumed values is called *image of f* or *proper range*:

$$f(D) = \{f(x);\ x \in D\}.$$

Example 2.3 A part of the graph of the quadratic function $f : \mathbb{R} \to \mathbb{R}$, $f(x) = x^2$
is shown in Fig. 2.2. If one chooses the domain to be $D = \mathbb{R}$, then the image is the
interval $f(D) = [0, \infty)$.

An important tool is the concept of *inverse functions*, whether to solve equations
or to find new types of functions. If and in which domain a given function has an
inverse depends on two main properties, the injectivity and the surjectivity, which
we investigate on their own first.

Definition 2.4 (a) A function $f : D \to \mathbb{R}$ is called *injective* or *one-to-one*, if different arguments always have different function values:

$$x_1 \neq x_2 \quad \Rightarrow \quad f(x_1) \neq f(x_2).$$

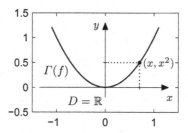

Fig. 2.2 Quadratic function

(b) A function $f : D \to B \subset \mathbb{R}$ is called *surjective* or *onto* from D to B, if each $y \in B$ appears as a function value:

$$\forall y \in B \ \exists x \in D : y = f(x).$$

(c) A function $f : D \to B$ is called *bijective*, if it is injective and surjective.

Figures 2.3 and 2.4 illustrate these notions.

Surjectivity can always be enforced by reducing the range B; for example, $f : D \to f(D)$ is always surjective. Likewise, injectivity can be obtained by restricting the domain to a subdomain.

If $f : D \to B$ is bijective, then for every $y \in B$ there exists *exactly one* $x \in D$ with $y = f(x)$. The mapping $y \mapsto x$ then defines the inverse of the mapping $x \mapsto y$.

Definition 2.5 If the function

$$f : D \to B : y = f(x),$$

is bijective, then the assignment

$$f^{-1} : B \to D : x = f^{-1}(y),$$

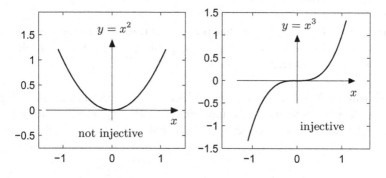

Fig. 2.3 Injectivity

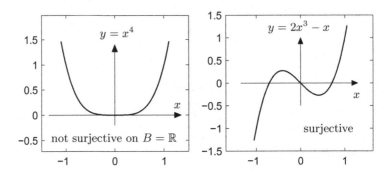

Fig. 2.4 Surjectivity

Fig. 2.5 Bijectivity and
inverse function

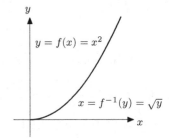

which maps each $y \in B$ to the unique $x \in D$ with $y = f(x)$ is called the *inverse function* of the function f.

Example 2.6 The quadratic function $f(x) = x^2$ is bijective from $D = [0, \infty)$ to $B = [0, \infty)$. In these intervals ($x \geq 0$, $y \geq 0$) one has

$$y = x^2 \quad \Leftrightarrow \quad x = \sqrt{y}.$$

Here $\sqrt{y}$ denotes the positive square root. Thus the inverse of the quadratic function on the above intervals is given by $f^{-1}(y) = \sqrt{y}$; see Fig. 2.5.

Once one has found the inverse function f^{-1}, it is usually written with variables $y = f^{-1}(x)$. This corresponds to flipping the graph of $y = f(x)$ about the diagonal $y = x$, as is shown in Fig. 2.6.

Experiment 2.7 The term inverse function is clearly illustrated by the MATLAB plot command. The graph of the inverse function can easily be plotted by interchanging the variables, which exactly corresponds to flipping the lists $y \leftrightarrow x$. For example,

Fig. 2.6 Inverse function
and reflection in the diagonal

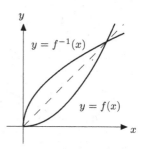

the graphs in Fig. 2.6 are obtained by

```
x = 0:0.01:1;
y = x.^2;
plot(x,y)
hold on
plot(y,x)
```

How the formatting, the dashed diagonal and the labelling are obtained can be learned from the M-file mat02_1.m.

2.2 Some Elementary Functions

The elementary functions are the powers and roots, exponential functions and logarithms, trigonometric functions and their inverse functions, as well as all functions which are obtained by combining these. We are going to discuss the most important basic types which have historically proven to be of importance for applications. The trigonometric functions will be dealt with in Chap. 3.

Linear functions (straight lines). A *linear function* $\mathbb{R} \to \mathbb{R}$ assigns each x-value a fixed multiple as y-value, i.e.,

$$y = kx.$$

Here

$$k = \frac{\text{increase in height}}{\text{increase in length}} = \frac{\Delta y}{\Delta x}$$

is the *slope* of the graph, which is a *straight line* through the origin. The connection between the slope and the angle between the straight line and x-axis is discussed in Sect. 3.1. Adding an *intercept* $d \in \mathbb{R}$ translates the straight line d units in y-direction (Fig. 2.7). The equation is then

$$y = kx + d.$$

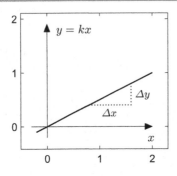 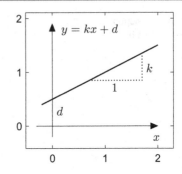

Fig. 2.7 Equation of a straight line

Quadratic parabolas. The quadratic function with domain $D = \mathbb{R}$ in its basic form
is given by

$$y = x^2.$$

Compression/stretching, horizontal and vertical translation are obtained via

$$y = \alpha x^2, \quad y = (x - \beta)^2, \quad y = x^2 + \gamma.$$

The effect of these transformations on the graph can be seen in Fig. 2.8.

$$\alpha > 1 \ldots \text{compression in } x\text{-direction}$$
$$0 < \alpha < 1 \ldots \text{stretching in } x\text{-direction}$$
$$\alpha < 0 \ldots \text{reflection in the } x\text{-axis}$$

$\beta > 0 \ldots$ translation to the right $\gamma > 0 \ldots$ translation upwards
$\beta < 0 \ldots$ translation to the left $\gamma < 0 \ldots$ translation downwards

The general quadratic function can be reduced to these cases by *completing the
square*:

$$\begin{aligned}
y &= ax^2 + bx + c \\
&= a\left(x + \frac{b}{2a}\right)^2 + c - \frac{b^2}{4a} \\
&= \alpha(x - \beta)^2 + \gamma.
\end{aligned}$$

Power functions. In the case of an integer exponent $n \in \mathbb{N}$ the following rules apply

$$x^n = x \cdot x \cdot x \cdot \cdots \cdot x \quad (n \text{ factors}), \quad x^1 = x,$$

$$x^0 = 1, \quad x^{-n} = \frac{1}{x^n} \quad (x \neq 0).$$

The behaviour of $y = x^3$ can be seen in the picture on the right-hand side of
Fig. 2.3, the one of $y = x^4$ in the picture on the left-hand side of Fig. 2.4. The graphs
for odd and even powers behave similarly.

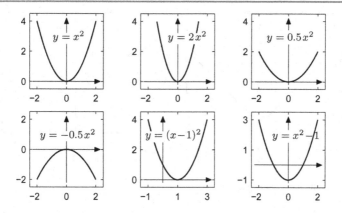

Fig. 2.8 Quadratic parabolas

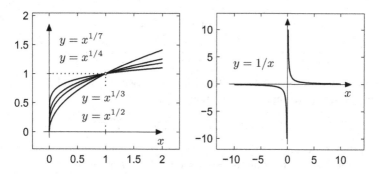

Fig. 2.9 Power functions with fractional and negative exponents

As an example of fractional exponents we consider the *root functions* $y = \sqrt[n]{x} = x^{1/n}$ for $n \in \mathbb{N}$ with domain $D = [0, \infty)$. Here $y = \sqrt[n]{x}$ is defined as the inverse function of the nth power, see Fig. 2.9 left. The graph of $y = x^{-1}$ with domain $D = \mathbb{R} \setminus \{0\}$ is pictured in Fig. 2.9 right.

Absolute value, sign and indicator function. The graph of the *absolute value function*

$$y = |x| = \begin{cases} x, & x \geq 0, \\ -x, & x < 0 \end{cases}$$

has a kink at the point $(0, 0)$, see Fig. 2.10 left.

The graph of the *sign function* or *signum function*

$$y = \operatorname{sign} x = \begin{cases} 1, & x > 0, \\ 0, & x = 0, \\ -1, & x < 0 \end{cases}$$

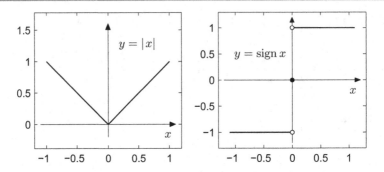

Fig. 2.10 Absolute value and sign

has a jump at $x = 0$ (Fig. 2.10 right). The *indicator function of a subset* $A \subset \mathbb{R}$ is defined as

$$\mathbb{1}_A(x) = \begin{cases} 1, & x \in A, \\ 0, & x \notin A. \end{cases}$$

Exponential functions and logarithms. *Integer powers* of a number $a > 0$ have just been defined. Fractional (rational) powers give

$$a^{1/n} = \sqrt[n]{a}, \qquad a^{m/n} = (\sqrt[n]{a})^m = \sqrt[n]{a^m}.$$

If r is an arbitrary real number then a^r is defined by its approximations $a^{m/n}$, where $\frac{m}{n}$ is the rational approximation to r obtained by decimal expansion.

Example 2.8 2^π is defined by the sequence

$$2^3, \ 2^{3.1}, \ 2^{3.14}, \ 2^{3.141}, \ 2^{3.1415}, \ \ldots,$$

where

$$2^{3.1} = 2^{31/10} = \sqrt[10]{2^{31}}; \quad 2^{3.14} = 2^{314/100} = \sqrt[100]{2^{314}}; \quad \ldots \text{ etc.}$$

This somewhat informal introduction of the exponential function should be sufficient to have some examples at hand for applications in the following sections. With the tools we have developed so far we cannot yet show that this process of approximation actually leads to a well-defined mathematical object. The success of this process is based on the *completeness* of the real numbers. This will be thoroughly discussed in Chap. 5.

From the definition above we obtain that the following rules of calculation are valid for rational exponents:

$$a^r a^s = a^{r+s}$$
$$(a^r)^s = a^{rs} = (a^s)^r$$
$$a^r b^r = (ab)^r$$

Fig. 2.11 Exponential functions

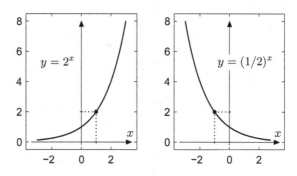

for $a, b > 0$ and arbitrary $r, s \in \mathbb{Q}$. The fact that these rules are also true for real-valued exponents $r, s \in \mathbb{R}$ can be shown by employing a limiting argument.

The *graph of the exponential function with base a*, the function $y = a^x$, increases for $a > 1$ and decreases for $a < 1$, see Fig. 2.11. Its *proper range* is $B = (0, \infty)$; the exponential function is *bijective* from $\mathbb{R}$ to $(0, \infty)$. Its inverse function is the *logarithm to the base a* (with domain $(0, \infty)$ and range $\mathbb{R}$):

$$y = a^x \quad \Leftrightarrow \quad x = \log_a y.$$

For example, $\log_{10} 2$ is the power by which 10 needs to be raised to obtain 2:

$$2 = 10^{\log_{10} 2}.$$

Other examples are, for instance:

$$2 = \log_{10}(10^2), \quad \log_{10} 10 = 1, \quad \log_{10} 1 = 0, \quad \log_{10} 0.001 = -3.$$

Euler's number[1] e is defined by

$$e = 1 + \frac{1}{1} + \frac{1}{2} + \frac{1}{6} + \frac{1}{24} + \dots$$
$$= 1 + \frac{1}{1!} + \frac{1}{2!} + \frac{1}{3!} + \frac{1}{4!} + \dots = \sum_{j=0}^{\infty} \frac{1}{j!}$$
$$\approx 2.718281828459045235360287471\dots$$

That this *summation of infinitely many numbers* can be defined rigorously will be proven in Chap. 5 by invoking the completeness of the real numbers. The logarithm to the base e is called *natural logarithm* and is denoted by log:

$$\log x = \log_e x$$

---

[1]L. Euler, 1707–1783.

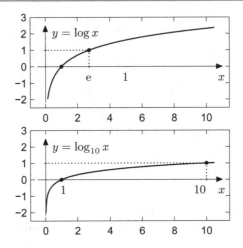

Fig. 2.12 Logarithms to the base e and to the base 10

In some books the natural logarithm is denoted by $\ln x$. We stick to the notation $\log x$ which is used, e.g., in MATLAB. The following rules are obtained directly by rewriting the rules for the exponential function:

$$u = e^{\log u}$$
$$\log(uv) = \log u + \log v$$
$$\log(u^z) = z \log u$$

for $u, v > 0$ and arbitrary $z \in \mathbb{R}$. In addition, it holds that

$$u = \log(e^u)$$

for all $u \in \mathbb{R}$, and $\log e = 1$. In particular it follows from the above that

$$\log \frac{1}{u} = -\log u, \quad \log \frac{v}{u} = \log v - \log u.$$

The graphs of $y = \log x$ and $y = \log_{10} x$ are shown in Fig. 2.12.

Hyperbolic functions and their inverses. Hyperbolic functions and their inverses will mainly be needed in Chap. 14 for the parametric representation of hyperbolas, in Chap. 10 for evaluating integrals and in Chap. 19 for explicitly solving some differential equations.

The *hyperbolic sine*, the *hyperbolic cosine* and the *hyperbolic tangent* are defined by

$$\sinh x = \frac{1}{2}\left(e^x - e^{-x}\right), \quad \cosh x = \frac{1}{2}\left(e^x + e^{-x}\right), \quad \tanh x = \frac{\sinh x}{\cosh x}$$

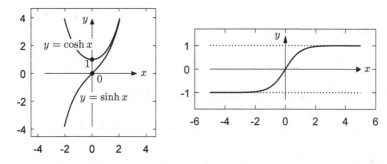

Fig. 2.13 Hyperbolic sine and cosine (left), and hyperbolic tangent (right)

for $x \in \mathbb{R}$. Their graphs are displayed in Fig. 2.13. An important property is the identity

$$\cosh^2 x - \sinh^2 x = 1,$$

which can easily be verified by inserting the defining expressions.

Figure 2.13 shows that the hyperbolic sine is invertible as a function from $\mathbb{R} \to \mathbb{R}$, the hyperbolic cosine is invertible as a function from $[0, \infty) \to [1, \infty)$, and the hyperbolic tangent is invertible as a function from $\mathbb{R} \to (-1, 1)$. The *inverse hyperbolic functions*, also known as *area functions*, are referred to as *inverse hyperbolic sine (cosine, tangent)* or *area hyperbolic sine (cosine, tangent)*. They can be expressed by means of logarithms as follows (see Exercise 15):

$$\operatorname{arsinh} x = \log\left(x + \sqrt{x^2 + 1}\right), \quad \text{for } x \in \mathbb{R},$$
$$\operatorname{arcosh} x = \log\left(x + \sqrt{x^2 - 1}\right), \quad \text{for } x \geq 1,$$
$$\operatorname{artanh} x = \frac{1}{2} \log \frac{1 + x}{1 - x}, \qquad \text{for } |x| < 1.$$

2.3 Exercises

1. How does the graph of an arbitrary function $y = f(x) : \mathbb{R} \to \mathbb{R}$ change under the transformations

$$y = f(ax), \qquad y = f(x - b), \qquad y = cf(x), \qquad y = f(x) + d$$

with $a, b, c, d \in \mathbb{R}$? Distinguish the following different cases for a:

$$a < -1, \qquad -1 \leq a < 0, \qquad 0 < a \leq 1, \qquad a > 1,$$

and for b, c, d the cases

$$b, c, d > 0, \qquad b, c, d < 0.$$

Sketch the resulting graphs.

2. Let the function $f : D \to \mathbb{R} : x \mapsto 3x^4 - 2x^3 - 3x^2 + 1$ be given. Using MATLAB plot the graphs of f for

$$D = [-1, 1.5], \qquad D = [-0.5, 0.5], \qquad D = [0.5, 1.5].$$

Explain the behaviour of the function for $D = \mathbb{R}$ and find

$$f([-1, 1.5]), \qquad f((-0.5, 0.5)), \qquad f((-\infty, 1]).$$

3. Which of the following functions are injective/surjective/bijective?

$$
\begin{aligned}
f : \mathbb{N} \to \mathbb{N} : &\quad n \mapsto n^2 - 6n + 10; \\
g : \mathbb{R} \to \mathbb{R} : &\quad x \mapsto |x + 1| - 3; \\
h : \mathbb{R} \to \mathbb{R} : &\quad x \mapsto x^3.
\end{aligned}
$$

Hint. Illustrative examples for the use of the MATLAB plot command may be found in the M-file mat02_2.m.

4. Sketch the graph of the function $y = x^2 - 4x$ and justify why it is bijective as a function from $D = (-\infty, 2]$ to $B = [-4, \infty)$. Compute its inverse function on the given domain.

5. Check that the following functions $D \to B$ are bijective in the given regions and compute the inverse function in each case:

$$
\begin{aligned}
y &= -2x + 3, &\quad D = \mathbb{R},\ B = \mathbb{R}; \\
y &= x^2 + 1, &\quad D = (-\infty, 0],\ B = [1, \infty); \\
y &= x^2 - 2x - 1, &\quad D = [1, \infty),\ B = [-2, \infty).
\end{aligned}
$$

6. Find the equation of the straight line through the points $(1, 1)$ and $(4, 3)$ as well as the equation of the quadratic parabola through the points $(-1, 6)$, $(0, 5)$ and $(2, 21)$.

7. Let the amount of a radioactive substance at time $t = 0$ be A grams. According to the law of radioactive decay, there remain $A \cdot q^t$ grams after t days. Compute q for radioactive iodine 131 from its half life (8 days) and work out after how many days $\frac{1}{100}$ of the original amount of iodine 131 is remaining.
Hint. The half life is the time span after which only half of the initial amount of radioactive substance is remaining.

8. Let I [Watt/cm^2] be the sound intensity of a sound wave that hits a detector surface. According to the Weber–Fechner law, its sound level L [Phon] is computed by

$$L = 10 \log_{10}(I/I_0)$$

where $I_0 = 10^{-16}$ W/cm^2. If the intensity I of a loudspeaker produces a sound level of 80 Phon, which level is then produced by an intensity of $2I$ by two loudspeakers?

9. For $x \in \mathbb{R}$ the floor function $\lfloor x \rfloor$ denotes the largest integer not greater than x, i.e.,

$$\lfloor x \rfloor = \max \{n \in \mathbb{N}; \, n \leq x\}.$$

Plot the following functions with domain $D = [0, 10]$ using the MATLAB command floor:

$$y = \lfloor x \rfloor, \qquad y = x - \lfloor x \rfloor, \qquad y = (x - \lfloor x \rfloor)^3, \qquad y = (\lfloor x \rfloor)^3.$$

Try to program correct plots in which the vertical connecting lines do not appear.

10. A function $f : D = \{1, 2, \ldots, N\} \to B = \{1, 2, \ldots, N\}$ is given by the list of its function values $y = (y_1, \ldots, y_N)$, $y_i = f(i)$. Write a MATLAB program which determines whether f is bijective. Test your program by generating random y-values using

(a) y = unirnd(N,1,N),   (b) y = randperm(N).

Hint. See the two M-files mat02_ex12a.m and mat02_ex12b.m or the Python-file python02_ex12.

11. Draw the graph of the function $f : \mathbb{R} \to \mathbb{R} : y = ax + \text{sign } x$ for different values of a. Distinguish between the cases $a > 0$, $a = 0$, $a < 0$. For which values of a is the function f injective and surjective, respectively?

12. Let $a > 0$, $b > 0$. Verify the *laws of exponents*

$$a^r a^s = a^{r+s}, \qquad (a^r)^s = a^{rs}, \qquad a^r b^r = (ab)^r$$

for rational $r = k/l$, $s = m/n$.

Hint. Start by verifying the laws for integer r and s (and arbitrary $a, b > 0$). To prove the first law for rational $r = k/l$, $s = m/n$, write

$$(a^{k/l} a^{m/n})^{ln} = (a^{k/l})^{ln}(a^{m/n})^{ln} = a^{kn} a^{lm} = a^{kn+lm}$$

using the third law for integer exponents and inspection; conclude that

$$a^{k/l} a^{m/n} = a^{(kn+lm)/ln} = a^{k/l+m/n}.$$

13. Using the arithmetics of exponentiation, verify the rules $\log(uv) = \log u + \log v$ and $\log u^z = z \log u$ for $u, v > 0$ and $z \in \mathbb{R}$.

Hint. Set $x = \log u$, $y = \log v$, so $uv = e^x e^y$. Use the laws of exponents and take the logarithm.

14. Verify the identity $\cosh^2 x - \sinh^2 x = 1$.

15. Show that $\operatorname{arsinh} x = \log(x + \sqrt{x^2 + 1})$ for $x \in \mathbb{R}$.

Hint. Set $y = \operatorname{arsinh} x$ and solve the identity $x = \sinh y = \frac{1}{2}(e^y - e^{-y})$ for y. Substitute $u = e^y$ to derive the quadratic equation $u^2 - 2xu - 1 = 0$ for u. Observe that $u > 0$ to select the appropriate root of this equation.

Trigonometry

3

Trigonometric functions play a major role in geometric considerations as well as in the modelling of oscillations. We introduce these functions at the right-angled triangle and extend them periodically to $\mathbb{R}$ using the unit circle. Furthermore, we will discuss the inverse functions of the trigonometric functions in this chapter. As an application we will consider the transformation between Cartesian and polar coordinates.

3.1 Trigonometric Functions at the Triangle

The definitions of the trigonometric functions are based on elementary properties of the right-angled triangle. Figure 3.1 shows a right-angled triangle. The sides adjacent to the right angle are called legs (or catheti), the opposite side hypotenuse.

One of the basic properties of the right-angled triangle is expressed by Pythagoras' theorem.[1]

Proposition 3.1 (Pythagoras) *In a right-angled triangle the sum of the squares of the legs equals the square of the hypotenuse. In the notation of Fig. 3.1 this says that* $a^2 + b^2 = c^2$.

Proof According to Fig. 3.2 one can easily see that

$$(a + b)^2 - c^2 = \text{area of the grey triangles} = 2ab.$$

From this it follows that $a^2 + b^2 - c^2 = 0$. □

---

[1]Pythagoras, approx. 570–501 B.C.

© Springer Nature Switzerland AG 2018
M. Oberguggenberger and A. Ostermann, *Analysis for Computer Scientists*,
Undergraduate Topics in Computer Science,
https://doi.org/10.1007/978-3-319-91155-7_3

Fig. 3.1 A right-angled
triangle with legs a, b and
hypotenuse c

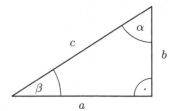

Fig. 3.2 Basic idea of the
proof of Pythagoras'
theorem

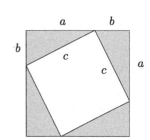

A fundamental fact is Thales' intercept theorem[2] which says that the ratios of
the sides in a triangle are scale invariant; i.e. they do not depend on the size of the
triangle.

In the situation of Fig. 3.3 Thales' theorem asserts that the following ratios are
valid:

$$\frac{a}{c} = \frac{a'}{c'}, \quad \frac{b}{c} = \frac{b'}{c'}, \quad \frac{a}{b} = \frac{a'}{b'}.$$

The reason for this is that by changing the scale (enlargement or reduction of the
triangle) all sides are changed by the same factor. One then concludes that the ratios
of the sides only depend on the angle α (and $\beta = 90° - \alpha$, respectively). This gives
rise to the following definition.

Fig. 3.3 Similar triangles

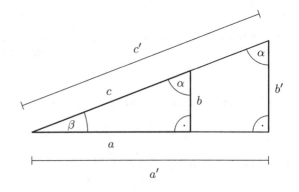

---

[2]Thales of Miletus, approx. 624–547 B.C.

Definition 3.2 (*Trigonometric functions*) For $0° \leq \alpha \leq 90°$ we define

$$\sin \alpha = \frac{a}{c} = \frac{\text{opposite leg}}{\text{hypotenuse}} \qquad \text{(sine)},$$

$$\cos \alpha = \frac{b}{c} = \frac{\text{adjacent leg}}{\text{hypotenuse}} \qquad \text{(cosine)},$$

$$\tan \alpha = \frac{a}{b} = \frac{\text{opposite leg}}{\text{adjacent leg}} \qquad \text{(tangent)},$$

$$\cot \alpha = \frac{b}{a} = \frac{\text{adjacent leg}}{\text{opposite leg}} \qquad \text{(cotangent)}.$$

Note that $\tan \alpha$ is not defined for $\alpha = 90°$ (since $b = 0$) and that $\cot \alpha$ is not defined for $\alpha = 0°$ (since $a = 0$). The identities

$$\tan \alpha = \frac{\sin \alpha}{\cos \alpha}, \quad \cot \alpha = \frac{\cos \alpha}{\sin \alpha}, \quad \sin \alpha = \cos \beta = \cos (90° - \alpha)$$

follow directly from the definition, the relationship

$$\sin^2 \alpha + \cos^2 \alpha = 1$$

is obtained using Pythagoras' theorem.

The trigonometric functions have many applications in mathematics. As a first example we derive the formula for the area of a general triangle; see Fig. 3.4. The sides of a triangle are usually labelled in counterclockwise direction using lowercase Latin letters, and the angles opposite the sides are labelled using the corresponding Greek letters. Because $F = \frac{1}{2}ch$ and $h = b \sin \alpha$ the formula for the area of a triangle can be written as

$$F = \frac{1}{2} bc \sin \alpha = \frac{1}{2} ac \sin \beta = \frac{1}{2} ab \sin \gamma.$$

So the area equals half the product of two sides times the sine of the enclosed angle. The last equality in the above formula is valid for reasons of symmetry. There γ denotes the angle opposite to the side c, in other words $\gamma = 180° - \alpha - \beta$.

As a second example we compute the slope of a straight line. Figure 3.5 shows a straight line $y = kx + d$. Its slope k is the change of the y-value per unit change in x. It is calculated from the triangle attached to the straight line in Fig. 3.5 as $k = \tan \alpha$.

Fig. 3.4 A general triangle

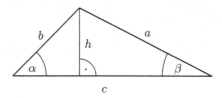

Fig. 3.5 Straight line with
slope k

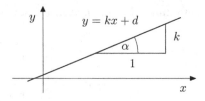

Fig. 3.6 Relationship
between degrees and radian
measure

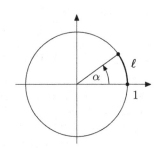

In order to have simple formulas such as

$$\frac{\mathrm{d}}{\mathrm{d}x}\sin x = \cos x,$$

one has to measure the angle in radian measure. The connection between degree and radian measure can be seen from the *unit circle* (i.e., the circle with centre 0 and radius 1); see Fig. 3.6.

The *radian measure* of the angle α (in degrees) is defined as the length ℓ of the corresponding arc of the unit circle with the sign of α. The arc length ℓ on the unit circle has no physical unit. However, one speaks about *radians* (rad) to emphasise the difference to degrees.

As is generally known the circumference of the unit circle is 2π with the constant

$$\pi = 3.141592653589793... \approx \frac{22}{7}.$$

For the conversion between the two measures we use that $360°$ corresponds to 2π in radian measure, for short $360° \leftrightarrow 2\pi$ [rad], so

$$\alpha° \leftrightarrow \frac{\pi}{180}\alpha\,[\text{rad}] \quad \text{and} \quad \ell\,[\text{rad}] \leftrightarrow \left(\frac{180}{\pi}\ell\right)°,$$

respectively. For example, $90° \leftrightarrow \frac{\pi}{2}$ and $-270° \leftrightarrow -\frac{3\pi}{2}$. Henceforth we always measure angles in radians.

3.2 Extension of the Trigonometric Functions to ℝ

For $0 \leq \alpha \leq \frac{\pi}{2}$ the values $\sin\alpha$, $\cos\alpha$, $\tan\alpha$ and $\cot\alpha$ have a simple interpretation on the unit circle; see Fig. 3.7. This representation follows from the fact that the hypotenuse of the defining triangle has length 1 on the unit circle.

One now extends the definition of the trigonometric functions for $0 \leq \alpha \leq 2\pi$ by continuation with the help of the unit circle. A general point P on the unit circle, which is defined by the angle α, is assigned the coordinates

$$P = (\cos\alpha, \sin\alpha),$$

see Fig. 3.8. For $0 \leq \alpha \leq \frac{\pi}{2}$ this is compatible with the earlier definition. For larger angles the sine and cosine functions are extended to the interval $[0, 2\pi]$ by this convention. For example, it follows from the above that

$$\sin\alpha = -\sin(\alpha - \pi), \qquad \cos\alpha = -\cos(\alpha - \pi)$$

for $\pi \leq \alpha \leq \frac{3\pi}{2}$, see Fig. 3.8.

For arbitrary values $\alpha \in \mathbb{R}$ one finally defines $\sin\alpha$ and $\cos\alpha$ by periodic continuation with period 2π. For this purpose one first writes $\alpha = x + 2k\pi$ with a unique $x \in [0, 2\pi)$ and $k \in \mathbb{Z}$. Then one sets

$$\sin\alpha = \sin(x + 2k\pi) = \sin x, \qquad \cos\alpha = \cos(x + 2k\pi) = \cos x.$$

Fig. 3.7 Definition of the trigonometric functions on the unit circle

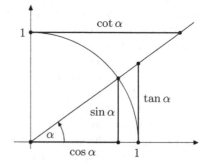

Fig. 3.8 Extension of the trigonometric functions on the unit circle

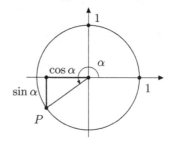

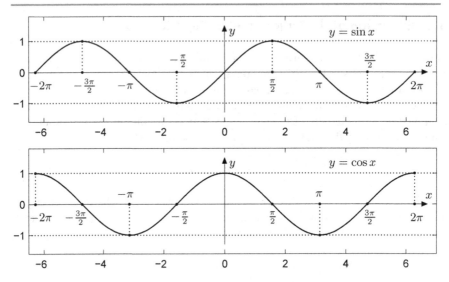

Fig. 3.9 The graphs of the sine and cosine functions in the interval $[-2\pi, 2\pi]$

With the help of the formulas

$$\tan \alpha = \frac{\sin \alpha}{\cos \alpha}, \qquad \cot \alpha = \frac{\cos \alpha}{\sin \alpha}$$

the tangent and cotangent functions are extended as well. Since the sine function equals zero for integer multiples of π, the cotangent is not defined for such arguments. Likewise the tangent is not defined for odd multiples of $\frac{\pi}{2}$.

The graphs of the functions $y = \sin x$, $y = \cos x$ are shown in Fig. 3.9. The domain of both functions is $D = \mathbb{R}$.

The graphs of the functions $y = \tan x$ and $y = \cot x$ are presented in Fig. 3.10. The domain D for the tangent is, as explained above, given by $D = \{x \in \mathbb{R} \ ; \ x \neq \frac{\pi}{2} + k\pi, \ k \in \mathbb{Z}\}$, the one for the cotangent is $D = \{x \in \mathbb{R} \ ; \ x \neq k\pi, \ k \in \mathbb{Z}\}$.

Many relations are valid between the trigonometric functions. For example, the following addition theorems, which can be proven by elementary geometrical considerations, are valid; see Exercise 3. The maple commands expand and combine use such identities to simplify trigonometric expressions.

Proposition 3.3 (Addition theorems) *For $x, y \in \mathbb{R}$ it holds that*

$$\sin (x + y) = \sin x \cos y + \cos x \sin y,$$
$$\cos (x + y) = \cos x \cos y - \sin x \sin y.$$

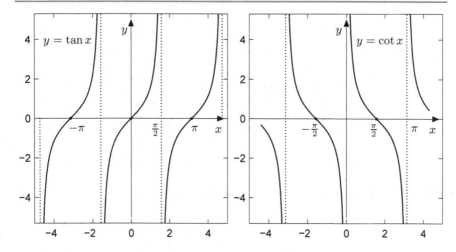

Fig. 3.10 The graphs of the tangent (left) and cotangent (right) functions

3.3 Cyclometric Functions

The cyclometric functions are inverse to the trigonometric functions in the appropriate bijectivity regions.

Sine and arcsine. The sine function is bijective from the interval $[-\frac{\pi}{2}, \frac{\pi}{2}]$ to the range $[-1, 1]$; see Fig. 3.9. This part of the graph is called *principal branch* of the sine. Its inverse function (Fig. 3.11) is called arcsine (or sometimes inverse sine)

$$\arcsin : [-1, 1] \to \left[-\frac{\pi}{2}, \frac{\pi}{2}\right].$$

According to the definition of the inverse function it follows that

$$\sin(\arcsin y) = y \quad \text{for all } y \in [-1, 1].$$

However, the converse formula is only valid for the principal branch; i.e.

$$\arcsin(\sin x) = x \quad \text{is only valid for } -\frac{\pi}{2} \le x \le \frac{\pi}{2}.$$

For example, $\arcsin(\sin 4) = -0.8584073... \neq 4$.

Cosine and arccosine. Likewise, the principal branch of the cosine is defined as restriction of the cosine to the interval $[0, \pi]$ with range $[-1, 1]$. The principal branch is bijective, and its inverse function (Fig. 3.12) is called arccosine (or sometimes inverse cosine)

$$\arccos : [-1, 1] \to [0, \pi].$$

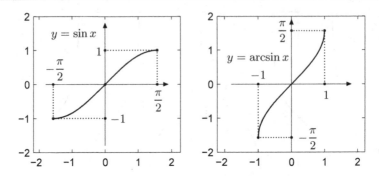

Fig. 3.11 The principal branch of the sine (left); the arcsine function (right)

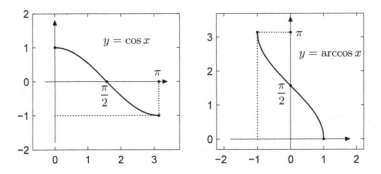

Fig. 3.12 The principal branch of the cosine (left); the arccosine function (right)

Tangent and arctangent. As can be seen in Fig. 3.10 the restriction of the tangent to the interval $(-\frac{\pi}{2}, \frac{\pi}{2})$ is bijective. Its inverse function is called arctangent (or inverse tangent)

$$\arctan : \mathbb{R} \rightarrow \left(-\frac{\pi}{2}, \frac{\pi}{2}\right).$$

To be precise this is again the principal branch of the inverse tangent (Fig. 3.13).

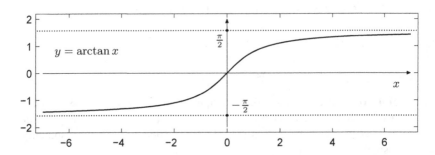

Fig. 3.13 The principal branch of the arctangent

Fig. 3.14 Plane polar coordinates

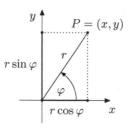

Application 3.4 (Polar coordinates in the plane) The polar coordinates (r, φ) of a point $P = (x, y)$ in the plane are obtained by prescribing its distance r from the origin and the angle φ with the positive x-axis (in counterclockwise direction); see Fig. 3.14.

The connection between Cartesian and polar coordinates is therefore described by

$$x = r \cos \varphi,$$
$$y = r \sin \varphi,$$

where $0 \le \varphi < 2\pi$ and $r \ge 0$. The range $-\pi < \varphi \le \pi$ is also often used.
In the converse direction the following conversion formulas are valid

$$r = \sqrt{x^2 + y^2},$$

$$\varphi = \arctan \frac{y}{x} \quad \text{(in the region } x > 0; \ -\tfrac{\pi}{2} < \varphi < \tfrac{\pi}{2}),$$

$$\varphi = \operatorname{sign} y \cdot \arccos \frac{x}{\sqrt{x^2 + y^2}} \quad \text{(if } y \neq 0 \text{ or } x > 0; \ -\pi < \varphi < \pi).$$

The reader is encouraged to verify these formulas with the help of maple.

3.4 Exercises

1. Using geometric considerations at suitable right-angled triangles, determine the values of the sine, cosine and tangent of the angles $\alpha = 45°$, $\beta = 60°$, $\gamma = 30°$. Extend your result for $\alpha = 45°$ to the angles $135°$, $225°$, $-45°$ with the help of the unit circle. What are the values of the angles under consideration in radian measure?
2. Using MATLAB write a function degrad.m which converts degrees to radian measure. The command degrad(180) should give π as a result. Furthermore, write a function mysin.m which calculates the sine of an angle in radian measure with the help of degrad.m.

Fig. 3.15 Proof of
Proposition 3.3

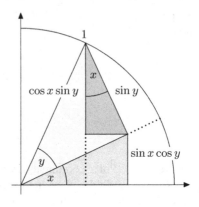

3. Prove the addition theorem of the sine function

$$\sin(x + y) = \sin x \cos y + \cos x \sin y.$$

Hint. If the angles x, y and their sum $x + y$ are between 0 and $\pi/2$ you can directly argue with the help of Fig. 3.15; the remaining cases can be reduced to this case.

4. Prove the *law of cosines*

$$a^2 = b^2 + c^2 - 2bc \cos \alpha$$

for the general triangle in Fig. 3.4.
Hint. The segment c is divided into two segments c_1 (left) and c_2 (right) by the height h. The following identities hold true by Pythagoras' theorem

$$a^2 = h^2 + c_2^2, \qquad b^2 = h^2 + c_1^2, \qquad c = c_1 + c_2.$$

Eliminating h gives $a^2 = b^2 + c^2 - 2cc_1$.

5. Compute the angles α, β, γ of the triangle with sides $a = 3$, $b = 4$, $c = 2$ and plot the triangle in maple.
Hint. Use the law of cosines from Exercise 4.

6. Prove the *law of sines*

$$\frac{a}{\sin \alpha} = \frac{b}{\sin \beta} = \frac{c}{\sin \gamma}$$

for the general triangle in Fig. 3.4.
Hint. The first identity follows from

$$\sin \alpha = \frac{h}{b}, \qquad \sin \beta = \frac{h}{a}.$$

Fig. 3.16 Right circular
truncated cone with unrolled
surface

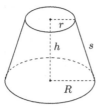
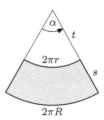

7. Compute the missing sides and angles of the triangle with data $b = 5$, $\alpha = 43°$,
   $\gamma = 62°$, and plot your solutions using MATLAB.
   Hint. Use the law of sines from Exercise 6.
8. With the help of MATLAB plot the following functions

   $$y = \cos(\arccos x), \qquad x \in [-1, 1];$$
   $$y = \arccos(\cos x), \qquad x \in [0, \pi];$$
   $$y = \arccos(\cos x), \qquad x \in [0, 4\pi].$$

   Why is $\arccos(\cos x) \neq x$ in the last case?
9. Plot the functions $y = \sin x$, $y = |\sin x|$, $y = \sin^2 x$, $y = \sin^3 x$, $y = \frac{1}{2}(|\sin x| - \sin x)$ and $y = \arcsin\left(\frac{1}{2}(|\sin x| - \sin x)\right)$ in the interval $[0, 6\pi]$. Explain your
   results.
   Hint. Use the MATLAB command `axis equal`.
10. Plot the graph of the function $f : \mathbb{R} \to \mathbb{R} : x \mapsto ax + \sin x$ for various values
    of a. For which values of a is the function f injective or surjective?
11. Show that the following formulas for the surface line s and the surface area M
    of a right circular truncated cone (see Fig. 3.16, left) hold true

    $$s = \sqrt{h^2 + (R - r)^2}, \qquad M = \pi(r + R)s.$$

    Hint. By unrolling the truncated cone a sector of an annulus with apex angle
    α is created; see Fig. 3.16, right. Therefore, the following relationships hold:
    $\alpha t = 2\pi r$, $\alpha(s + t) = 2\pi R$ and $M = \frac{1}{2}\alpha\left((s + t)^2 - t^2\right)$.
12. The *secant* and *cosecant functions* are defined as the reciprocals of the cosine
    and the sine functions, respectively,

    $$\sec \alpha = \frac{1}{\cos \alpha}, \qquad \csc \alpha = \frac{1}{\sin \alpha}.$$

    Due to the zeros of the cosine and the sine function, the secant is not defined for
    odd multiples of $\frac{\pi}{2}$, and the cosecant is not defined for integer multiples of π.
    (a) Prove the identities $1 + \tan^2 \alpha = \sec^2 \alpha$ and $1 + \cot^2 \alpha = \csc^2 \alpha$.
    (b) With the help of MATLAB plot the graph of the functions $y = \sec x$ and
        $y = \csc x$ for x between -2π and 2π.

Complex Numbers

4

Complex numbers are not just useful when solving polynomial equations but play an important role in many fields of mathematical analysis. With the help of complex functions transformations of the plane can be expressed, solution formulas for differential equations can be obtained, and matrices can be classified. Not least, fractals can be defined by complex iteration processes. In this section we introduce complex numbers and then discuss some elementary complex functions, like the complex exponential function. Applications can be found in Chaps. 9 (fractals), 20 (systems of differential equations) and in Appendix B (normal form of matrices).

4.1 The Notion of Complex Numbers

The set of *complex numbers* $\mathbb{C}$ represents an extension of the real numbers, in which the polynomial $z^2 + 1$ has a root. Complex numbers can be introduced as pairs (a, b) of real numbers for which addition and multiplication are defined as follows:

$$(a, b) + (c, d) = (a + c, b + d),$$
$$(a, b) \cdot (c, d) = (ac - bd, ad + bc).$$

The real numbers are considered as the subset of all pairs of the form $(a, 0), a \in \mathbb{R}$. Squaring the pair $(0, 1)$ shows that

$$(0, 1) \cdot (0, 1) = (-1, 0).$$

The square of $(0, 1)$ thus corresponds to the real number -1. Therefore, $(0, 1)$ provides a root for the polynomial $z^2 + 1$. This root is denoted by i; in other words

$$i^2 = -1.$$

© Springer Nature Switzerland AG 2018
M. Oberguggenberger and A. Ostermann, *Analysis for Computer Scientists*,
Undergraduate Topics in Computer Science,
https://doi.org/10.1007/978-3-319-91155-7_4

Using this notation and rewriting the pairs (a, b) in the form $a + ib$, one obtains a computationally more convenient representation of the set of complex numbers:

$$\mathbb{C} = \{a + ib \; ; \; a \in \mathbb{R}, b \in \mathbb{R}\}.$$

The rules of calculation with pairs (a, b) then simply amount to the common calculations with the expressions $a + ib$ like *with terms* with the additional rule that $i^2 = -1$:

$$(a + ib) + (c + id) = a + c + i(b + d),$$
$$(a + ib)(c + id) = ac + ibc + iad + i^2bd$$
$$= ac - bd + i(ad + bc).$$

So, for example,

$$(2 + 3i)(-1 + i) = -5 - i.$$

Definition 4.1 For the complex number $z = x + iy$,

$$x = \operatorname{Re} z, \quad y = \operatorname{Im} z$$

denote the *real part* and the *imaginary part* of z, respectively. The real number

$$|z| = \sqrt{x^2 + y^2}$$

is the *absolute value* (or modulus) of z, and

$$\bar{z} = x - iy$$

is the *complex conjugate* to z.

A simple calculation shows that

$$z\bar{z} = (x + iy)(x - iy) = x^2 + y^2 = |z|^2,$$

which means that $z\bar{z}$ is always a real number. From this we obtain the rule for calculating with fractions

$$\frac{u + iv}{x + iy} = \left(\frac{u + iv}{x + iy}\right)\left(\frac{x - iy}{x - iy}\right) = \frac{(u + iv)(x - iy)}{x^2 + y^2} = \frac{ux + vy}{x^2 + y^2} + i\frac{vx - uy}{x^2 + y^2}.$$

It is achieved by expansion with the complex conjugate of the denominator. Apparently one can therefore divide by any complex number not equal to zero, and the set $\mathbb{C}$ forms a *field*.

Experiment 4.2 Type in MATLAB: z = complex(2,3) (equivalently z = 2+3*i or z = 2+3*j) as well as w = complex(-1,1) and try out the commands z * w, z/w as well as real(z), imag(z), conj(z), abs(z).

Clearly every negative real x has two square roots in $\mathbb{C}$, namely $i\sqrt{|x|}$ and $-i\sqrt{|x|}$. More than that the *fundamental theorem of algebra* says that $\mathbb{C}$ is *algebraically closed*. Thus every polynomial equation

$$\alpha_n z^n + \alpha_{n-1} z^{n-1} \cdots + \alpha_1 z + \alpha_0 = 0$$

with coefficients $\alpha_j \in \mathbb{C}$, $\alpha_n \neq 0$ has n complex solutions (counted with their multiplicity).

Example 4.3 (Taking the square root of complex numbers)   The equation $z^2 = a + ib$ can be solved by the ansatz

$$(x + iy)^2 = a + ib$$

so

$$x^2 - y^2 = a, \; 2xy = b.$$

If one uses the second equation to express y through x and substitutes this into the first equation, one obtains the *quartic* equation

$$x^4 - ax^2 - b^2/4 = 0.$$

Solving this by substitution $t = x^2$ one obtains the two real solutions. In the case of $b = 0$, either x or y equals zero depending on the sign of a.

The complex plane. A geometric representation of the complex numbers is obtained by identifying $z = x + iy \in \mathbb{C}$ with the point $(x, y) \in \mathbb{R}^2$ in the coordinate plane (Fig. 4.1). Geometrically $|z| = \sqrt{x^2 + y^2}$ is the distance of point (x, y) from the origin; the complex conjugate $\bar{z} = x - iy$ is obtained by reflection in the x-axis.

The *polar representation* of a complex number $z = x + iy$ is obtained like in Application 3.4 by

$$r = |z|, \quad \varphi = \arg z.$$

Fig. 4.1 Complex plane

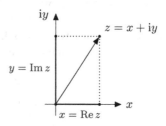

The angle φ to the positive x-axis is called *argument* of the complex number, whereupon the choice of the interval $-\pi < \varphi \leq \pi$ defines the *principal value* $\operatorname{Arg} z$ of the argument. Thus

$$z = x + iy = r(\cos\varphi + i\sin\varphi).$$

The multiplication of two complex numbers $z = r(\cos\varphi + i\sin\varphi)$, $w = s(\cos\psi + i\sin\psi)$ in polar representation corresponds to the product of the absolute values and the sum of the angles:

$$zw = rs\big(\cos(\varphi + \psi) + i\sin(\varphi + \psi)\big),$$

which follows from the addition formulas for sine and cosine:

$$\sin(\varphi + \psi) = \sin\varphi\cos\psi + \cos\varphi\sin\psi,$$
$$\cos(\varphi + \psi) = \cos\varphi\cos\psi - \sin\varphi\sin\psi,$$

see Proposition 3.3.

4.2 The Complex Exponential Function

An important tool for the representation of complex numbers and functions, but also for the real trigonometric functions, is given by the *complex exponential function*. For $z = x + iy$ this function is defined by

$$e^z = e^x(\cos y + i\sin y).$$

The complex exponential function maps $\mathbb{C}$ to $\mathbb{C} \setminus \{0\}$. We will study its mapping behaviour below. It is an *extension* of the real exponential function; i.e. if $z = x \in \mathbb{R}$, then $e^z = e^x$. This is in accordance with the previously defined real-valued exponential function. We also use the notation $\exp(z)$ for e^z.

The addition theorems for sine and cosine imply the usual rules of calculation

$$e^{z+w} = e^z e^w, \quad e^0 = 1, \quad (e^z)^n = e^{nz},$$

valid for $z, w \in \mathbb{C}$ and $n \in \mathbb{Z}$. In contrast to the case when z is a real number, the last rule (for raising to powers) is generally not true, if n is not an integer.

Exponential function and polar coordinates. According to the definition the exponential function of a purely imaginary number $i\varphi$ equals

$$e^{i\varphi} = \cos\varphi + i\sin\varphi,$$
$$|e^{i\varphi}| = \sqrt{\cos^2\varphi + \sin^2\varphi} = 1.$$

Fig. 4.2 The unit circle in the complex plane

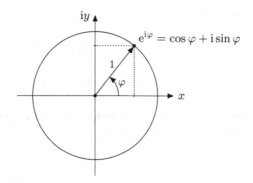

Thus the complex numbers

$$\{e^{i\varphi} \; ; \; -\pi < \varphi \le \pi\}$$

lie on the unit circle (Fig. 4.2).

For example, the following identities hold:

$$e^{i\pi/2} = i, \quad e^{i\pi} = -1, \quad e^{2i\pi} = 1, \quad e^{2ki\pi} = 1 \; (k \in \mathbb{Z}).$$

Using $r = |z|$, $\varphi = \text{Arg}\, z$ results in the especially simple form of the polar representation

$$z = r e^{i\varphi}.$$

Taking roots is accordingly simple.

Example 4.4 (Taking square roots in complex polar coordinates) If $z^2 = r e^{i\varphi}$, then one obtains the two solutions $\pm\sqrt{r}\, e^{i\varphi/2}$ for z. For example, the problem

$$z^2 = 2i = 2\, e^{i\pi/2}$$

has the two solutions

$$z = \sqrt{2}\, e^{i\pi/4} = 1 + i$$

and

$$z = -\sqrt{2}\, e^{i\pi/4} = -1 - i.$$

Euler's formulas. By addition and subtraction, respectively, of the relations

$$e^{i\varphi} = \cos\varphi + i\sin\varphi,$$
$$e^{-i\varphi} = \cos\varphi - i\sin\varphi,$$

one obtains at once Euler's formulas

$$\cos \varphi = \frac{1}{2}\left(e^{i\varphi} + e^{-i\varphi}\right),$$

$$\sin \varphi = \frac{1}{2i}\left(e^{i\varphi} - e^{-i\varphi}\right).$$

They permit a representation of the real trigonometric functions by means of the complex exponential function.

4.3 Mapping Properties of Complex Functions

In this section we study the mapping properties of complex functions. More precisely, we ask how their effect can be described geometrically. Let

$$f : D \subset \mathbb{C} \to \mathbb{C} : z \mapsto w = f(z)$$

be a complex function, defined on a subset D of the complex plane. The effect of the function f can best be visualised by plotting two complex planes next to each other, the z-plane and the w-plane, and studying the images of rays and circles under f.

Example 4.5 The complex quadratic function maps $D = \mathbb{C}$ to $\mathbb{C} : w = z^2$. Using polar coordinates one obtains

$$z = x + iy = r\,e^{i\varphi} \quad \Rightarrow \quad w = u + iv = r^2 e^{2i\varphi}.$$

From this representation it can be seen that the complex quadratic function maps a circle of radius r in the z-plane onto a circle of radius r^2 in the w-plane. Further, it maps half-rays

$$\{z = re^{i\psi} : r > 0\}$$

with the angle of inclination ψ onto half-rays with angle of inclination 2ψ (Fig. 4.3).

Particularly important are the mapping properties of the complex exponential function $w = e^z$ because they form the basis for the definition of the complex logarithm and the root functions. If $z = x + iy$ then $e^z = e^x(\cos y + i \sin y)$. We always have that $e^x > 0$; furthermore $\cos y + i \sin y$ defines a point on the complex unit circle which is unique for $-\pi < y \le \pi$. If x moves along the real line then the points $e^x(\cos y + i \sin y)$ form a half-ray with angle y, as can be seen in Fig. 4.4. Conversely, if x is fixed and y varies between $-\pi$ and π one obtains the circle with radius e^x in the w-plane. For example, the dotted circle (Fig. 4.4, right) is the image of the dotted straight line (Fig. 4.4, left) under the exponential function.

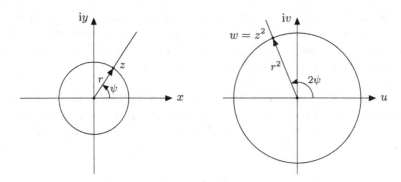

Fig. 4.3 The complex quadratic function

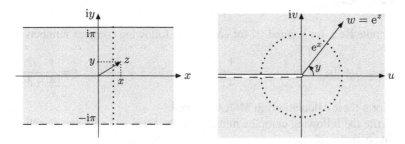

Fig. 4.4 The complex exponential function

From what has just been said it follows that the exponential function is bijective on the domain

$$D = \{z = x + iy \; ; \; x \in \mathbb{R}, -\pi < y \le \pi\} \to B = \mathbb{C} \setminus \{0\}.$$

It thus maps the strip of width 2π onto the complex plane without zero. The argument of e^z exhibits a jump along the negative u-axis as indicated in Fig. 4.4 (right). Within the domain D the exponential function has an inverse function, the *principal branch* of the complex logarithm. From the representation $w = e^z = e^x e^{iy}$ one derives at once the relation $x = \log |w|$, $y = \text{Arg } w$. Thus the principal value of the complex logarithm of the complex number w is given by

$$z = \text{Log } w = \log |w| + i \, \text{Arg } w$$

and in polar coordinates

$$\text{Log} \left(r \, e^{i\varphi} \right) = \log r + i\varphi, \quad -\pi < \varphi \le \pi,$$

respectively.

With the help of the principal value of the complex logarithm, the principal values of the nth complex root function can be defined by $\sqrt[n]{z} = \exp\left(\frac{1}{n} \text{Log}(z)\right).$

Experiment 4.6 Open the applet *2D visualisation of complex functions* and investigate how the power functions $w = z^n$, $n \in \mathbb{N}$, map circles and rays of the complex plane. Set the pattern *polar coordinates* and experiment with different sectors (intervals of the argument $[\alpha, \beta]$ with $0 \leq \alpha < \beta \leq 2\pi$).

Experiment 4.7 Open the applet *2D visualisation of complex functions* and investigate how the exponential function $w = e^z$ maps horizontal and vertical straight lines of the complex plane. Set the pattern *grid* and experiment with different strips, for example $1 \leq \operatorname{Re} z \leq 2, -2 \leq \operatorname{Im} z \leq 2$.

4.4 Exercises

1. Compute $\operatorname{Re} z$, $\operatorname{Im} z$, $\bar{z}$ and $|z|$ for each of the following complex numbers z:

$$z = 3 + 2i, \quad z = -i, \quad z = \frac{1+i}{2-i}, \quad z = 3 - i + \frac{1}{3-i}, \quad z = \frac{1-2i}{4-3i}.$$

   Perform these calculations in MATLAB as well.

2. Rewrite the following complex numbers in the form $z = re^{i\varphi}$ and sketch them in the complex plane:

$$z = -1 - i, \quad z = -5, \quad z = 3i, \quad z = 2 - 2i, \quad z = 1 - i\sqrt{3}.$$

   What are the values of φ in radian measure?

3. Compute the two complex solutions of the equation

$$z^2 = 2 + 2i$$

   with the help of the ansatz $z = x + iy$ and equating the real and the imaginary parts. Test and explain the MATLAB commands

   ```
   roots([2,0,-2-2*i])
   sqrt(2+2*i)
   ```

4. Compute the two complex solutions of the equation

$$z^2 = 2 + 2i$$

   in the form $z = re^{i\varphi}$ from the polar representation of $2 + 2i$.

5. Compute the four complex solutions of the quartic equation

$$z^4 - 2z^2 + 2 = 0$$

   by hand and with MATLAB (command `roots`).

6. Let $z = x + iy$, $w = u + iv$. Check the formula $e^{z+w} = e^z e^w$ by using the definition and applying the addition theorems for the trigonometric functions.

7. Compute $z = \text{Log } w$ for

$$w = 1 + i, \quad w = -5i, \quad w = -1.$$

Sketch w and z in the complex plane and verify your results with the help of the relation $w = e^z$ and with MATLAB (command `log`).

8. The complex sine and cosine functions are defined by

$$\sin z = \frac{1}{2i}\left(e^{iz} - e^{-iz}\right), \quad \cos z = \frac{1}{2}\left(e^{iz} + e^{-iz}\right)$$

for $z \in \mathbb{C}$.

(a) Show that both functions are periodic with period 2π, that is $\sin(z + 2\pi) = \sin z$, $\cos(z + 2\pi) = \cos z$.

(b) Verify that, for $z = x + iy$,
$\sin z = \sin x \cosh y + i \cos x \sinh y$, $\quad \cos z = \cos x \cosh y - i \sin x \sinh y$.

(c) Show that $\sin z = 0$ if and only if $z = k\pi$, $k \in \mathbb{Z}$, and $\cos z = 0$ if and only if $z = (k + \frac{1}{2})\pi$, $k \in \mathbb{Z}$.

Sequences and Series

5

The concept of a limiting process *at infinity* is one of the central ideas of mathematical analysis. It forms the basis for all its essential concepts, like continuity, differentiability, series expansions of functions, integration, etc. The transition from a *discrete* to a *continuous* setting constitutes the modelling strength of mathematical analysis. Discrete models of physical, technical or economic processes can often be better and more easily understood, provided that the number of their *atoms*— their discrete building blocks—is sufficiently big, if they are approximated by a continuous model with the help of a limiting process. The transition from difference equations for biological growth processes in discrete time to differential equations in continuous time are examples for that, as is the description of share prices by stochastic processes in continuous time. The majority of models in physics are *field models*, that is, they are expressed in a continuous space and time structure. Even though the models are *discretised* again in numerical approximations, the continuous model is still helpful as a background, for example for the derivation of error estimates.

The following sections are dedicated to the specification of the idea of limiting processes. This chapter starts by studying infinite sequences and series, gives some applications and covers the corresponding notion of a limit. One of the achievements which we especially emphasise is the completeness of the real numbers. It guarantees the existence of limits for arbitrary monotonically increasing bounded sequences of numbers, the existence of zeros of continuous functions, of maxima and minima of differentiable functions, of integrals, etc. It is an indispensable building block of mathematical analysis.

5.1 The Notion of an Infinite Sequence

Definition 5.1 Let X be a set. An *(infinite) sequence with values in X* is a mapping from $\mathbb{N}$ to X.

© Springer Nature Switzerland AG 2018
M. Oberguggenberger and A. Ostermann, *Analysis for Computer Scientists*,
Undergraduate Topics in Computer Science,
https://doi.org/10.1007/978-3-319-91155-7_5

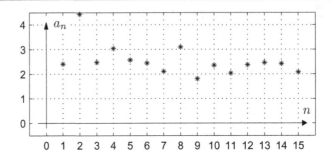

Fig. 5.1 Graph of a sequence

Thus each natural number n *(the index)* is mapped to an element a_n of X *(the nth term of the sequence)*. We express this by using the notation

$$(a_n)_{n \geq 1} = (a_1, a_2, a_3, \ldots).$$

In the case of $X = \mathbb{R}$ one speaks of *real-valued* sequences, if $X = \mathbb{C}$ of *complex-valued* sequences, if $X = \mathbb{R}^m$ of *vector-valued* sequences. In this section we only discuss real-valued sequences.

Sequences can be added

$$(a_n)_{n \geq 1} + (b_n)_{n \geq 1} = (a_n + b_n)_{n \geq 1}$$

and multiplied by a scalar factor

$$\lambda(a_n)_{n \geq 1} = (\lambda a_n)_{n \geq 1}.$$

These operations are performed componentwise and endow the set of all real-valued sequences with the structure of a vector space. The *graph of a sequence* is visualised by plotting the points $(n, a_n), n = 1, 2, 3, \ldots$ in a coordinate system, see Fig. 5.1.

Experiment 5.2 The M-file `mat05_1a.m` offers the possibility to study various examples of sequences which are increasing/decreasing, bounded/unbounded, oscillating, convergent. For a better visualisation the discrete points of the graph of the sequence are often connected by line segments (exclusively for graphical purpose)—this is implemented in the M-file `mat05_1b.m`. Open the applet *Sequences* and use it to illustrate the sequences given in the M-file `mat05_1a.m`.

Sequences can either be defined *explicitly* by a formula, for instance

$$a_n = 2^n,$$

or *recursively* by giving a starting value and a rule how to calculate a term from the preceding one,

$$a_1 = 1, \quad a_{n+1} = 2a_n.$$

The recursion can also involve several previous terms at a time.

Example 5.3 A discrete population model which goes back to Verhulst[1] (limited growth) describes the population x_n at the point in time n (using time intervals of length 1) by the recursive relation

$$x_{n+1} = x_n + \beta x_n (L - x_n).$$

Here β is a growth factor and L the limiting population, i.e. the population which is not exceeded in the long-term (short-term overruns are possible, however, lead to immediate decay of the population). Additionally one has to prescribe the initial population $x_1 = A$. According to the model the population increase $x_{n+1} - x_n$ during one time interval is proportional to the existing population and to the difference to the population limit. The M-file mat05_2.m contains a MATLAB function, called as

$$x = \text{mat05\_2(A,beta,N)}$$

which computes and plots the first N terms of the sequence $x = (x_1, \ldots, x_N)$. The initial value is A, the growth rate β; L was set to $L = 1$. Experiments with $A = 0.1$, $N = 50$ and $\beta = 0.5$, $\beta = 1$, $\beta = 2$, $\beta = 2.5$, $\beta = 3$ show convergent, oscillating and chaotic behaviour of the sequence, respectively.

Below we develop some concepts which help to describe the behaviour of sequences.

Definition 5.4 A sequence $(a_n)_{n \geq 1}$ is called *monotonically increasing*, if

$$n \leq m \quad \Rightarrow \quad a_n \leq a_m;$$

$(a_n)_{n \geq 1}$ is called *monotonically decreasing*, if

$$n \leq m \quad \Rightarrow \quad a_n \geq a_m;$$

$(a_n)_{n \geq 1}$ is called *bounded from above*, if

$$\exists T \in \mathbb{R} \ \forall n \in \mathbb{N} : a_n \leq T.$$

We will show in Proposition 5.13 below that the set of upper bounds of a bounded sequence has a smallest element. This least upper bound T_0 is called the *supremum* of the sequence and denoted by

$$T_0 = \sup_{n \in \mathbb{N}} a_n.$$

---

[1]P.-F. Verhulst, 1804–1849.

The supremum is characterised by the following two conditions:

(a) $a_n \leq T_0$ for all $n \in \mathbb{N}$;

(b) if T is a real number and $a_n \leq T$ for all $n \in \mathbb{N}$, then $T \geq T_0$.

Note that the supremum itself does not have to be a term of the sequence. However, if this is the case, it is called *maximum* of the sequence and denoted by

$$T_0 = \max_{n \in \mathbb{N}} a_n.$$

A sequence has a maximum T_0 if the following two conditions are fulfilled:

(a) $a_n \leq T_0$ for all $n \in \mathbb{N}$;

(b) there exists at least one $m \in \mathbb{N}$ such that $a_m = T_0$.

In the same way, a sequence $(a_n)_{n \geq 1}$ is called *bounded from below*, if

$$\exists S \in \mathbb{R} \ \forall n \in \mathbb{N} : S \leq a_n.$$

The greatest lower bound is called *infimum* (or *minimum*, if it is attained by a term of the sequence).

Experiment 5.5 Investigate the sequences produced by the M-file `mat05_1a.m` with regard to the concepts developed above.

As mentioned in the introduction to this chapter, the concept of *convergence* is a central concept of mathematical analysis. Intuitively it states that the terms of the sequence $(a_n)_{n \geq 1}$ approach a *limit* a with growing index n. For example, in Fig. 5.2 with $a = 0.8$ one has

$$|a - a_n| < 0.2 \ \text{from} \ n = 6, \quad |a - a_n| < 0.05 \ \text{from} \ n = 21.$$

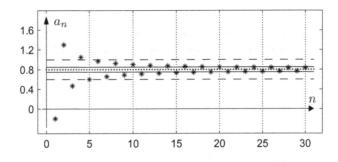

Fig. 5.2 Convergence of a sequence

For a precise definition of the concept of convergence we first introduce the notion of an ε-*neighbourhood* of a point $a \in \mathbb{R}$ ($\varepsilon > 0$):

$$U_\varepsilon(a) = \{x \in \mathbb{R} \; ; \; |a - x| < \varepsilon\} = (a - \varepsilon, a + \varepsilon).$$

We say that a sequence $(a_n)_{n \geq 1}$ *settles* in a neighbourhood $U_\varepsilon(a)$, if from a certain index $n(\varepsilon)$ on all subsequent terms a_n of the sequence lie in $U_\varepsilon(a)$.

Definition 5.6 The sequence $(a_n)_{n \geq 1}$ *converges to a limit* a if it settles in each ε-neighbourhood of a.

These facts can be expressed in quantifier notation as follows:

$$\forall \varepsilon > 0 \; \exists n(\varepsilon) \in \mathbb{N} \; \forall n \geq n(\varepsilon) \; : \; |a - a_n| < \varepsilon.$$

If a sequence $(a_n)_{n \geq 1}$ converges to a limit a, one writes

$$a = \lim_{n \to \infty} a_n \quad \text{or} \quad a_n \to a \text{ as } n \to \infty.$$

In the example of Fig. 5.2 the limit a is indicated as a dotted line, the neighbourhood $U_{0.2}(a)$ as a strip with a dashed boundary line and the neighbourhood $U_{0.05}(a)$ as a strip with a solid boundary line.

In the case of convergence the limit can be interchanged with addition, multiplication and division (with the exception of zero), as expected.

Proposition 5.7 (Rules of calculation for limits) *If the sequences $(a_n)_{n \geq 1}$ and $(b_n)_{n \geq 1}$ are convergent then the following rules hold:*

$$\lim_{n \to \infty} (a_n + b_n) = \lim_{n \to \infty} a_n + \lim_{n \to \infty} b_n$$
$$\lim_{n \to \infty} (\lambda a_n) = \lambda \lim_{n \to \infty} a_n \quad (\textit{for } \lambda \in \mathbb{R})$$
$$\lim_{n \to \infty} (a_n b_n) = (\lim_{n \to \infty} a_n)(\lim_{n \to \infty} b_n)$$
$$\lim_{n \to \infty} (a_n/b_n) = (\lim_{n \to \infty} a_n)/(\lim_{n \to \infty} b_n) \quad (\textit{if } \lim_{n \to \infty} b_n \neq 0)$$

Proof The verification of these trivialities is left to the reader as an exercise. The proofs are not deep, but one has to carefully pick the right approach in order to verify the conditions of Definition 5.6. In order to illustrate at least once how such proofs are done, we will show the statement about multiplication. Assume that

$$\lim_{n \to \infty} a_n = a \quad \text{and} \quad \lim_{n \to \infty} b_n = b.$$

Let $\varepsilon > 0$. According to Definition 5.6 we have to find an index $n(\varepsilon) \in \mathbb{N}$ satisfying

$$|ab - a_n b_n| < \varepsilon$$

for all $n \geq n(\varepsilon)$. Due to the convergence of the sequence $(a_n)_{n \geq 1}$ we can first find an $n_1(\varepsilon) \in \mathbb{N}$ so that $|a - a_n| \leq 1$ for all $n \geq n_1(\varepsilon)$. For these n it also applies that

$$|a_n| = |a_n - a + a| \leq 1 + |a|.$$

Furthermore, we can find $n_2(\varepsilon) \in \mathbb{N}$ and $n_3(\varepsilon) \in \mathbb{N}$ which guarantee that

$$|a - a_n| < \frac{\varepsilon}{2 \max(|b|, 1)} \quad \text{and} \quad |b - b_n| < \frac{\varepsilon}{2(1 + |a|)}$$

for all $n \geq n_2(\varepsilon)$ and $n \geq n_3(\varepsilon)$, respectively. It thus follows that

$$|ab - a_n b_n| = |(a - a_n)b + a_n(b - b_n)| \leq |a - a_n||b| + |a_n||b - b_n|$$
$$\leq |a - a_n||b| + (|a| + 1)|b - b_n| \leq \frac{\varepsilon}{2} + \frac{\varepsilon}{2} \leq \varepsilon$$

for all $n \geq n(\varepsilon)$ with $n(\varepsilon) = \max(n_1(\varepsilon), n_2(\varepsilon), n_3(\varepsilon))$. This is the statement that was to be proven. $\square$

The important ideas of the proof were: Splitting in two summands with the help of the triangle inequality (see Exercise 2 of Chap. 1); bounding $|a_n|$ by $1 + |a|$ using the assumed convergence; upper bounds for the terms $|a - a_n|$ and $|b - b_n|$ by fractions of ε (again possible due to the convergence) so that the summands together stay less than ε. All elementary proofs of convergence in mathematical analysis proceed in a similar way.

Real-valued sequences with terms that increase to infinity with growing index n have no limit in the sense of the definition given above. However, it is practical to assign them the symbol ∞ as an *improper limit*.

Definition 5.8 A sequence $(a_n)_{n \geq 1}$ has the *improper limit* ∞ if it has the property of unlimited increase

$$\forall T \in \mathbb{R} \ \exists n(T) \in \mathbb{N} \ \forall n \geq n(T) : a_n \geq T.$$

In this case one writes

$$\lim_{n \to \infty} a_n = \infty.$$

In the same way one defines

$$\lim_{n \to \infty} b_n = -\infty, \quad \text{if} \quad \lim_{n \to \infty} (-b_n) = \infty.$$

Example 5.9 We consider the geometric sequence $(q^n)_{n \geq 1}$. It obviously holds that

$$\lim_{n \to \infty} q^n = 0, \quad \text{if } |q| < 1,$$

$$\lim_{n \to \infty} q^n = \infty, \quad \text{if } q > 1,$$

$$\lim_{n \to \infty} q^n = 1, \quad \text{if } q = 1.$$

For $q \leq -1$ the sequence has no limit (neither proper nor improper).

5.2 The Completeness of the Set of Real Numbers

As remarked in the introduction to this chapter, the completeness of the set of real numbers is one of the pillars of real analysis. The property of completeness can be expressed in different ways. We will use a simple formulation which is particularly helpful in many applications.

Proposition 5.10 (Completeness of the set of real numbers) *Each monotonically increasing sequence of real numbers that is bounded from above has a limit (in $\mathbb{R}$).*

Proof Let $(a_n)_{n \geq 1}$ be a monotonically increasing, bounded sequence. First we prove the theorem in the case that all terms a_n are non-negative. We write the terms as decimal numbers

$$a_n = A^{(n)}.\alpha_1^{(n)}\alpha_2^{(n)}\alpha_3^{(n)} \ldots$$

with $A^{(n)} \in \mathbb{N}_0$, $\alpha_j^{(n)} \in \{0, 1, \ldots, 9\}$. By assumption there is a bound $T \geq 0$ so that $a_n \leq T$ for all n. Therefore, also $A^{(n)} \leq T$ for all n. But the sequence $(A^{(n)})_{n \geq 1}$ is a monotonically increasing, bounded sequence of integers and therefore must eventually reach its least upper bound A (and stay there). In other words, there exists $n_0 \in \mathbb{N}$ such that

$$A^{(n)} = A \quad \text{for all } n \geq n_0.$$

Thus we have found the integer part of the limit a to be constructed:

$$a = A. \ldots$$

Let now $\alpha_1 \in \{0, \ldots, 9\}$ be the least upper bound for $\alpha_1^{(n)}$. As the sequence is monotonically increasing there is again an $n_1 \in \mathbb{N}$ with

$$\alpha_1^{(n)} = \alpha_1 \quad \text{for all } n \geq n_1$$

and consequently

$$a = A.\alpha_1 \ldots$$

Let now $\alpha_2 \in \{0, \ldots, 9\}$ be the least upper bound for $\alpha_2^{(n)}$. There is an $n_2 \in \mathbb{N}$ with

$$\alpha_2^{(n)} = \alpha_2 \quad \text{for all } n \geq n_2$$

and consequently

$$a = A.\alpha_1\alpha_2 \ldots$$

Successively one defines a real number

$$a = A.\alpha_1\alpha_2\alpha_3\alpha_4 \ldots$$

in that way. It remains to show that $a = \lim_{n \to \infty} a_n$. Let $\varepsilon > 0$. We first choose $j \in \mathbb{N}$ so that $10^{-j} < \varepsilon$. For $n \geq n_j$

$$a - a_n = 0.000\ldots0\,\alpha_{j+1}^{(n)}\,\alpha_{j+2}^{(n)} \ldots,$$

since the first j digits after the decimal point in a coincide with those of a_n provided $n \geq n_j$. Therefore,

$$|a - a_n| \leq 10^{-j} < \varepsilon \quad \text{for} \quad n \geq n_j.$$

With $n(\varepsilon) = n_j$ the condition required in Definition 5.6 is fulfilled.

If the sequence $(a_n)_{n \geq 1}$ also has negative terms, it can be transformed to a sequence with non-negative terms by adding the absolute value of the first term which results in the sequence $(|a_1| + a_n)_{n \geq 1}$. Using the obvious rule $\lim(c + a_n) = c + \lim a_n$ allows one to apply the first part of the proof.   □

Remark 5.11 The set of rational numbers is not complete. For example, the decimal expansion of $\sqrt{2}$,

$$(1, 1.4, 1.41, 1.414, 1.4142, \ldots)$$

is a monotonically increasing, bounded sequence of rational numbers (an upper bound is, e.g. $T = 1.5$, since $1.5^2 > 2$), but the limit $\sqrt{2}$ does not belong to $\mathbb{Q}$ (as it is an irrational number).

Example 5.12 (Arithmetic of real numbers) Due to Proposition 5.10 the arithmetical operations on the real numbers introduced in Sect. 1.2 can be legitimised a posteriori. Let us look, for instance, at the addition of two non-negative real numbers $a = A.\alpha_1\alpha_2 \ldots$ and $b = B.\beta_1\beta_2 \ldots$ with $A, B \in \mathbb{N}_0$, $\alpha_j, \beta_j \in \{0, 1, \ldots, 9\}$. By truncating them after the nth decimal place we obtain two approximating sequences of rational numbers $a_n = A.\alpha_1\alpha_2\ldots\alpha_n$ and $b_n = B.\beta_1\beta_2\ldots\beta_n$ with

$$a = \lim_{n\to\infty} a_n, \quad b = \lim_{n\to\infty} b_n.$$

The sum of two approximations $a_n + b_n$ is defined by the addition of rational numbers in an elementary way. The sequence $(a_n + b_n)_{n\geq 1}$ is evidently monotonically increasing and bounded from above, for instance, by $A + B + 2$. According to Proposition 5.10 this sequence has a limit and this limit *defines* the sum of the real numbers

$$a + b = \lim_{n\to\infty} (a_n + b_n).$$

In this way the addition of real numbers is rigorously justified. In a similar way one can proceed with multiplication. Finally, Proposition 5.7 allows one to prove the usual rules for addition and multiplication.

Consider a sequence with upper bound T. Each real number $T_1 > T$ is also an upper bound. We can now show that there always exists a smallest upper bound. A bounded sequence thus actually has a supremum as claimed earlier.

Proposition 5.13 *Each sequence $(a_n)_{n\geq 1}$ of real numbers which is bounded from above has a supremum.*

Proof Let $T_n = \max\{a_1,\ldots,a_n\}$ be the maximum of the first n terms of the sequence. These maxima on their part define a sequence $(T_n)_{n\geq 1}$ which is bounded from above by the same bounds as $(a_n)_{n\geq 1}$ but is additionally monotonically increasing. According to the previous proposition it has a limit T_0. We are going to show that this limit is the supremum of the original sequence. Indeed, as $T_n \leq T_0$ for all n, we have $a_n \leq T_0$ for all n as well. Assume that the sequence $(a_n)_{n\geq 1}$ had a smaller upper bound $T < T_0$, i.e. $a_n \leq T$ for all n. This in turn implies $T_n \leq T$ for all n and contradicts the fact that $T_0 = \lim T_n$. Therefore, T_0 is the least upper bound. $\square$

Application 5.14 We are now in a position to show that the construction of the exponential function for real exponents given informally in Sect. 2.2 is justified. Let $a > 0$ be a basis for the power a^r to be defined with real exponent $r \in \mathbb{R}$. It is sufficient to treat the case $r > 0$ (for negative r, the expression a^r is defined by the reciprocal of $a^{|r|}$). We write r as the limit of a monotonically increasing sequence $(r_n)_{n\geq 1}$ of rational numbers by choosing for r_n the decimal representation of r, truncated at the nth digit. The rules of calculation for rational exponents imply the inequality $a^{r_{n+1}} - a^{r_n} = a^{r_n}\left(a^{r_{n+1}-r_n} - 1\right) \geq 0$. This shows that the sequence $(a^{r_n})_{n\geq 1}$ is monotonically increasing. It is also bounded from above, for instance, by a^q, if q is a rational number bigger than r. According to Proposition 5.10 this sequence has a limit. It defines a^r.

Application 5.15 Let $a > 0$. Then $\lim_{n\to\infty} \sqrt[n]{a} = 1$.

In the proof we can restrict ourselves to the case $0 < a < 1$ since otherwise the argument can be used for $1/a$. One can easily see that the sequence $(\sqrt[n]{a})_{n\geq 1}$ is monotonically increasing; it is also bounded from above by 1. Therefore, it has a limit b. Suppose that $b < 1$. From $\sqrt[n]{a} \leq b$ we infer that $a \leq b^n \to 0$ for $n \to \infty$, which contradicts the assumption $a > 0$. Consequently $b = 1$.

5.3 Infinite Series

Sums of the form

$$\sum_{k=1}^{\infty} a_k = a_1 + a_2 + a_3 + \cdots$$

with infinitely many summands can be given a meaning under certain conditions. The starting point of our considerations is a sequence of coefficients $(a_k)_{k\geq 1}$ of real numbers. The nth *partial sum* is defined as

$$S_n = \sum_{k=1}^{n} a_k = a_1 + a_2 + \cdots + a_n,$$

thus

$$S_1 = a_1,$$
$$S_2 = a_1 + a_2,$$
$$S_3 = a_1 + a_2 + a_3, \quad \text{etc.}$$

As needed we also use the notation $S_n = \sum_{k=0}^{n} a_k$ without further comment if the sequence $a_0, a_1, a_2, a_3, \ldots$ starts with the index $k = 0$.

Definition 5.16 The sequence of the partial sums $(S_n)_{n\geq 1}$ is called a *series*. If the limit $S = \lim_{n\to\infty} S_n$ exists, then the series is called *convergent*, otherwise *divergent*.

In the case of convergence one writes

$$S = \sum_{k=1}^{\infty} a_k = \lim_{n\to\infty} \left(\sum_{k=1}^{n} a_k \right).$$

In this way the summation problem is reduced to the question of convergence of the sequence of the partial sums.

Experiment 5.17 The M-file `mat05_3.m`, when called as `mat05_3(N,Z)`, generates the first N partial sums with time delay Z [seconds] of five series, i.e. it computes S_n for $1 \leq n \leq N$ in each case:

$$\text{Series 1}: \quad S_n = \sum_{k=1}^{n} k^{-0.99} \qquad\qquad \text{Series 2}: \quad S_n = \sum_{k=1}^{n} k^{-1}$$

$$\text{Series 3}: \quad S_n = \sum_{k=1}^{n} k^{-1.01} \qquad\qquad \text{Series 4}: \quad S_n = \sum_{k=1}^{n} k^{-2}$$

$$\text{Series 5}: \quad S_n = \sum_{k=1}^{n} \frac{1}{k!}$$

Experiment with increasing values of N and try to see which series shows convergence or divergence.

In the experiment the convergence of Series 5 seems obvious, while the observations for the other series are rather not as conclusive. Actually, Series 1 and 2 are divergent while the others are convergent. This shows the need for analytical tools in order to be able to decide the question of convergence. However, we first look at a few examples.

Example 5.18 (Geometric series)   In this example we are concerned with the series $\sum_{k=0}^{\infty} q^k$ with real factor $q \in \mathbb{R}$. For the partial sums we deduce that

$$S_n = \sum_{k=0}^{n} q^k = \frac{1 - q^{n+1}}{1 - q}.$$

Indeed, by subtraction of the two lines

$$\begin{aligned} S_n &= 1 + q + q^2 + \cdots + q^n, \\ qS_n &= \quad\ \ q + q^2 + q^3 + \cdots + q^{n+1} \end{aligned}$$

one obtains the formula $(1 - q)S_n = 1 - q^{n+1}$ from which the result follows.
The case $|q| < 1$: As $q^{n+1} \to 0$ the series converges with value

$$S = \lim_{n \to \infty} \frac{1 - q^{n+1}}{1 - q} = \frac{1}{1 - q}.$$

The case $|q| > 1$: For $q > 1$ the partial sum $S_n = (q^{n+1} - 1)/(q - 1) \to \infty$ and the series diverges. In the case of $q < -1$ the partial sums $S_n = (1 - (-1)^{n+1}|q|^{n+1})/(1 - q)$ are unbounded and oscillate. They thus diverge as well.

The case $|q| = 1$: For $q = 1$ we have $S_n = 1 + 1 + \cdots + 1 = n + 1$ which tends to infinity; for $q = -1$, the partial sums S_n oscillate between 1 and 0. In both cases the series diverges.

Example 5.19 The nth partial sum of the series $\sum_{k=1}^{\infty} \frac{1}{k(k+1)}$ is

$$
\begin{aligned}
S_n &= \sum_{k=1}^{n} \frac{1}{k(k+1)} = \sum_{k=1}^{n} \left(\frac{1}{k} - \frac{1}{k+1} \right) \\
&= 1 - \frac{1}{2} + \frac{1}{2} - \frac{1}{3} + \frac{1}{3} - \frac{1}{4} + \cdots - \frac{1}{n} + \frac{1}{n} - \frac{1}{n+1} = 1 - \frac{1}{n+1}.
\end{aligned}
$$

It is called a *telescopic sum*. The series converges to

$$
S = \sum_{k=1}^{\infty} \frac{1}{k(k+1)} = \lim_{n \to \infty} \left(1 - \frac{1}{n+1} \right) = 1.
$$

Example 5.20 (Harmonic series) We consider the series $\sum_{k=1}^{\infty} \frac{1}{k}$. By combining blocks of two, four, eight, sixteen, etc., elements, one obtains the grouping

$$
\begin{aligned}
& 1 + \tfrac{1}{2} + \left(\tfrac{1}{3} + \tfrac{1}{4} \right) + \left(\tfrac{1}{5} + \tfrac{1}{6} + \tfrac{1}{7} + \tfrac{1}{8} \right) + \left(\tfrac{1}{9} + \cdots + \tfrac{1}{16} \right) + \left(\tfrac{1}{17} + \cdots \right) + \cdots \\
\geq \; & 1 + \tfrac{1}{2} + \left(\tfrac{1}{4} + \tfrac{1}{4} \right) + \left(\tfrac{1}{8} + \tfrac{1}{8} + \tfrac{1}{8} + \tfrac{1}{8} \right) + \left(\tfrac{1}{16} + \cdots + \tfrac{1}{16} \right) + \left(\tfrac{1}{32} + \cdots \right) + \cdots \\
= \; & 1 + \tfrac{1}{2} + \tfrac{1}{2} + \tfrac{1}{2} + \tfrac{1}{2} + \tfrac{1}{2} + \cdots \to \infty.
\end{aligned}
$$

The partial sums tend to infinity, therefore, the series diverges.

There are a number of criteria which allow one to decide whether a series converges or diverges. Here we only discuss two simple comparison criteria, which suffice for our purpose. For further considerations we refer to the literature, for instance [3, Chap. 9.2].

Proposition 5.21 (Comparison criteria) *Let $0 \leq a_k \leq b_k$ for all $k \in \mathbb{N}$ or at least for all k greater than or equal to a certain k_0. Then we have:*

(a) If the series $\sum_{k=1}^{\infty} b_k$ is convergent then the series $\sum_{k=1}^{\infty} a_k$ converges, too.

(b) If the series $\sum_{k=1}^{\infty} a_k$ is divergent then the series $\sum_{k=1}^{\infty} b_k$ diverges, too.

Proof (a) The partial sums fulfill $S_n = \sum_{k=1}^{n} a_k \leq \sum_{k=1}^{\infty} b_k = T$ and $S_n \leq S_{n+1}$, hence are bounded and monotonically increasing. According to Proposition 5.10 the limit of the partial sums exists.

(b) This time, we have for the partial sums

$$
T_n = \sum_{k=1}^{n} b_k \geq \sum_{k=1}^{n} a_k \to \infty,
$$

since the latter are positive and divergent. $\square$

Under the condition $0 \leq a_k \leq b_k$ of the proposition one says that $\sum_{k=1}^{\infty} b_k$ dominates $\sum_{k=1}^{\infty} a_k$. A series thus converges if it is dominated by a convergent series; it diverges if it dominates a divergent series.

Example 5.22 The series $\sum_{k=1}^{\infty} \frac{1}{k^2}$ is convergent. For the proof we use that

$$\sum_{k=1}^{n} \frac{1}{k^2} = 1 + \sum_{j=1}^{n-1} \frac{1}{(j+1)^2} \quad \text{and} \quad a_j = \frac{1}{(j+1)^2} \leq \frac{1}{j(j+1)} = b_j.$$

Example 5.19 shows that $\sum_{j=1}^{\infty} b_j$ converges. Proposition 5.21 then implies convergence of the original series.

Example 5.23 The series $\sum_{k=1}^{\infty} k^{-0.99}$ diverges. This follows from the fact that $k^{-1} \leq k^{-0.99}$. Therefore, the series $\sum_{k=1}^{\infty} k^{-0.99}$ dominates the harmonic series which itself is divergent, see Example 5.20.

Example 5.24 In Chap. 2 Euler's number

$$e = \sum_{j=0}^{\infty} \frac{1}{j!} = 1 + 1 + \frac{1}{2} + \frac{1}{6} + \frac{1}{24} + \frac{1}{120} + \cdots$$

was introduced. We can now show that this definition makes sense, i.e. the series converges. For $j \geq 4$ it is obvious that

$$j! = 1 \cdot 2 \cdot 3 \cdot 4 \cdot 5 \cdots \cdot j \geq 2 \cdot 2 \cdot 2 \cdot 2 \cdot 2 \cdots \cdot 2 = 2^j.$$

Thus the geometric series $\sum_{j=0}^{\infty} (\frac{1}{2})^j$ is a dominating convergent series.

Example 5.25 The decimal notation of a positive real number

$$a = A.\alpha_1\alpha_2\alpha_3 \ldots$$

with $A \in \mathbb{N}_0$, $\alpha_k \in \{0, \ldots, 9\}$ can be understood as a representation by the series

$$a = A + \sum_{k=1}^{\infty} \alpha_k 10^{-k}.$$

The series converges since $A + 9 \sum_{k=1}^{\infty} 10^{-k}$ is a dominating convergent series.

5.4   Supplement: Accumulation Points of Sequences

Occasionally we need sequences which themselves do not converge but have convergent *subsequences*. The notions of *accumulation points, limit superior* and *limit inferior* are connected with this concept.

Definition 5.26 A number b is called *accumulation point* of a sequence $(a_n)_{n\geq1}$ if each neighbourhood $U_\varepsilon(b)$ of b contains infinitely many terms of the sequence:

$$\forall \varepsilon > 0 \ \forall n \in \mathbb{N} \ \exists m = m(n, \varepsilon) \geq n : \ |b - a_m| < \varepsilon.$$

Figure 5.3 displays the sequence

$$a_n = \arctan n + \cos(n\pi/2) + \frac{1}{n}\sin(n\pi/2).$$

It has three accumulation points, namely $b_1 = \pi/2 + 1 \approx 2.57$, $b_2 = \pi/2 \approx 1.57$ and $b_3 = \pi/2 - 1 \approx 0.57$.

If a sequence is convergent with limit a then a is the unique accumulation point. Accumulation points of a sequence can also be characterised with the help of the concept of subsequences.

Definition 5.27 If $1 \leq n_1 < n_2 < n_3 < \cdots$ is a strictly monotonically increasing sequence of integers (indices) then

$$(a_{n_j})_{j\geq1}$$

is called a *subsequence* of the sequence $(a_n)_{n\geq1}$.

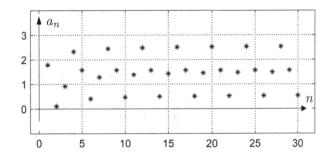

Fig. 5.3 Accumulation points of a sequence

Example 5.28 We start with the sequence $a_n = \frac{1}{n}$. If we take, for instance, $n_j = j^2$ then we obtain the sequence $a_{n_j} = \frac{1}{j^2}$ as subsequence:

$$(a_n)_{n \geq 1} = (1, \tfrac{1}{2}, \tfrac{1}{3}, \tfrac{1}{4}, \tfrac{1}{5}, \tfrac{1}{6}, \tfrac{1}{7}, \tfrac{1}{8}, \tfrac{1}{9}, \tfrac{1}{10}, \dots),$$
$$(a_{n_j})_{j \geq 1} = (1, \tfrac{1}{4}, \tfrac{1}{9}, \dots).$$

Proposition 5.29 *A number b is an accumulation point of the sequence $(a_n)_{n \geq 0}$ if and only if b is the limit of a convergent subsequence $(a_{n_j})_{j \geq 1}$.*

Proof Let b be an accumulation point of the sequence $(a_n)_{n \geq 0}$. Step by step we will construct a strictly monotonically increasing sequence of indices $(n_j)_{j \geq 1}$ so that

$$|b - a_{n_j}| < \frac{1}{j}$$

is fulfilled for all $j \in \mathbb{N}$. According to Definition 5.26 for $\varepsilon_1 = 1$ we have

$$\forall n \in \mathbb{N} \, \exists m \geq n \, : \, |b - a_m| < \varepsilon_1.$$

We choose $n = 1$ and denote the smallest $m \geq n$ which fulfills this condition by n_1. Thus

$$|b - a_{n_1}| < \varepsilon_1 = 1.$$

For $\varepsilon_2 = \frac{1}{2}$ one again obtains according to Definition 5.26:

$$\forall n \in \mathbb{N} \, \exists m \geq n \, : \, |b - a_m| < \varepsilon_2.$$

This time we choose $n = n_1 + 1$ and denote the smallest $m \geq n_1 + 1$ which fulfills this condition by n_2. Thus

$$|b - a_{n_2}| < \varepsilon_2 = \frac{1}{2}.$$

It is clear how one has to proceed. Once n_j is constructed one sets $\varepsilon_{j+1} = 1/(j+1)$ and uses Definition 5.26 according to which

$$\forall n \in \mathbb{N} \, \exists m \geq n \, : \, |b - a_m| < \varepsilon_{j+1}.$$

We choose $n = n_j + 1$ and denote the smallest $m \geq n_j + 1$ which fulfills this condition by n_{j+1}. Thus

$$|b - a_{n_{j+1}}| < \varepsilon_{j+1} = \frac{1}{j+1}.$$

This procedure guarantees on the one hand that the sequence of indices $(n_j)_{j\geq 1}$ is strictly monotonically increasing and on the other hand that the desired inequality is fulfilled for all $j \in \mathbb{N}$. In particular, $(a_{n_j})_{j\geq 1}$ is a subsequence that converges to b.

Conversely, it is obvious that the limit of a convergent subsequence is an accumulation point of the original sequence. $\square$

In the proof of the proposition we have used the method of *recursive definition* of a sequence, namely the subsequence $(a_{n_j})_{j\geq 1}$.

We next want to show that each bounded sequence has at least one accumulation point—or equivalently—a convergent subsequence. This result bears the names of Bolzano[2] and Weierstrass[3] and is an important technical tool for proofs in many areas of analysis.

Proposition 5.30 (Theorem of Bolzano–Weierstrass) *Every bounded sequence* $(a_n)_{n\geq 1}$ *has (at least) one accumulation point.*

Proof Due to the boundedness of the sequence there are bounds $b < c$ so that all terms of the sequence a_n lie between b and c. We bisect the interval $[b, c]$. Then in at least one of the two half-intervals $[b, (b + c)/2]$ or $[(b + c)/2, c]$ there have to be infinitely many terms of the sequence. We choose such a half-interval and call it $[b_1, c_1]$. This interval is also bisected; in one of the two halves again there have to be infinitely many terms of the sequence. We call this quarter-interval $[b_2, c_2]$. Continuing this way we obtain a sequence of intervals $[b_n, c_n]$ of length $2^{-n}(c - b)$ each of which contains infinitely many terms of the sequence. Obviously the b_n are monotonically increasing and bounded, therefore converge to a limit b. Since each interval $[b - 2^{-n}, b + 2^{-n}]$ by construction contains infinitely many terms of the sequence, b is an accumulation point of the sequence. $\square$

If the sequence $(a_n)_{n\geq 1}$ is bounded then the set of its accumulation points is also bounded and hence has a supremum. This supremum is itself an accumulation point of the sequence (which can be shown by constructing a suitable convergent subsequence) and thus forms the largest accumulation point.

Definition 5.31 The largest accumulation point of a bounded sequence is called *limit superior* and is denoted by $\overline{\lim}_{n\to\infty} a_n$ or $\limsup_{n\to\infty} a_n$. The smallest accumulation point is called *limit inferior* with the corresponding notation $\underline{\lim}_{n\to\infty} a_n$ or $\liminf_{n\to\infty} a_n$.

---

[2]B. Bolzano, 1781–1848.
[3]K. Weierstrass, 1815–1897.

The relationships

$$\limsup_{n\to\infty} a_n = \lim_{n\to\infty}\left(\sup_{m\ge n} a_m\right), \qquad \liminf_{n\to\infty} a_n = \lim_{n\to\infty}\left(\inf_{m\ge n} a_m\right)$$

follow easily from the definition and justify the notation.

For example, the sequence $(a_n)_{n\ge 1}$ from Fig. 5.3 has $\limsup_{n\to\infty} a_n = \pi/2 + 1$ and $\liminf_{n\to\infty} a_n = \pi/2 - 1$.

5.5 Exercises

1. Find a law of formation for the sequences below and check for monotonicity, boundedness and convergence:

$$-3,\ -2,\ -1,\ 0,\ \tfrac{1}{4},\ \tfrac{3}{9},\ \tfrac{5}{16},\ \tfrac{7}{25},\ \tfrac{9}{36},\ \ldots;$$
$$0,\ -1,\ \tfrac{1}{2},\ -2,\ \tfrac{1}{4},\ -3,\ \tfrac{1}{8},\ -4,\ \tfrac{1}{16},\ \ldots.$$

2. Verify that the sequence $a_n = \frac{n^2}{1+n^2}$ converges to 1.
   Hint. Given $\varepsilon > 0$, find $n(\varepsilon)$ such that

   $$\left|\frac{n^2}{1+n^2} - 1\right| < \varepsilon$$

   for all $n \ge n(\varepsilon)$.

3. Determine a recursion formula that provides the terms of the geometric sequence $a_n = q^n$, $n \ge 0$ successively. Write a MATLAB program that calculates the first N terms of the geometric sequence for an arbitrary $q \in \mathbb{R}$.
   Check the convergence behaviour for different values of q and plot the results. Do the same with the help of the applet *Sequences*.

4. Investigate whether the following sequences converge and, in case of convergence, compute the limit:

   $$a_n = \frac{n}{n+1} - \frac{n+1}{n}, \qquad b_n = -n + \frac{1}{n}, \qquad c_n = \left(-\frac{1}{n}\right)^n,$$

   $$d_n = n - \frac{n^2 + 3n + 1}{n}, \qquad e_n = \frac{1}{2}\left(e^n + e^{-n}\right), \qquad f_n = \cos(n\pi).$$

5. Investigate whether the following sequences have a limit or an accumulation point. Compute, if existent, $\lim$, $\liminf$, $\limsup$, $\inf$, $\sup$:

   $$a_n = \frac{n+7}{n^3 + n + 1}, \qquad b_n = \frac{1 - 3n^2}{7n + 5}, \qquad c_n = \frac{e^n - e^{-n}}{e^n + e^{-n}},$$

   $$d_n = 1 + (-1)^n, \qquad e_n = \frac{1 + (-1)^n}{n}, \qquad f_n = \left(1 + (-1)^n\right)(-1)^{n/2}.$$

6. Open the applet *Sequences*, visualise the sequences from Exercises 4 and 5 and discuss their behaviour by means of their graphs.

7. The population model of Verhulst from Example 5.3 can be described in appropriate units in simplified form by the recursive relationship

$$x_{n+1} = rx_n(1 - x_n), \quad n = 0, 1, 2, 3, \ldots$$

with an initial value x_0 and a parameter r. We presume in this sequence that $0 \le x_0 \le 1$ and $0 \le r \le 4$ (since all x_n then stay in the interval $[0, 1]$). Write a MATLAB program which calculates for given r, x_0, N the first N terms of the sequence $(x_n)_{n \ge 1}$. With the help of your program (and some numerical values for r, x_0, N) check the following statements:

(a) For $0 \le r \le 1$ the sequence x_n converges to 0.

(b) For $1 < r < 2\sqrt{2}$ the sequence x_n tends to a positive limit.

(c) For $3 < r < 1 + \sqrt{6}$ the sequence x_n eventually oscillates between two different positive values.

(d) For $3.75 < r \le 4$ the sequence x_n behaves *chaotically*.

Illustrate these assertions also with the applet *Sequences*.

8. The sequence $(a_n)_{n \ge 1}$ is given recursively by

$$a_1 = A, \quad a_{n+1} = \frac{1}{2}a_n^2 - \frac{1}{2}.$$

Which starting values $A \in \mathbb{R}$ are fixed points of the recursion, i.e. it holds $A = a_1 = a_2 = \ldots$? Investigate for which starting values $A \in \mathbb{R}$ the sequence converges or diverges, respectively. You can use the applet *Sequences* for that. Try to locate the regions of convergence and divergence as precisely as possible.

9. Write a MATLAB program which, for given $\alpha \in [0, 1]$ and $N \in \mathbb{N}$, calculates the first N terms of the sequence

$$x_n = n\alpha - \lfloor n\alpha \rfloor, \quad n = 1, 2, 3, \ldots, N$$

($\lfloor n\alpha \rfloor$ denotes the largest integer smaller than $n\alpha$). With the help of your program, investigate the behaviour of the sequence for a rational $\alpha = \frac{p}{q}$ and for an irrational α (or at least a very precise rational approximation to an irrational α) by plotting the terms of the sequence and by visualising their distribution in a histogram. Use the MATLAB commands `floor` and `hist`.

10. Give formal proofs for the remaining rules of calculation of Proposition 5.7, i.e. for addition and division by modifying the proof for the multiplication rule.

11. Check the following series for convergence with the help of the comparison criteria:

$$\sum_{k=1}^{\infty} \frac{1}{k(k+2)}, \quad \sum_{k=1}^{\infty} \frac{1}{\sqrt{k}}, \quad \sum_{k=1}^{\infty} \frac{1}{k^3}.$$

12. Check the following series for convergence:

$$\sum_{k=1}^{\infty} \frac{2+k^2}{k^4}, \quad \sum_{k=1}^{\infty} \left(\frac{1}{2}\right)^{2k}, \quad \sum_{k=1}^{\infty} \frac{2}{k!}.$$

13. Try to find out how the partial sums S_n of the series in Exercises 11 and 12 can be calculated with the help of a recursion and then study their behaviour with the applet *Sequences*.

14. Prove the convergence of the series

$$\sum_{k=0}^{\infty} \frac{2^k}{k!}.$$

Hint. Use the fact that $j! \geq 4^j$ is fulfilled for $j \geq 9$ (why)? From this it follows that $2^j/j! \leq 1/2^j$. Now apply the appropriate comparison criterion.

15. Prove the *ratio test* for series with positive terms $a_k > 0$: If there exists a number $q, 0 < q < 1$ such that the quotients satisfy

$$\frac{a_{k+1}}{a_k} \leq q$$

for all $k \in \mathbb{N}_0$, then the series $\sum_{k=0}^{\infty} a_k$ converges.
Hint. From the assumption it follows that $a_1 \leq a_0 q, a_2 \leq a_1 q \leq a_0 q^2$ and thus successively $a_k \leq a_0 q^k$ for all k. Now use the comparison criteria and the convergence of the geometric series with $q < 1$.

Limits and Continuity of Functions

6

In this section we extend the notion of the limit of a sequence to the concept of the limit of a function. Hereby we obtain a tool which enables us to investigate the behaviour of graphs of functions in the neighbourhood of chosen points. Moreover, limits of functions form the basis of one of the central themes in mathematical analysis, namely differentiation (Chap. 7). In order to derive certain differentiation formulas some elementary limits are needed, for instance, limits of trigonometric functions. The property of continuity of a function has far-reaching consequences like, for instance, the *intermediate value theorem*, according to which a continuous function which changes its sign in an interval has a zero. Not only does this theorem allow one to show the solvability of equations, it also provides numerical procedures to approximate the solutions. Further material on continuity can be found in Appendix C.

6.1 The Notion of Continuity

We start with the investigation of the behaviour of graphs of real functions

$$f : (a, b) \to \mathbb{R}$$

while approaching a point x in the open interval (a, b) or a boundary point of the closed interval $[a, b]$. For that we need the notion of a *zero sequence*, i.e. a sequence of real numbers $(h_n)_{n \geq 1}$ with $\lim_{n \to \infty} h_n = 0$.

© Springer Nature Switzerland AG 2018
M. Oberguggenberger and A. Ostermann, *Analysis for Computer Scientists*,
Undergraduate Topics in Computer Science,
https://doi.org/10.1007/978-3-319-91155-7_6

Definition 6.1 (Limits and continuity)

(a) The function f has a *limit M* at a point $x \in (a, b)$, if

$$\lim_{n \to \infty} f(x + h_n) = M$$

for all zero sequences $(h_n)_{n \geq 1}$ with $h_n \neq 0$. In this case one writes

$$M = \lim_{h \to 0} f(x + h) = \lim_{\xi \to x} f(\xi)$$

or

$$f(x + h) \to M \text{ as } h \to 0.$$

(b) The function f has a *right-hand limit R* at the point $x \in [a, b)$, if

$$\lim_{n \to \infty} f(x + h_n) = R$$

for all zero sequences $(h_n)_{n \geq 1}$ with $h_n > 0$, with the corresponding notation

$$R = \lim_{h \to 0+} f(x + h) = \lim_{\xi \to x+} f(\xi).$$

(c) The function f has a *left-hand limit L* at the point $x \in (a, b]$, if:

$$\lim_{n \to \infty} f(x + h_n) = L$$

for all zero sequences $(h_n)_{n \geq 1}$ with $h_n < 0$. Notations:

$$L = \lim_{h \to 0-} f(x + h) = \lim_{\xi \to x-} f(\xi).$$

(d) If f has a limit M at $x \in (a, b)$ which coincides with the value of the function, i.e. $f(x) = M$, then f is called *continuous at the point x*.

(e) If f is continuous at every $x \in (a, b)$, then f is said to be *continuous on the open interval* (a, b). If in addition f has right- and left-hand limits at the endpoints a and b, it is called *continuous on the closed interval* $[a, b]$.

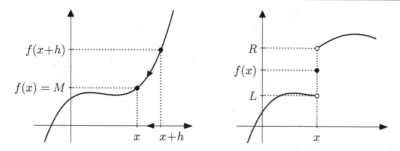

Fig. 6.1 Limit and continuity; left- and right-hand limits

Figure 6.1 illustrates the idea of approaching a point x for $h \to 0$ as well as possible differences between left-hand and right-hand limits and the value of the function.

If a function f is continuous at a point x, the function evaluation can be interchanged with the limit:

$$\lim_{\xi \to x} f(\xi) = f(x) = f(\lim_{\xi \to x} \xi).$$

The following examples show some further possibilities how a function can behave in the neighbourhood of a point: Jump discontinuity with left- and right-hand limits, vertical asymptote, oscillations with non-vanishing amplitude and ever-increasing frequency.

Example 6.2 The quadratic function $f(x) = x^2$ is continuous at every $x \in \mathbb{R}$ since

$$f(x + h_n) - f(x) = (x + h_n)^2 - x^2 = 2xh_n + h_n^2 \to 0$$

as $n \to \infty$ for any zero sequence $(h_n)_{n \geq 1}$. Therefore

$$\lim_{h \to 0} f(x + h) = f(x).$$

Likewise the continuity of the power functions $x \mapsto x^m$ for $m \in \mathbb{N}$ can be shown.

Example 6.3 The absolute value function $f(x) = |x|$ and the third root $g(x) = \sqrt[3]{x}$ are everywhere continuous. The former has a kink at $x = 0$, the latter a vertical tangent; see Fig. 6.2.

Example 6.4 The sign function $f(x) = \operatorname{sign} x$ has different left- and right-hand limits $L = -1$, $R = 1$ at $x = 0$. In particular, it is discontinuous at that point. At all other points $x \neq 0$ it is continuous; see Fig. 6.3.

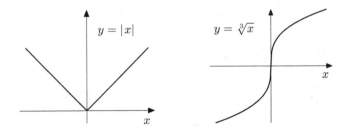

Fig. 6.2 Continuity and kink or vertical tangent

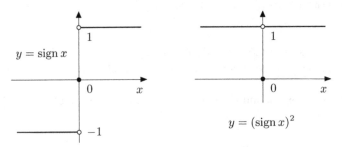

Fig. 6.3 Discontinuities: jump discontinuity and exceptional value

Example 6.5 The square of the sign function

$$g(x) = (\text{sign } x)^2 = \begin{cases} 1, & x \neq 0 \\ 0, & x = 0 \end{cases}$$

has equal left- and right-hand limits at $x = 0$. However, they are different from the value of the function (see Fig. 6.3):

$$\lim_{\xi \to 0} g(\xi) = 1 \neq 0 = g(0).$$

Therefore, g is discontinuous at $x = 0$.

Example 6.6 The functions $f(x) = \frac{1}{x}$ and $g(x) = \tan x$ have vertical asymptotes at $x = 0$ and $x = \frac{\pi}{2} + k\pi$, $k \in \mathbb{Z}$, respectively, and in particular no left- or right-hand limit at these points. At all other points, however, they are continuous. We refer to Figs. 2.9 and 3.10.

Example 6.7 The function $f(x) = \sin \frac{1}{x}$ has no left- or right-hand limit at $x = 0$ but oscillates with non-vanishing amplitude (Fig. 6.4). Indeed, one obtains different limits for different zero sequences. For example, for

$$h_n = \frac{1}{n\pi}, \quad k_n = \frac{1}{\pi/2 + 2n\pi}, \quad l_n = \frac{1}{3\pi/2 + 2n\pi}$$

Fig. 6.4 No limits, oscillation with non-vanishing amplitude

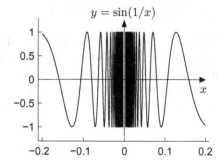

Fig. 6.5 Continuity, oscillation with vanishing amplitude

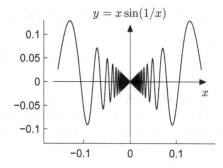

the respective limits are

$$\lim_{n\to\infty} f(h_n) = 0, \quad \lim_{n\to\infty} f(k_n) = 1, \quad \lim_{n\to\infty} f(l_n) = -1.$$

All other values in the interval $[-1, 1]$ can also be obtained as limits with the help of suitable zero sequences.

Example 6.8 The function $g(x) = x \sin \frac{1}{x}$ can be continuously extended by $g(0) = 0$ at $x = 0$; it oscillates with vanishing amplitude (Fig. 6.5). Indeed,

$$|g(h_n) - g(0)| = |h_n \sin \tfrac{1}{h_n} - 0| \le |h_n| \to 0$$

for all zero sequences $(h_n)_{n\ge 1}$, thus $\lim_{h\to 0} h \sin \frac{1}{h} = 0$.

Experiment 6.9 Open the M-files mat06_1.m and mat06_2.m, and study the graphs of the functions in Figs. 6.4 and 6.5 with the use of the zoom tool in the figure window. How can you improve the accuracy of the visualisation in the neighbourhood of $x = 0$?

6.2 Trigonometric Limits

Comparing the areas in Fig.6.6 shows that the area of the grey triangle with sides $\cos x$ and $\sin x$ is smaller than the area of the sector which in turn is smaller or equal to the area of the big triangle with sides 1 and $\tan x$.

The area of a sector in the unit circle (with angle x in radian measure) equals $x/2$ as is well-known. In summary we obtain the inequalities

$$\frac{1}{2} \sin x \cos x \le \frac{x}{2} \le \frac{1}{2} \tan x$$

or after division by $\sin x$ and taking the reciprocal

$$\cos x \le \frac{\sin x}{x} \le \frac{1}{\cos x},$$

valid for all x with $0 < |x| < \pi/2$.

With the help of these inequalities we can compute several important limits. From an elementary geometric consideration, one obtains

$$|\cos x| \ge \frac{1}{2} \qquad \text{for} \quad -\frac{\pi}{3} \le x \le \frac{\pi}{3},$$

and together with the previous inequalities

$$|\sin h_n| \le \frac{|h_n|}{|\cos h_n|} \le 2 |h_n| \to 0$$

for all zero sequences $(h_n)_{n\ge 1}$. This means that

$$\lim_{h \to 0} \sin h = 0.$$

Fig. 6.6 Illustration of trigonometric inequalities

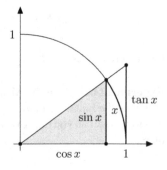

The sine function is therefore continuous at zero. From the continuity of the square function and the root function as well as the fact that $\cos h$ equals the *positive* square root of $1 - \sin^2 h$ for small h it follows that

$$\lim_{h \to 0} \cos h = \lim_{h \to 0} \sqrt{1 - \sin^2 h} = 1.$$

With this the continuity of the sine function at every point $x \in \mathbb{R}$ can be proven:

$$\lim_{h \to 0} \sin(x + h) = \lim_{h \to 0} \left(\sin x \cos h + \cos x \sin h \right) = \sin x.$$

The inequality illustrated at the beginning of the section allows one to deduce one of the most important trigonometric limits. It forms the basis of the differentiation rules for trigonometric functions.

Proposition 6.10 $\lim_{x \to 0} \frac{\sin x}{x} = 1.$

Proof We combine the above result $\lim_{x \to 0} \cos x = 1$ with the inequality deduced earlier and obtain

$$1 = \lim_{x \to 0} \cos x \leq \lim_{x \to 0} \frac{\sin x}{x} \leq \lim_{x \to 0} \frac{1}{\cos x} = 1,$$

and therefore $\lim_{x \to 0} \frac{\sin x}{x} = 1$. $\qquad\qquad\square$

6.3 Zeros of Continuous Functions

Figure 6.7 shows the graph of a function that is continuous on a closed interval $[a, b]$ and that is negative at the left endpoint and positive at the right endpoint. Geometrically the graph has to intersect the x-axis at least once since it has no jumps due to the continuity. This means that f has to have at least one zero in (a, b). This is a criterion that guarantees the existences of a solution to the equation $f(x) = 0$. A first rigorous proof of this intuitively evident statement goes back to Bolzano.

Proposition 6.11 (Intermediate value theorem) *Let $f : [a, b] \to \mathbb{R}$ be continuous and $f(a) < 0$, $f(b) > 0$. Then there exists a point $\xi \in (a, b)$ with $f(\xi) = 0$.*

Proof The proof is based on the successive bisection of the intervals and the completeness of the set of real numbers. One starts with the interval $[a, b]$ and sets $a_1 = a$, $b_1 = b$.

Fig. 6.7 The intermediate
value theorem

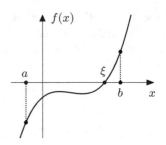

Step 1: Compute $y_1 = f\left(\frac{a_1+b_1}{2}\right)$.

If $y_1 > 0$: set $a_2 = a_1, b_2 = \frac{a_1+b_1}{2}$.

If $y_1 < 0$: set $a_2 = \frac{a_1+b_1}{2}, b_2 = b_1$.

If $y_1 = 0$: termination, $\xi = \frac{a_1+b_1}{2}$ is a zero.

By construction $f(a_2) < 0$, $f(b_2) > 0$ and the interval length is halved:

$$b_2 - a_2 = \frac{1}{2}(b_1 - a_1).$$

Step 2: Compute $y_2 = f\left(\frac{a_2+b_2}{2}\right)$.

If $y_2 > 0$: set $a_3 = a_2, b_3 = \frac{a_2+b_2}{2}$.

If $y_2 < 0$: set $a_3 = \frac{a_2+b_2}{2}, b_3 = b_2$.

If $y_2 = 0$: termination, $\xi = \frac{a_2+b_2}{2}$ is a zero.

Further iterations lead to a monotonically increasing sequence

$$a_1 \leq a_2 \leq a_3 \leq \cdots \leq b$$

which is bounded from above. According to Proposition 5.10 the limit $\xi = \lim_{n\to\infty} a_n$ exists.

On the other hand $|a_n - b_n| \leq |a - b|/2^{n-1} \to 0$, therefore $\lim_{n\to\infty} b_n = \xi$ as well. If ξ has not appeared after a finite number of steps as either a_k or b_k then for all $n \in \mathbb{N}$:

$$f(a_n) < 0, \quad f(b_n) > 0.$$

From the continuity of f it follows that

$$f(\xi) = \lim_{n\to\infty} f(a_n) \leq 0, \qquad f(\xi) = \lim_{n\to\infty} f(b_n) \geq 0$$

which implies $f(\xi) = 0$, as claimed. $\square$

The proof provides at the same time a numerical method to compute zeros of functions, the *bisection method*. Although it converges rather slowly, it is easily implementable and universally applicable—also for non-differentiable, continuous functions. For differentiable functions, however, considerably faster algorithms exist. The order of convergence and the discussion of faster procedures will be taken up in Sect. 8.2.

Example 6.12 Calculation of $\sqrt{2}$ as the root of $f(x) = x^2 - 2 = 0$ in the interval $[1, 2]$ using the bisection method:

Start:	$f(1) = -1 < 0,\ f(2) = 2 > 0;$	$a_1 = 1,\ b_1 = 2$
Step 1:	$f(1.5) = 0.25 > 0;$	$a_2 = 1,\ b_2 = 1.5$
Step 2:	$f(1.25) = -0.4375 < 0;$	$a_3 = 1.25,\ b_3 = 1.5$
Step 3:	$f(1.375) = -0.109375 < 0;$	$a_4 = 1.375,\ b_4 = 1.5$
Step 4:	$f(1.4375) = 0.066406\ldots > 0;$	$a_5 = 1.375,\ b_5 = 1.4375$
Step 5:	$f(1.40625) = -0.022461\ldots < 0;$	$a_6 = 1.40625,\ b_6 = 1.4375$
etc.		

After 5 steps the first decimal place is ascertained:

$$1.40625 < \sqrt{2} < 1.4375$$

Experiment 6.13 Sketch the graph of the function $y = x^3 + 3x^2 - 2$ on the interval $[-3, 2]$, and try to first estimate graphically one of the roots by successive bisection. Execute the interval bisection with the help of the applet *Bisection method*. Assure yourself of the plausibility of the intermediate value theorem using the applet *Animation of the intermediate value theorem*.

As an important application of the intermediate value theorem we now show that images of intervals under continuous functions are again intervals. For the different types of intervals which appear in the following proposition we refer to Sect. 1.2; for the notion of the proper range to Sect. 2.1.

Proposition 6.14 *Let $I \subset \mathbb{R}$ be an interval (open, half-open or closed, bounded or improper) and $f : I \to \mathbb{R}$ a continuous function with proper range $J = f(I)$. Then J is also an interval.*

Proof As subsets of the real line, intervals are characterised by the following property: With any two points all intermediate points are contained in it as well. Let $y_1, y_2 \in J$, $y_1 < y_2$, and η be an intermediate point, i.e. $y_1 < \eta < y_2$. Since $f : I \to J$ is surjective there are $x_1, x_2 \in I$ such that $y_1 = f(x_1)$ and $y_2 = f(x_2)$. We consider the case $x_1 < x_2$. Since $f(x_1) - \eta < 0$ and $f(x_2) - \eta > 0$ it follows from the intermediate value theorem applied on the interval $[x_1, x_2]$ that there exists a point $\xi \in (x_1, x_2)$ with $f(\xi) - \eta = 0$, thus $f(\xi) = \eta$. Hence η is attained as a value of the function and therefore lies in $J = f(I)$. $\square$

Proposition 6.15 *Let $I = [a, b]$ be a closed, bounded interval and $f : I \to \mathbb{R}$ a continuous function. Then the proper range $J = f(I)$ is also a closed, bounded interval.*

Proof According to Proposition 6.14 the range J is an interval. Let d be the least upper bound (possibly $d = \infty$). We take a sequence of values $y_n \in J$ which converges to d. The values y_n are function values of certain arguments $x_n \in I = [a, b]$. The sequence $(x_n)_{n \geq 1}$ is bounded and, according to Proposition 5.30, has an accumulation point $x_0, a \leq x_0 \leq b$. Thus a subsequence $(x_{n_j})_{j \geq 1}$ exists which converges to x_0 (see Sect. 5.4). From the continuity of the function f it follows that

$$d = \lim_{j \to \infty} y_{n_j} = \lim_{j \to \infty} f(x_{n_j}) = f(x_0).$$

This shows that the upper endpoint of the interval J is finite and is attained as function value. The same argument is applied to the lower boundary c; the range J is therefore a closed, bounded interval $[c, d]$. □

From the proof of the proposition it is clear that d is the largest and c the smallest value of the function f on the interval $[a, b]$. We thus obtain the following important consequence.

Corollary 6.16 *Each continuous function defined on a closed interval $I = [a, b]$ attains its maximum and minimum there.*

6.4 Exercises

1. (a) Investigate the behaviour of the functions

$$\frac{x + x^2}{|x|}, \quad \frac{\sqrt{1 + x} - 1}{x}, \quad \frac{x^2 + \sin x}{\sqrt{1 - \cos^2 x}}$$

in a neighbourhood of $x = 0$ by plotting their graphs for arguments in $[-2, -\frac{1}{100}) \cup (\frac{1}{100}, 2]$.

   (b) Find out by inspection of the graphs whether there are left- or right-hand limits at $x = 0$. Which value do they have? Explain your results by rearranging the expressions in (a).

   Hint. Some guidance for part (a) can be found in the M-file mat06_ex1.m. Expand the middle term in (b) with $\sqrt{1 + x} + 1$.

2. Do the following functions have a limit at the given points? If so, what is its value?
   (a) $y = x^3 + 5x + 10$, $x = 1$.
   (b) $y = \frac{x^2-1}{x^2+x}$, $x = 0$, $x = 1$, $x = -1$.
   (c) $y = \frac{1-\cos x}{x^2}$, $x = 0$.
      Hint. Expand with $(1 + \cos x)$.
   (d) $y = \operatorname{sign} x \cdot \sin x$, $x = 0$.
   (e) $y = \operatorname{sign} x \cdot \cos x$, $x = 0$.
3. Let $f_n(x) = \arctan nx$, $g_n(x) = (1 + x^2)^{-n}$. Compute the limits

$$f(x) = \lim_{n \to \infty} f_n(x), \qquad g(x) = \lim_{n \to \infty} g_n(x)$$

   for each $x \in \mathbb{R}$, and sketch the graphs of the thereby defined functions f and g. Are they continuous? Plot f_n and g_n using MATLAB, and investigate the behaviour of the graphs for $n \to \infty$.
   Hint. An advice can be found in the M-file mat06_ex3.m.
4. With the help of zero sequences, carry out a formal proof of the fact that the absolute value function and the third root function of Example 6.3 are continuous.
5. Argue with the help of the intermediate value theorem that $p(x) = x^3 + 5x + 10$ has a zero in the interval $[-2, 1]$. Compute this zero up to four decimal places using the applet *Bisection method*.
6. Compute all zeros of the following functions in the given interval with accuracy 10^{-3}, using the applet *Bisection method*.

$$f(x) = x^4 - 2, \qquad I = \mathbb{R};$$
$$g(x) = x - \cos x, \qquad I = \mathbb{R};$$
$$h(x) = \sin \tfrac{1}{x}, \qquad I = \left[\tfrac{1}{20}, \tfrac{1}{10}\right].$$

7. Write a MATLAB program which locates—with the help of the bisection method—the zero of an arbitrary polynomial

$$p(x) = x^3 + c_1 x^2 + c_2 x + c_3$$

   of degree three. Your program should automatically provide starting values a, b with $p(a) < 0$, $p(b) > 0$ (why do such values always exist?). Test your program by choosing the coefficient vector (c_1, c_2, c_3) randomly, for example by using c = 1000*rand(1,3).
   Hint. A solution is suggested in the M-file mat06_ex7.m.

The Derivative of a Function

7

Starting from the problem to define the tangent to the graph of a function, we introduce the derivative of a function. Two points on the graph can always be joined by a secant, which is a good model for the tangent whenever these points are close to each other. In a limiting process, the secant (discrete model) is replaced by the tangent (continuous model). Differential calculus, which is based on this limiting process, has become one of the most important building blocks of mathematical modelling.

In this section we discuss the derivative of important elementary functions as well as general differentiation rules. Thanks to the meticulous implementation of these rules, expert systems such as maple have become helpful tools in mathematical analysis. Furthermore, we will discuss the interpretation of the derivative as linear approximation and as rate of change. These interpretations form the basis of numerous applications in science and engineering.

The concept of the numerical derivative follows the opposite direction. The continuous model is discretised, and the derivative is replaced by a difference quotient. We carry out a detailed error analysis which allows us to find an optimal approximation. Further, we will illustrate the relevance of symmetry in numerical procedures.

7.1 Motivation

Example 7.1 (The free fall according to Galilei[1]) Imagine an object, which released at time $t = 0$, falls down under the influence of gravity. We are interested in the position $s(t)$ of the object at time $t \geq 0$ as well as in its velocity $v(t)$, see Fig. 7.1. Due to the definition of velocity as change in travelled distance divided by change

---

[1]G. Galilei, 1564–1642.

© Springer Nature Switzerland AG 2018
M. Oberguggenberger and A. Ostermann, *Analysis for Computer Scientists*,
Undergraduate Topics in Computer Science,
https://doi.org/10.1007/978-3-319-91155-7_7

Fig. 7.1 The free fall

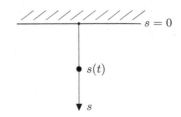

in time, the object has the *average velocity*

$$v_{\text{average}} = \frac{s(t + \Delta t) - s(t)}{\Delta t}$$

in the time interval $[t, t + \Delta t]$. In order to obtain the *instantaneous velocity* $v = v(t)$ we take the limit $\Delta t \to 0$ in the above formula and arrive at

$$v(t) = \lim_{\Delta t \to 0} \frac{s(t + \Delta t) - s(t)}{\Delta t}.$$

Galilei discovered through his experiments that the travelled distance in free fall increases quadratically with the time passed, i.e. the law

$$s(t) = \frac{g}{2} t^2$$

with $g \approx 9.81$ m/s^2 holds. Thus we obtain the expression

$$v(t) = \lim_{\Delta t \to 0} \frac{\frac{g}{2}(t + \Delta t)^2 - \frac{g}{2} t^2}{\Delta t} = \frac{g}{2} \lim_{\Delta t \to 0} (2t + \Delta t) = gt$$

for the instantaneous velocity. The velocity is hence proportional to the time passed.

Example 7.2 (The tangent problem) Consider a real function f and two different points $P = (x_0, f(x_0))$ and $Q = (x, f(x))$ on the graph of the function. The uniquely defined straight line through these two points is called *secant* of the function f through P and Q, see Fig. 7.2. The slope of the secant is given by the *difference quotient*

$$\frac{\Delta y}{\Delta x} = \frac{f(x) - f(x_0)}{x - x_0}.$$

As x tends to x_0, the secant graphically turns into the tangent, provided the limit exists. Motivated by this idea we define the slope

$$k = \lim_{x \to x_0} \frac{f(x) - f(x_0)}{x - x_0} = \lim_{h \to 0} \frac{f(x_0 + h) - f(x_0)}{h}$$

of the function f at x_0. If this limit exists, we call the straight line

$$y = k \cdot (x - x_0) + f(x_0)$$

the *tangent* to the graph of the function at the point $(x_0, f(x_0))$.

Fig. 7.2 Slope of the secant

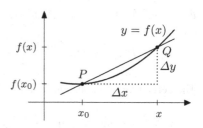

Experiment 7.3 Plot the function $f(x) = x^2$ on the interval $[0, 2]$ in MATLAB. Draw the straight lines through the points $(1, 1)$, $(2, z)$ for various values of z. Adjust z until you find the tangent to the graph of the function f at $(1, 1)$ and read off its slope.

7.2 The Derivative

Motivated by the above applications we are going to define the derivative of a real-valued function.

Definition 7.4 (Derivative) Let $I \subset \mathbb{R}$ be an open interval, $f : I \to \mathbb{R}$ a real-valued function and $x_0 \in I$.

(a) The function f is called *differentiable* at x_0 if the difference quotient

$$\frac{\Delta y}{\Delta x} = \frac{f(x) - f(x_0)}{x - x_0}$$

has a (finite) limit for $x \to x_0$. In this case one writes

$$f'(x_0) = \lim_{x \to x_0} \frac{f(x) - f(x_0)}{x - x_0} = \lim_{h \to 0} \frac{f(x_0 + h) - f(x_0)}{h}$$

and calls the limit *derivative of f at the point x_0*.

(b) The function f is called *differentiable* (in the interval I) if $f'(x)$ exists for all $x \in I$. In this case the function

$$f' : I \to \mathbb{R} : x \mapsto f'(x)$$

is called the *derivative of f*. The process of computing f' from f is called *differentiation*.

In place of $f'(x)$ one often writes $\frac{\mathrm{d}f}{\mathrm{d}x}(x)$ or $\frac{\mathrm{d}}{\mathrm{d}x} f(x)$, respectively. The following examples show how the derivative of a function is obtained by means of the limiting process above.

Example 7.5 (The constant function $f(x) = c$)

$$f'(x) = \lim_{h \to 0} \frac{f(x+h) - f(x)}{h} = \lim_{h \to 0} \frac{c - c}{h} = \lim_{h \to 0} \frac{0}{h} = 0.$$

The derivative of a constant function is zero.

Example 7.6 (The affine function $g(x) = ax + b$)

$$g'(x) = \lim_{h \to 0} \frac{g(x+h) - g(x)}{h} = \lim_{h \to 0} \frac{ax + ah + b - ax - b}{h} = \lim_{h \to 0} a = a.$$

The derivative is the slope a of the straight line $y = ax + b$.

Example 7.7 (The derivative of the quadratic function $y = x^2$)

$$y' = \lim_{h \to 0} \frac{(x+h)^2 - x^2}{h} = \lim_{h \to 0} \frac{2hx + h^2}{h} = \lim_{h \to 0} (2x + h) = 2x.$$

Similarly, one can show for the power function (with $n \in \mathbb{N}$):

$$f(x) = x^n \quad \Rightarrow \quad f'(x) = n \cdot x^{n-1}.$$

Example 7.8 (The derivative of the square root function $y = \sqrt{x}$ for $x > 0$)

$$y' = \lim_{\xi \to x} \frac{\sqrt{\xi} - \sqrt{x}}{\xi - x} = \lim_{\xi \to x} \frac{\sqrt{\xi} - \sqrt{x}}{(\sqrt{\xi} - \sqrt{x})(\sqrt{\xi} + \sqrt{x})} = \lim_{\xi \to x} \frac{1}{\sqrt{\xi} + \sqrt{x}} = \frac{1}{2\sqrt{x}}.$$

Example 7.9 (Derivatives of the sine and cosine functions) We first recall from Proposition 6.10 that

$$\lim_{t \to 0} \frac{\sin t}{t} = 1.$$

Due to

$$(\cos t - 1)(\cos t + 1) = -\sin^2 t$$

it also holds that

$$\frac{\cos t - 1}{t} = -\underbrace{\sin t}_{\to 0} \cdot \underbrace{\frac{\sin t}{t}}_{\to 1} \cdot \underbrace{\frac{1}{\cos t + 1}}_{\to 1/2} \to 0 \quad \text{for } t \to 0,$$

and thus

$$\lim_{t \to 0} \frac{\cos t - 1}{t} = 0.$$

Due to the addition theorems (Proposition 3.3) we get with the preparations from above

$$\sin' x = \lim_{h \to 0} \frac{\sin(x + h) - \sin x}{h} = \lim_{h \to 0} \frac{\sin x \cos h + \cos x \sin h - \sin x}{h}$$

$$= \lim_{h \to 0} \sin x \cdot \frac{\cos h - 1}{h} + \lim_{h \to 0} \cos x \cdot \frac{\sin h}{h}$$

$$= \sin x \cdot \underbrace{\lim_{h \to 0} \frac{\cos h - 1}{h}}_{= 0} + \cos x \cdot \underbrace{\lim_{h \to 0} \frac{\sin h}{h}}_{= 1}$$

$$= \cos x.$$

This proves the formula $\sin' x = \cos x$. Likewise it can be shown that $\cos' x = -\sin x$.

Example 7.10 (The derivative of the exponential function with base e) Rearranging terms in the series expansion of the exponential function (Proposition C.12) we obtain

$$\frac{e^h - 1}{h} = \sum_{k=0}^{\infty} \frac{h^k}{(k + 1)!} = 1 + \frac{h}{2} + \frac{h^2}{6} + \frac{h^3}{24} + \cdots$$

From that one infers

$$\left| \frac{e^h - 1}{h} - 1 \right| \leq |h| \left(\frac{1}{2} + \frac{|h|}{6} + \frac{|h|^3}{24} + \cdots \right) \leq |h| e^{|h|}.$$

Letting $h \to 0$ hence gives the important limit

$$\lim_{h \to 0} \frac{e^h - 1}{h} = 1.$$

The existence of the limit

$$\lim_{h \to 0} \frac{e^{x+h} - e^x}{h} = e^x \cdot \lim_{h \to 0} \frac{e^h - 1}{h} = e^x$$

shows that the exponential function is differentiable and that $(e^x)' = e^x$.

Example 7.11 (New representation of Euler's number) By substituting $y = e^h - 1$, $h = \log(y + 1)$ in the above limit one obtains

$$\lim_{y \to 0} \frac{y}{\log(y + 1)} = 1$$

and in this way

$$\lim_{y \to 0} \log(1 + \alpha y)^{1/y} = \lim_{y \to 0} \frac{\log(1 + \alpha y)}{y} = \alpha \lim_{y \to 0} \frac{\log(1 + \alpha y)}{\alpha y} = \alpha.$$

Due to the continuity of the exponential function it further follows that

$$\lim_{y \to 0} (1 + \alpha y)^{1/y} = e^{\alpha}.$$

In particular, for $y = 1/n$, we obtain a new representation of the exponential function

$$e^{\alpha} = \lim_{n \to \infty} \left(1 + \frac{\alpha}{n}\right)^n.$$

For $\alpha = 1$ the identity

$$e = \lim_{n \to \infty} \left(1 + \frac{1}{n}\right)^n = \sum_{k=0}^{\infty} \frac{1}{k!} = 2.718281828459...$$

follows.

Example 7.12 Not every continuous function is differentiable. For instance, the function

$$f(x) = |x| = \begin{cases} x, & x \geq 0 \\ -x, & x \leq 0 \end{cases}$$

is not differentiable at the vertex $x = 0$, see Fig. 7.3, left picture. However, it is differentiable for $x \neq 0$ with

$$(|x|)' = \begin{cases} 1, & \text{if } x > 0 \\ -1, & \text{if } x < 0. \end{cases}$$

The function $g(x) = \sqrt[3]{x}$ is not differentiable at $x = 0$ either. The reason for that is the vertical tangent, see Fig. 7.3, right picture.

There are even continuous functions that are nowhere differentiable. It is possible to write down such functions in the form of certain intricate infinite series. However, an analogous example of a (continuous) *curve in the plane* which is nowhere differentiable is the boundary of *Koch's snowflake*, which can be constructed in a simple geometric manner, see Examples 9.9 and 14.17.

Fig. 7.3 Functions that are not differentiable at $x = 0$

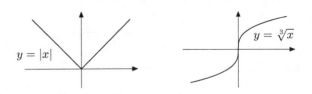

Definition 7.13 If the function f' is again differentiable then

$$f''(x) = \frac{d^2}{dx^2}f(x) = \frac{d^2 f}{dx^2}(x) = \lim_{h \to 0} \frac{f'(x+h) - f'(x)}{h}$$

is called the *second derivative* of f with respect to x. Likewise higher derivatives are defined recursively as

$$f'''(x) = \left(f''(x)\right)' \quad \text{or} \quad \frac{d^3}{dx^3}f(x) = \frac{d}{dx}\left(\frac{d^2}{dx^2}f(x)\right), \quad \text{etc.}$$

Differentiating with maple. Using maple one can differentiate expressions as well as functions. If the expression g is of the form

```
g := x^2 - a*x;
```

then the corresponding function f is defined by

```
f := x -> x^2 - a*x;
```

The evaluation of functions generates expressions, for example f(t) produces the expression $t^2 - at$. Conversely, expressions can be converted to functions using unapply

```
h := unapply(g,x);
```

The derivative of expressions can be obtained using diff, those of functions using D. Examples can be found in the maple worksheet mp07_1.mws.

7.3 Interpretations of the Derivative

We introduced the derivative geometrically as the slope of the tangent, and we saw that the tangent to a graph of a differentiable function f at the point $(x_0, f(x_0))$ is given by

$$y = f'(x_0)(x - x_0) + f(x_0).$$

Example 7.14 Let $f(x) = x^4 + 1$ with derivative $f'(x) = 4x^3$.

(i) The tangent to the graph of f at the point $(0, 1)$ is

$$y = f'(0) \cdot (x - 0) + f(0) = 1$$

and thus horizontal.

(ii) The tangent to the graph of f at the point $(1, 2)$ is

$$y = f'(1)(x - 1) + 2 = 4(x - 1) + 2 = 4x - 2.$$

The derivative allows further interpretations.

Interpretation as linear approximation. We start off by emphasising that every differentiable function f can be written in the form

$$f(x) = f(x_0) + f'(x_0)(x - x_0) + R(x, x_0),$$

where the remainder $R(x, x_0)$ has the property

$$\lim_{x \to x_0} \frac{R(x, x_0)}{x - x_0} = 0.$$

This follows immediately from

$$R(x, x_0) = f(x) - f(x_0) - f'(x_0)(x - x_0)$$

by dividing by $x - x_0$, since

$$\frac{f(x) - f(x_0)}{x - x_0} \to f'(x_0) \quad \text{as } x \to x_0.$$

Application 7.15 As we have just seen, a differentiable function f is characterised by the property that

$$f(x) = f(x_0) + f'(x_0)(x - x_0) + R(x, x_0),$$

where the remainder term $R(x, x_0)$ tends faster to zero than $x - x_0$. Taking the limit $x \to x_0$ in this equation shows in particular that *every differentiable function is continuous*.

Application 7.16 Let g be the function given by

$$g(x) = k \cdot (x - x_0) + f(x_0).$$

Its graph is the straight line with slope k passing through the point $(x_0, f(x_0))$. Since

$$\frac{f(x) - g(x)}{x - x_0} = \frac{f(x) - f(x_0) - k \cdot (x - x_0)}{x - x_0} = f'(x_0) - k + \underbrace{\frac{R(x, x_0)}{x - x_0}}_{\to 0}$$

as $x \to x_0$, the tangent with $k = f'(x_0)$ is the straight line which approximates the graph best. One therefore calls

$$g(x) = f(x_0) + f'(x_0) \cdot (x - x_0)$$

the *linear approximation* to f at x_0. For x close to x_0 one can consider $g(x)$ as a good approximation to $f(x)$. In applications the (possibly complicated) function f is often replaced by its linear approximation g which is easier to handle.

Example 7.17 Let $f(x) = \sqrt{x} = x^{1/2}$. Consequently,

$$f'(x) = \frac{1}{2}x^{-\frac{1}{2}} = \frac{1}{2\sqrt{x}}.$$

We want to find the linear approximation to the function f at $x_0 = a$. According to the formula above it holds that

$$\sqrt{x} \approx g(x) = \sqrt{a} + \frac{1}{2\sqrt{a}}(x - a)$$

for x close to a, or, alternatively with $h = x - a$,

$$\sqrt{a + h} \approx \sqrt{a} + \frac{1}{2\sqrt{a}} h \qquad \text{for small } h.$$

If we now substitute $a = 1$ and $h = 0.1$, we obtain the approximation

$$\sqrt{1.1} \approx 1 + \frac{0.1}{2} = 1.05.$$

The first digits of the actual value are 1.0488...

Physical interpretation as rate of change. In physical applications the derivative often plays the role of a rate of change. A well-known example from everyday life is the *velocity*, see Sect. 7.1. Consider a particle which is moving along a straight line. Let $s(t)$ be the position where the particle is at time t. The average velocity is given by the quotient

$$\frac{s(t) - s(t_0)}{t - t_0} \qquad \text{(difference in displacement divided by difference in time)}.$$

In the limit $t \to t_0$ the average velocity turns into the *instantaneous velocity*

$$v(t_0) = \frac{ds}{dt}(t_0) = \dot{s}(t_0) = \lim_{t \to t_0} \frac{s(t) - s(t_0)}{t - t_0}.$$

Note that one often writes $\dot{f}(t)$ instead of $f'(t)$ if the time t is the argument of the function f. In particular, in physics the *dot notation* is most commonly used.

Likewise one obtains the acceleration by differentiating the velocity

$$a(t) = \dot{v}(t) = \ddot{s}(t).$$

The notion of velocity is also used in the modelling of other processes that vary over time, e.g. for growth or decay.

7.4 Differentiation Rules

In this section $I \subset \mathbb{R}$ denotes an open interval. We first note that differentiation is a *linear* process.

Proposition 7.18 (Linearity of the derivative) *Let $f, g : I \to \mathbb{R}$ be two functions which are differentiable at $x \in I$ and take $c \in \mathbb{R}$. Then the functions $f + g$ and $c \cdot f$ are differentiable at x as well and*

$$\big(f(x) + g(x)\big)' = f'(x) + g'(x),$$
$$\big(cf(x)\big)' = cf'(x).$$

Proof The result follows from the corresponding rules for limits. The first statement is true because

$$\frac{f(x+h) + g(x+h) - (f(x) + g(x))}{h} = \underbrace{\frac{f(x+h) - f(x)}{h}}_{\to\, f'(x)} + \underbrace{\frac{g(x+h) - g(x)}{h}}_{\to\, g'(x)}$$

as $h \to 0$. The second statement follows similarly.   □

Linearity together with the differentiation rule $(x^m)' = m\,x^{m-1}$ for powers implies that every polynomial is differentiable. Let

$$p(x) = a_n x^n + a_{n-1} x^{n-1} + \cdots + a_1 x + a_0.$$

Then its derivative has the form

$$p'(x) = na_n x^{n-1} + (n-1)a_{n-1} x^{n-2} + \cdots + a_1.$$

For example, $(3x^7 - 4x^2 + 5x - 1)' = 21x^6 - 8x + 5$.

The following two rules allow one to determine the derivative of products and quotients of functions from their factors.

Proposition 7.19 (Product rule) *Let $f, g : I \to \mathbb{R}$ be two functions which are differentiable at $x \in I$. Then the function $f \cdot g$ is differentiable at x and*

$$\big(f(x) \cdot g(x)\big)' = f'(x) \cdot g(x) + f(x) \cdot g'(x).$$

Proof This fact follows again from the corresponding rules for limits

$$\frac{f(x+h) \cdot g(x+h) - f(x) \cdot g(x)}{h}$$

$$= \frac{f(x+h) \cdot g(x+h) - f(x) \cdot g(x+h)}{h} + \frac{f(x) \cdot g(x+h) - f(x) \cdot g(x)}{h}$$

$$= \underbrace{\frac{f(x+h) - f(x)}{h}}_{\to f'(x)} \cdot \underbrace{g(x+h)}_{\to g(x)} + f(x) \cdot \underbrace{\frac{g(x+h) - g(x)}{h}}_{\to g'(x)}$$

as $h \to 0$. The required continuity of g at x is a consequence of Application 7.15. $\square$

Proposition 7.20 (Quotient rule) *Let $f, g : I \to \mathbb{R}$ be two functions differentiable at $x \in I$ and $g(x) \neq 0$. Then the quotient $\frac{f}{g}$ is differentiable at the point x and*

$$\left(\frac{f(x)}{g(x)} \right)' = \frac{f'(x) \cdot g(x) - f(x) \cdot g'(x)}{g(x)^2}.$$

In particular,

$$\left(\frac{1}{g(x)} \right)' = - \frac{g'(x)}{(g(x))^2}.$$

The proof is similar to the one for the product rule and can be found in [3, Chap. 3.1], for example.

Example 7.21 An application of the quotient rule to $\tan x = \dfrac{\sin x}{\cos x}$ shows that

$$\tan' x = \frac{\cos^2 x + \sin^2 x}{\cos^2 x} = \frac{1}{\cos^2 x} = 1 + \tan^2 x.$$

Complicated functions can often be written as a composition of simpler functions. For example, the function

$$h : [2, \infty) \to \mathbb{R} : x \mapsto h(x) = \sqrt{\log(x-1)}$$

can be interpreted as $h(x) = f(g(x))$ with

$$f : [0, \infty) \to \mathbb{R} : y \mapsto \sqrt{y}, \qquad g : [2, \infty) \to [0, \infty) : x \mapsto \log(x-1).$$

One denotes the composition of the functions f and g by $h = f \circ g$. The following proposition shows how such compound functions can be differentiated.

Proposition 7.22 (Chain rule) *The composition of two differentiable functions g : $I \to B$ and $f : B \to \mathbb{R}$ is also differentiable and*

$$\frac{d}{dx} f(g(x)) = f'(g(x)) \cdot g'(x).$$

In shorthand notation the rule is

$$(f \circ g)' = (f' \circ g) \cdot g'.$$

Proof We write

$$\frac{1}{h}\Big(f(g(x+h)) - f(g(x))\Big) = \frac{f(g(x+h)) - f(g(x))}{g(x+h) - g(x)} \cdot \frac{g(x+h) - g(x)}{h}$$

$$= \frac{f(g(x)+k) - f(g(x))}{k} \cdot \frac{g(x+h) - g(x)}{h},$$

where, due to the interpretation as a linear approximation (see Sect. 7.3), the expression

$$k = g(x+h) - g(x)$$

is of the form

$$k = g'(x)h + R(x+h, x)$$

and tends to zero itself as $h \to 0$. It follows that

$$\frac{d}{dx} f(g(x)) = \lim_{h \to 0} \frac{1}{h}\Big(f(g(x+h)) - f(g(x))\Big)$$

$$= \lim_{h \to 0} \left(\frac{f(g(x)+k) - f(g(x))}{k} \cdot \frac{g(x+h) - g(x)}{h} \right)$$

$$= f'(g(x)) \cdot g'(x)$$

and hence the assertion of the proposition.   □

The differentiation of a composite function $h(x) = f(g(x))$ is consequently performed in three steps:

1. Identify the *outer* function f and the *inner* function g with $h(x) = f(g(x))$.
2. Differentiate the outer function f at the point $g(x)$, i.e. compute $f'(y)$ and then substitute $y = g(x)$. The result is $f'(g(x))$.
3. Inner derivative: Differentiate the inner function g and multiply it with the result of step 2. One obtains $h'(x) = f'(g(x)) \cdot g'(x)$.

In the case of three or more compositions, the above rules have to be applied recursively.

Example 7.23 (a) Let $h(x) = (\sin x)^3$. We identify the outer function $f(y) = y^3$ and the inner function $g(x) = \sin x$. Then

$$h'(x) = 3(\sin x)^2 \cdot \cos x.$$

(b) Let $h(x) = e^{-x^2}$. We identify $f(y) = e^y$ and $g(x) = -x^2$. Thus

$$h'(x) = e^{-x^2} \cdot (-2x).$$

The last rule that we will discuss concerns the differentiation of the inverse of a differentiable function.

Proposition 7.24 (Inverse function rule) *Let $f : I \to J$ be bijective, differentiable and $f'(y) \neq 0$ for all $y \in I$. Then $f^{-1} : J \to I$ is also differentiable and*

$$\frac{d}{dx} f^{-1}(x) = \frac{1}{f'(f^{-1}(x))}.$$

In shorthand notation this rule is

$$\left(f^{-1}\right)' = \frac{1}{f' \circ f^{-1}}.$$

Proof We set $y = f^{-1}(x)$ and $\eta = f^{-1}(\xi)$. Due to the continuity of the inverse function (see Proposition C.3) we have that $\eta \to y$ as $\xi \to x$. It thus follows that

$$\frac{d}{dx} f^{-1}(x) = \lim_{\xi \to x} \frac{f^{-1}(\xi) - f^{-1}(x)}{\xi - x} = \lim_{\eta \to y} \frac{\eta - y}{f(\eta) - f(y)}$$

$$= \lim_{\eta \to y} \left(\frac{f(\eta) - f(y)}{\eta - y} \right)^{-1} = \frac{1}{f'(y)} = \frac{1}{f'(f^{-1}(x))}$$

and hence the statement of the proposition. $\qquad\qquad\square$

Figure 7.4 shows the geometric background of the inverse function rule: The slope of a straight line in x-direction is the inverse of the slope in y-direction.

If it is known beforehand that the inverse function is differentiable then its derivative can also be obtained in the following way. One differentiates the identity

$$x = f(f^{-1}(x))$$

with respect to x using the chain rule. This yields

$$1 = f'(f^{-1}(x)) \cdot (f^{-1})'(x)$$

and one obtains the inverse rule by division by $f'(f^{-1}(x))$.

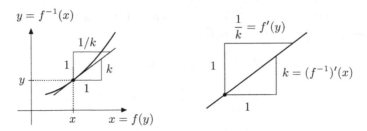

Fig. 7.4 Derivative of the inverse function with detailed view of the slopes

Example 7.25 (Derivative of the logarithm) Since $y = \log x$ is the inverse function to $x = e^y$, it follows from the inverse function rule that

$$(\log x)' = \frac{1}{e^{\log x}} = \frac{1}{x}$$

for $x > 0$. Furthermore

$$\log |x| = \begin{cases} \log x, & x > 0, \\ \log(-x), & x < 0, \end{cases}$$

and thus

$$(\log |x|)' = \begin{cases} (\log x)' = \dfrac{1}{x}, & x > 0, \\ (\log(-x))' = \dfrac{1}{(-x)} \cdot (-1) = \dfrac{1}{x}, & x < 0. \end{cases}$$

Altogether one obtains the formula

$$(\log |x|)' = \frac{1}{x} \quad \text{for} \quad x \neq 0.$$

For logarithms to the base a one has

$$\log_a x = \frac{\log x}{\log a}, \quad \text{thus} \quad (\log_a x)' = \frac{1}{x \log a}.$$

Example 7.26 (Derivatives of general power functions) From $x^\alpha = e^{\alpha \log x}$ we infer by the chain rule that

$$(x^\alpha)' = e^{\alpha \log x} \cdot \frac{\alpha}{x} = x^\alpha \cdot \frac{\alpha}{x} = \alpha \, x^{\alpha-1}.$$

Example 7.27 (Derivative of the general exponential function) For $a > 0$ we have $a^x = e^{x \log a}$. An application of the chain rule shows that

$$(a^x)' = (e^{x \log a})' = e^{x \log a} \cdot \log a = a^x \log a.$$

Example 7.28 For $x > 0$ we have $x^x = e^{x \log x}$ and thus

$$\left(x^x\right)' = e^{x \log x} \left(\log x + \frac{x}{x}\right) = x^x \left(\log x + 1\right).$$

Example 7.29 (Derivatives of cyclometric functions) We recall the differentiation rules for the trigonometric functions on their principal branches:

$$\begin{aligned}
(\sin x)' &= \cos x = \sqrt{1 - \sin^2 x}, & -\tfrac{\pi}{2} \leq x \leq \tfrac{\pi}{2}, \\
(\cos x)' &= -\sin x = -\sqrt{1 - \cos^2 x}, & 0 \leq x \leq \pi, \\
(\tan x)' &= 1 + \tan^2 x, & -\tfrac{\pi}{2} < x < \tfrac{\pi}{2}.
\end{aligned}$$

The inverse function rule thus yields

$$\begin{aligned}
(\arcsin x)' &= \frac{1}{\sqrt{1 - \sin^2(\arcsin x)}} = \frac{1}{\sqrt{1 - x^2}}, & -1 < x < 1, \\
(\arccos x)' &= \frac{-1}{\sqrt{1 - \cos^2(\arccos x)}} = -\frac{1}{\sqrt{1 - x^2}}, & -1 < x < 1, \\
(\arctan x)' &= \frac{1}{1 + \tan^2(\arctan x)} = \frac{1}{1 + x^2}, & -\infty < x < \infty.
\end{aligned}$$

Example 7.30 (Derivatives of hyperbolic and inverse hyperbolic functions) The derivative of the hyperbolic sine is readily computed by invoking the defining formula:

$$(\sinh x)' = \left(\frac{1}{2}\left(e^x - e^{-x}\right)\right)' = \frac{1}{2}\left(e^x + e^{-x}\right) = \cosh x.$$

The derivative of the hyperbolic cosine is obtained in the same way; for differentiating the hyperbolic tangent, the quotient rule is to be applied (see Exercise 3):

$$(\cosh x)' = \sinh x, \quad (\tanh x)' = 1 - \tanh^2 x.$$

The derivative of the inverse hyperbolic sine can be computed by means of the inverse function rule:

$$(\operatorname{arsinh} x)' = \frac{1}{\cosh(\operatorname{arsinh} x)} = \frac{1}{\sqrt{1 + \sinh^2(\operatorname{arsinh} x)}} = \frac{1}{\sqrt{1 + x^2}}$$

for $x \in \mathbb{R}$, where we have used the identity $\cosh^2 x - \sinh^2 x = 1$. In a similar way, the derivatives of the other inverse hyperbolic functions can be computed on their respective domains (Exercise 3):

$$\begin{aligned}
(\operatorname{arcosh} x)' &= \frac{1}{\sqrt{x^2 - 1}}, & x > 1, \\
(\operatorname{artanh} x)' &= \frac{1}{1 - x^2}, & -1 < x < 1.
\end{aligned}$$

Table 7.1 Derivatives of the elementary functions ($\alpha \in \mathbb{R}$, $a > 0$)

$f(x)$	1	x^α	e^x	a^x	$\log|x|$	$\log_a x$
$f'(x)$	0	$\alpha x^{\alpha-1}$	e^x	$a^x \log a$	$\dfrac{1}{x}$	$\dfrac{1}{x \log a}$
$f(x)$	$\sin x$	$\cos x$	$\tan x$	$\arcsin x$	$\arccos x$	$\arctan x$
$f'(x)$	$\cos x$	$-\sin x$	$1 + \tan^2 x$	$\dfrac{1}{\sqrt{1-x^2}}$	$\dfrac{-1}{\sqrt{1-x^2}}$	$\dfrac{1}{1+x^2}$
$f(x)$	$\sinh x$	$\cosh x$	$\tanh x$	$\operatorname{arsinh} x$	$\operatorname{arcosh} x$	$\operatorname{artanh} x$
$f'(x)$	$\cosh x$	$\sinh x$	$1 - \tanh^2 x$	$\dfrac{1}{\sqrt{1+x^2}}$	$\dfrac{1}{\sqrt{x^2-1}}$	$\dfrac{1}{1-x^2}$

The derivatives of the most important elementary functions are collected in Table 7.1. The formulas are valid on the respective domains.

7.5 Numerical Differentiation

In applications it often happens that a function can be evaluated for arbitrary arguments, but no analytic formula is known which represents the function. This situation, for example, arises if the dependent variable is determined using a measuring instrument, e.g. the temperature at a given point as a function of time.

The definition of the derivative as a limit of difference quotients suggests that the derivative of such functions can be approximated by an appropriate difference quotient

$$f'(a) \approx \frac{f(a+h) - f(a)}{h}.$$

The question is how small h should be chosen. In order to decide this we will first carry out a numerical experiment.

Experiment 7.31 Use the above formula to approximate the derivative $f'(a)$ of $f(x) = e^x$ at $a = 1$. Consider different values of h, for example for $h = 10^{-j}$ with $j = 0, 1, \ldots, 16$. One expects a value close to $e = 2.71828\ldots$ as result. Typical outcomes of such an experiment are listed in Table 7.2.

One sees that the error initially decreases with h, but increases again for smaller h. The reason lies in the representation of numbers on a computer. The experiment was carried out in IEEE double precision which corresponds to a relative machine accuracy of eps $\approx 10^{-16}$. The experiment shows that the best result is obtained for

$$h \approx \sqrt{\text{eps}} \approx 10^{-8}.$$

Table 7.2 Numerical differentiation of the exponential function at $a = 1$ using a *one-sided* difference quotient. The numerical results and errors are given as functions of h

h	Value	Error
1.000E-000	4.67077427047160	1.95249244201256E-000
1.000E-001	2.85884195487388	1.40560126414838E-001
1.000E-002	2.73191865578714	1.36368273280976E-002
1.000E-003	2.71964142253338	1.35959407433051E-003
1.000E-004	2.71841774708220	1.35918623152431E-004
1.000E-005	2.71829541994577	1.35914867218645E-005
1.000E-006	2.71828318752147	1.35906242526573E-006
1.000E-007	2.71828196740610	1.38947053418548E-007
1.000E-008	2.71828183998415	1.15251088672608E-008
1.000E-009	2.71828219937549	3.70916445113778E-007
1.000E-010	2.71828349976758	1.67130853068187E-006
1.000E-011	2.71829650802524	1.46795661959409E-005
1.000E-012	2.71866817252997	3.86344070924416E-004
1.000E-013	2.71755491373926	-7.26914719783700E-004
1.000E-014	2.73058485544819	1.23030269891471E-002
1.000E-015	3.16240089670572	4.44119068246674E-001
1.000E-016	1.44632569809566	-1.27195613036338E-000

This behaviour can be explained by using *Taylor expansion*. In Chap. 12 we will derive the formula

$$f(a + h) = f(a) + hf'(a) + \frac{h^2}{2}f''(\xi),$$

where ξ denotes an appropriate point between a and $a + h$. (The value of ξ is usually not known.) Thus, after rearranging, we get

$$f'(a) = \frac{f(a + h) - f(a)}{h} - \frac{h}{2}f''(\xi).$$

The *discretisation error*, i.e. the error which arises from replacing the derivative by the difference quotient, is proportional to h and decreases *linearly* with h. This behaviour can also be seen in the numerical experiment for h between 10^{-2} and 10^{-8}.

For very small h *rounding errors* additionally come into play. As we have seen in Sect. 1.4 the calculation of $f(a)$ on a computer yields

$$\mathrm{rd}(f(a)) = f(a) \cdot (1 + \varepsilon) = f(a) + \varepsilon f(a)$$

Fig. 7.5 Approximation of
the tangent by a symmetric
secant

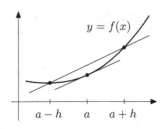

with $|\varepsilon| \leq$ eps. The rounding error turns out to be proportional to eps/h and increases dramatically for small h. This behaviour can be seen in the numerical experiment for h between 10^{-8} and 10^{-16}.

The result of the numerical derivative using the *one-sided difference quotient*

$$f'(a) \approx \frac{f(a+h) - f(a)}{h}$$

is then most precise if discretisation and rounding error have approximately the same magnitude, so if

$$h \approx \frac{\text{eps}}{h} \quad \text{or} \quad h \approx \sqrt{\text{eps}} \approx 10^{-8}.$$

In order to calculate the derivative of $f'(a)$ one can also use a secant placed *symmetrically* around $(a, f(a))$, i.e.

$$f'(a) = \lim_{h \to 0} \frac{f(a+h) - f(a-h)}{2h}$$

This suggests the *symmetric* formula

$$f'(a) \approx \frac{f(a+h) - f(a-h)}{2h}.$$

This approximation is called *symmetric difference quotient* (Fig. 7.5).

To analyse the accuracy of the approximation, we need the Taylor series from Chap. 12:

$$f(a+h) = f(a) + hf'(a) + \frac{h^2}{2}f''(a) + \frac{h^3}{6}f'''(a) + \cdots$$

If one replaces h by $-h$ in this formula

$$f(a-h) = f(a) - hf'(a) + \frac{h^2}{2}f''(a) - \frac{h^3}{6}f'''(a) + \cdots$$

Table 7.3 Numerical differentiation of the exponential function at $a = 1$ using a *symmetric* difference quotient. The numerical results and errors are given as functions of h

h	Value	Error
1.000E-000	3.19452804946533	4.76246221006280E-001
1.000E-001	2.72281456394742	4.53273548837307E-003
1.000E-002	2.71832713338270	4.53049236583958E-005
1.000E-003	2.71828228150582	4.53046770765297E-007
1.000E-004	2.71828183298958	4.53053283777649E-009
1.000E-005	2.71828182851255	5.35020916458961E-011
1.000E-006	2.71828182834134	-1.17704512803130E-010
1.000E-007	2.71828182903696	5.77919490041268E-010
1.000E-008	2.71828181795317	-1.05058792776447E-008
1.000E-009	2.71828182478364	-3.67540575751946E-009
1.000E-010	2.71828199164235	1.63183308643511E-007
1.000E-011	2.71829103280427	9.20434522511116E-006
1.000E-012	2.71839560410381	1.13775644761560E-004

and takes the difference, one obtains

$$f(a+h) - f(a-h) = 2hf'(a) + 2\frac{h^3}{6}f'''(a) + \cdots$$

and furthermore

$$f'(a) = \frac{f(a+h) - f(a-h)}{2h} - \frac{h^2}{6}f'''(a) + \cdots$$

In this case the discretisation error is hence proportional to h^2, while the rounding error is still proportional to $\texttt{eps}/h$.

The symmetric procedure thus delivers the best results for

$$h^2 \approx \frac{\texttt{eps}}{h} \quad \text{or} \quad h \approx \sqrt[3]{\texttt{eps}},$$

respectively. We repeat Experiment 7.31 with $f(x) = e^x$, $a = 1$ and $h = 10^{-j}$ for $j = 0, \ldots, 12$. The results are listed in Table 7.3.

As expected one obtains the best result for $h \approx 10^{-5}$. The obtained approximation is more precise than that of Table 7.2. Since symmetric procedures generally give better results, *symmetry* is an important concept in numerical mathematics.

Numerical differentiation of noisy functions. In practice it often occurs that a function which has to be differentiated consists of *discrete* data that are additionally perturbed by a noise. The noise represents small measuring errors and behaves statistically like random numbers.

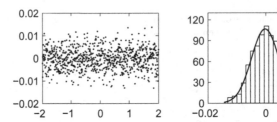

Fig. 7.6 The left picture shows random noise which masks the data. The noise is modelled by 801 normally distributed random numbers. The frequencies of the chosen random numbers can be seen in the histogram in the right picture. For comparison, the (scaled) density of the corresponding normal distribution is given there as well

Example 7.32 Digitising a line of a picture by $J + 1$ pixels produces a function

$$f : \{0, 1, \ldots, J\} \to \mathbb{R} : j \mapsto f(j) = f_j = \text{brightness of the } j\text{th pixel.}$$

In order to find an edge in the picture, where the brightness locally changes very rapidly, this function has to be differentiated.

We consider a concrete example. Suppose that the picture information consists of the function

$$g : [a, b] \to \mathbb{R} : x \mapsto g(x) = -2x^3 + 4x$$

with $a = -2$ and $b = 2$. Let Δx be the distance between two pixels and

$$J = \frac{b - a}{\Delta x}$$

denote the total number of pixels minus 1. We choose $\Delta x = 1/200$ and thus obtain $J = 800$. The actual brightness of the jth pixel would then be

$$g_j = g(a + j \Delta x), \qquad 0 \le j \le J.$$

However, due to measuring errors the measuring instrument supplies

$$f_j = g_j + \varepsilon_j,$$

where ε_j are random numbers. We choose normally distributed random numbers with expected value 0 and variance $2.5 \cdot 10^{-5}$ for ε_j, see Fig. 7.6. For an exact definition of the notions of expected value and variance we refer to the literature, for instance [18].

These random numbers can be generated in MATLAB using the command

```
randn(1,801)*sqrt(2.5e-5).
```

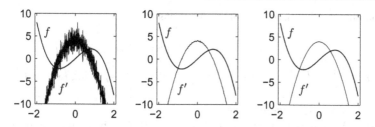

Fig. 7.7 Numerically obtained derivative of a noisy function f, consisting of 801 data values (left); derivative of the same function after filtering using a Gaussian filter (middle) and after smoothing using splines (right)

Differentiating f using the previous rules generates

$$f'_j \approx \frac{f_j - f_{j-1}}{\Delta x} = \frac{g_j - g_{j-1}}{\Delta x} + \frac{\varepsilon_j - \varepsilon_{j-1}}{\Delta x}$$

and the part with g gives the desired value of the derivative, namely

$$\frac{g_j - g_{j-1}}{\Delta x} = \frac{g(a + j\Delta x) - g(a + j\Delta x - \Delta x)}{\Delta x} \approx g'(a + j\Delta x).$$

The sequence of random numbers results in a *non-differentiable* graph. The expression

$$\frac{\varepsilon_j - \varepsilon_{j-1}}{\Delta x}$$

is proportional to $J \cdot \max_{0 \le j \le J} |\varepsilon_j|$. The errors become dominant for large J, see Fig. 7.7, left picture.

To still obtain reliable results, the data have to be smoothed before differentiating. The simplest method is a so-called *convolution* with a *Gaussian filter* which amounts to a weighted averaging of the data (Fig. 7.7, middle). Alternatively one can also use *splines* for smoothing, for example the routine csaps in MATLAB. For the right picture in Fig. 7.7 this method has been used.

Experiment 7.33 Generate Fig. 7.7 using the MATLAB program mat07_1.m and investigate the influence of the choice of random numbers and the smoothing parameter in csaps on the result.

7.6 Exercises

1. Compute the first derivative of the functions

$$f(x) = x^3, \quad g(t) = \frac{1}{t^2}, \quad h(x) = \cos x, \quad k(x) = \frac{1}{\sqrt{x}}, \quad \ell(t) = \tan t$$

using the definition of the derivative as a limit.

2. Compute the first derivative of the functions

$$a(x) = \frac{x^2-1}{x^2+2x+1}, \qquad b(x) = (x^3-1)\sin^2 x, \quad c(t) = \sqrt{1+t^2}\arctan t,$$

$$d(t) = t^2 e^{\cos(t^2+1)}, \quad e(x) = x^{2\sin x}, \qquad\qquad f(s) = \log\left(s + \sqrt{1+s^2}\right).$$

Check your results with maple.

3. Derive the remaining formulas in Example 7.30. Start by computing the derivatives of the hyperbolic cosine and hyperbolic tangent. Use the inverse function rule to differentiate the inverse hyperbolic cosine and inverse hyperbolic tangent.

4. Compute an approximation of $\sqrt{34}$ by replacing the function $f(x) = \sqrt{x}$ at $x = 36$ by its linear approximation. How accurate is your result?

5. Find the equation of the tangent line to the graph of the function $y = f(x)$ through the point $(x_0, f(x_0))$, where

$$f(x) = \frac{x}{2} + \frac{x}{\log x} \quad \text{and} \quad \text{(a)}\ x_0 = e; \quad \text{(b)}\ x_0 = e^2.$$

6. Sand runs from a conveyor belt onto a heap with a velocity of 2 m^3/min. The sand forms a cone-shaped pile whose height equals $\frac{4}{3}$ of the radius. With which velocity does the radius grow if the sand cone has a diameter of 6 m?

Hint. Determine the volume V as a function of the radius r, consider V and r as functions of time t and differentiate the equation with respect to t. Compute $\dot{r}$.

7. Use the Taylor series

$$y(x+h) = y(x) + hy'(x) + \frac{h^2}{2}y''(x) + \frac{h^3}{6}y'''(x) + \frac{h^4}{24}y^{(4)}(x) + \cdots$$

to derive the formula

$$y''(x) = \frac{y(x+h) - 2y(x) + y(x-h)}{h^2} - \frac{h^2}{12}y^{(4)}(x) + \cdots$$

and read off from that a numerical method for calculating the second derivative. The discretisation error is proportional to h^2, and the rounding error is proportional to eps/h^2. By equating the discretisation and the rounding error deduce the optimal step size h. Check your considerations by performing a numerical experiment in MATLAB, computing the second derivative of $y(x) = e^{2x}$ at the point $x = 1$.

8. Write a MATLAB program which numerically differentiates a given function on a given interval and plots the function and its first derivative. Test your program on the functions

$$f(x) = \cos x, \qquad 0 \le x \le 6\pi,$$

and

$$g(x) = e^{-\cos(3x)}, \qquad 0 \le x \le 2.$$

9. Show that the nth derivative of the power function $y = x^n$ equals $n!$ for $n \geq 1$. Verify that the derivative of order $n + 1$ of a polynomial $p(x) = a_n x^n + a_{n-1} x^{n-1} + \cdots + a_1 x + a_0$ of degree n equals zero.

10. Compute the second derivative of the functions

$$f(x) = e^{-x^2}, \quad g(x) = \log\left(x + \sqrt{1 + x^2}\right), \quad h(x) = \log\frac{x + 1}{x - 1}.$$

Applications of the Derivative

8

This chapter is devoted to some applications of the derivative which form part of the basic skills in modelling. We start with a discussion of features of graphs. More precisely, we use the derivative to describe geometric properties like maxima, minima and monotonicity. Even though plotting functions with MATLAB or maple is simple, understanding the connection with the derivative is important, for example, when a function with given properties is to be chosen from a particular class of functions.

In the following section we discuss Newton's method and the concept of order of convergence. Newton's method is one of the most important tools for computing zeros of functions. It is nearly universally in use.

The final section of this chapter is devoted to an elementary method from data analysis. We show how to compute a regression line through the origin. There are many areas of application that involve linear regression. This topic will be developed in more detail in Chap. 18.

8.1 Curve Sketching

In the following we investigate some geometric properties of graphs of functions using the derivative: maxima and minima, intervals of monotonicity and convexity. We further discuss the mean value theorem which is an important technical tool for proofs.

Definition 8.1 A function $f : [a, b] \to \mathbb{R}$ has

(a) a *global maximum* at $x_0 \in [a, b]$ if

$$f(x) \leq f(x_0) \text{ for all } x \in [a, b];$$

© Springer Nature Switzerland AG 2018
M. Oberguggenberger and A. Ostermann, *Analysis for Computer Scientists*,
Undergraduate Topics in Computer Science,
https://doi.org/10.1007/978-3-319-91155-7_8

Fig. 8.1 Minima and
maxima of a function

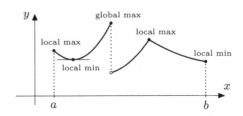

(b) a *local maximum* at $x_0 \in [a, b]$, if there exists a neighbourhood $U_\varepsilon(x_0)$ so that

$$f(x) \leq f(x_0) \text{ for all } x \in U_\varepsilon(x_0) \cap [a, b].$$

The maximum is called *strict* if the strict inequality $f(x) < f(x_0)$ holds in (a) or
(b) for $x \neq x_0$.

The definition for *minimum* is analogous by inverting the inequalities. Maxima
and minima are subsumed under the term *extrema*. Figure 8.1 shows some possible
situations. Note that the function there does not have a global minimum on the chosen
interval.

For points x_0 in the open interval (a, b) one has a simple necessary condition for
extrema of differentiable functions:

Proposition 8.2 *Let $x_0 \in (a, b)$ and f be differentiable at x_0. If f has a local maximum or minimum at x_0 then $f'(x_0) = 0$.*

Proof Due to the differentiability of f we have

$$f'(x_0) = \lim_{h \to 0+} \frac{f(x_0 + h) - f(x_0)}{h} = \lim_{h \to 0-} \frac{f(x_0 + h) - f(x_0)}{h}.$$

In the case of a maximum the slope of the secant satisfies the inequalities

$$\frac{f(x_0 + h) - f(x_0)}{h} \leq 0, \quad \text{if } h > 0,$$

$$\frac{f(x_0 + h) - f(x_0)}{h} \geq 0, \quad \text{if } h < 0.$$

Consequently the limit $f'(x_0)$ has to be greater than or equal to zero as well as
smaller than or equal to zero, thus necessarily $f'(x_0) = 0$. $\qquad\square$

The function $f(x) = x^3$, whose derivative vanishes at $x = 0$, shows that the condition of the proposition is not sufficient for the existence of a maximum or minimum.

The geometric content of the proposition is that in the case of differentiability the
graph of the function has a horizontal tangent at a maximum or minimum. A point
$x_0 \in (a, b)$ where $f'(x_0) = 0$ is called a *stationary point*.

Fig. 8.2 The mean value theorem

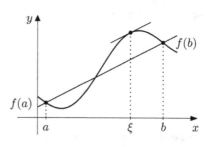

Remark 8.3 The proposition shows that the following point sets have to be checked in order to determine the maxima and minima of a function $f: [a, b] \to \mathbb{R}$:

(a) the boundary points $x_0 = a$, $x_0 = b$;

(b) points $x_0 \in (a, b)$ at which f is not differentiable;

(c) points $x_0 \in (a, b)$ at which f is differentiable and $f'(x_0) = 0$.

The following proposition is a useful technical tool for proofs. One of its applications lies in estimating the error of numerical methods. Similarly to the intermediate value theorem, the proof is based on the completeness of the real numbers. We are not going to present it here but instead refer to the literature, for instance [3, Chap. 3.2].

Proposition 8.4 (Mean value theorem) *Let f be continuous on $[a, b]$ and differentiable on (a, b). Then there exists a point $\xi \in (a, b)$ such that*

$$\frac{f(b) - f(a)}{b - a} = f'(\xi).$$

Geometrically this means that the tangent at ξ has the same slope as the secant through $(a, f(a))$, $(b, f(b))$. Figure 8.2 illustrates this fact.

We now turn to the description of the behaviour of the slope of differentiable functions.

Definition 8.5 A function $f: I \to \mathbb{R}$ is called *monotonically increasing*, if

$$x_1 < x_2 \quad \Rightarrow \quad f(x_1) \le f(x_2)$$

for all $x_1, x_2 \in I$. It is called *strictly monotonically increasing*, if

$$x_1 < x_2 \quad \Rightarrow \quad f(x_1) < f(x_2).$$

A function f is said to be (strictly) monotonically decreasing, if $-f$ is (strictly) monotonically increasing.

Examples of strictly monotonically increasing functions are the power functions $x \mapsto x^n$ with odd powers n; a monotonically, but not strictly monotonically increasing function is the sign function $x \mapsto \text{sign } x$, for instance. The behaviour of the slope of a differentiable function can be described by the sign of the first derivative.

Fig. 8.3 Local maximum

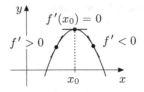

Proposition 8.6 *For differentiable functions* $f: (a, b) \to \mathbb{R}$ *the following implications hold:*

(a) $\quad \begin{array}{l} f' \geq 0 \ on \ (a, b) \quad \Leftrightarrow \ f \ is \ monotonically \ increasing; \\ f' > 0 \ on \ (a, b) \quad \Rightarrow \ f \ is \ strictly \ monotonically \ increasing. \end{array}$

(b) $\quad \begin{array}{l} f' \leq 0 \ on \ (a, b) \quad \Leftrightarrow \ f \ is \ monotonically \ decreasing; \\ f' < 0 \ on \ (a, b) \quad \Rightarrow \ f \ is \ strictly \ monotonically \ decreasing. \end{array}$

Proof (a) According to the mean value theorem we have $f(x_2) - f(x_1) = f'(\xi) \cdot (x_2 - x_1)$ for a certain $\xi \in (a, b)$. If $x_1 < x_2$ and $f'(\xi) \geq 0$ then $f(x_2) - f(x_1) \geq 0$. If $f'(\xi) > 0$ then $f(x_2) - f(x_1) > 0$. Conversely

$$f'(x) = \lim_{h \to 0} \frac{f(x + h) - f(x)}{h} \geq 0,$$

if f is increasing. The proof for (b) is similar. $\square$

Remark 8.7 The example $f(x) = x^3$ shows that f can be strictly monotonically increasing even if $f' = 0$ at isolated points.

Proposition 8.8 (Criterion for local extrema) *Let f be differentiable on (a, b), $x_0 \in (a, b)$ and $f'(x_0) = 0$. Then*

(a) $\quad \left. \begin{array}{l} f'(x) > 0 \quad for \ x < x_0 \\ f'(x) < 0 \quad for \ x > x_0 \end{array} \right\} \quad \Rightarrow \quad f \ has \ a \ local \ maximum \ in \ x_0,$

(b) $\quad \left. \begin{array}{l} f'(x) < 0 \quad for \ x < x_0 \\ f'(x) > 0 \quad for \ x > x_0 \end{array} \right\} \quad \Rightarrow \quad f \ has \ a \ local \ minimum \ in \ x_0.$

Proof The proof follows from the previous proposition which characterises the monotonic behaviour as shown in Fig. 8.3. $\square$

Remark 8.9 (Convexity and concavity of a function graph) If $f'' > 0$ holds in an interval then f' is monotonically increasing there. Thus the graph of f is *curved to the left* or *convex*. On the other hand, if $f'' < 0$, then f' is monotonically decreasing and the graph of f is *curved to the right* or *concave* (see Fig. 8.4). A quantitative description of the curvature of the graph of a function will be given in Sect. 14.2.

Fig. 8.4 Convexity/
concavity and second
derivative

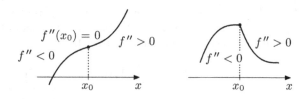

Let x_0 be a point where $f'(x_0) = 0$. If f' does not change its sign at x_0, then x_0 is an *inflection point*. Here f changes from positive to negative curvature or vice versa.

Proposition 8.10 (Second derivative criterion for local extrema) *Let f be twice continuously differentiable on (a, b), $x_0 \in (a, b)$ and $f'(x_0) = 0$.*

(a) If $f''(x_0) > 0$ then f has a local minimum at x_0.
(b) If $f''(x_0) < 0$ then f has a local maximum at x_0.

Proof (a) Since f'' is continuous, $f''(x) > 0$ for all x in a neighbourhood of x_0. According to Proposition 8.6, f' is strictly monotonically increasing in this neighbourhood. Because of $f'(x_0) = 0$ this means that $f'(x_0) < 0$ for $x < x_0$ and $f'(x) > 0$ for $x > x_0$; according to the criterion for local extrema, x_0 is a minimum. The assertion (b) can be shown similarly.   □

Remark 8.11 If $f''(x_0) = 0$ there can either be an inflection point or a minimum or maximum. The functions $f(x) = x^n$, $n = 3, 4, 5, \ldots$ supply a typical example. In fact, they have for n even a global minimum at $x = 0$, and an inflection point for n odd. More general functions can easily be assessed using Taylor expansion. An extreme value criterion based on this expansion will be discussed in Application 12.14.

One of the applications of the previous propositions is *curve sketching*, which is the detailed investigation of the properties of the graph of a function using differential calculus. Even though graphs can easily be plotted in MATLAB or maple it is still often necessary to check the graphical output at certain points using analytic methods.

Experiment 8.12 Plot the function

$$y = x(\operatorname{sign} x - 1)(x + 1)^3 + \big(\operatorname{sign}(x - 1) + 1\big)\big((x - 2)^4 - 1/2\big)$$

on the interval $-2 \le x \le 3$ and try to read off the local and global extrema, the inflection points and the monotonic behaviour. Check your observations using the criteria discussed above.

A further application of the previous propositions consists in finding *extrema*, i.e. solving one-dimensional *optimisation problems*. We illustrate this topic using a standard example.

Example 8.13 Which rectangle with a given perimeter has the largest area? To answer this question we denote the lengths of the sides of the rectangle by x and y. Then the perimeter and the area are given by

$$U = 2x + 2y, \qquad F = xy.$$

Since U is fixed, we obtain $y = U/2 - x$, and from that

$$F = x(U/2 - x),$$

where x can vary in the domain $0 \le x \le U/2$. We want to find the maximum of the function F on the interval $[0, U/2]$. Since F is differentiable, we only have to investigate the boundary points and the stationary points. At the boundary points $x = 0$ and $x = U/2$ we have $F(0) = 0$ and $F(U/2) = 0$. The stationary points are obtained by setting the derivative to zero

$$F'(x) = U/2 - 2x = 0,$$

which brings us to $x = U/4$ with the function value $F(U/4) = U^2/16$.

As result we get that the maximum area is obtained at $x = U/4$, thus in the case of a square.

8.2   Newton's Method

With the help of differential calculus efficient numerical methods for computing zeros of differentiable functions can be constructed. One of the basic procedures is *Newton's method*[1] which will be discussed in this section for the case of real-valued functions $f: D \subset \mathbb{R} \to \mathbb{R}$.

First we recall the *bisection method* discussed in Sect. 6.3. Consider a continuous, real-valued function f on an interval $[a, b]$ with

$$f(a) < 0, \ f(b) > 0 \ \text{ or } \ f(a) > 0, \ f(b) < 0.$$

With the help of continued bisection of the interval, one obtains a zero ξ of f satisfying

$$a = a_1 \le a_2 \le a_3 \le \cdots \le \xi \le \cdots \le b_3 \le b_2 \le b_1 = b,$$

where

$$|b_{n+1} - a_{n+1}| = \frac{1}{2}|b_n - a_n| = \frac{1}{4}|b_{n-1} - a_{n-1}| = \ldots = \frac{1}{2^n}|b_1 - a_1|.$$

---

[1]I. Newton, 1642–1727.

If one stops after n iterations and chooses a_n or b_n as approximation for ξ then one gets a guaranteed error bound

$$|\text{error}| \leq \varphi(n) = |b_n - a_n|.$$

Note that we have

$$\varphi(n + 1) = \frac{1}{2}\,\varphi(n).$$

The error thus decays with each iteration by (at least) a constant factor $\frac{1}{2}$, and one calls the method *linearly convergent*. More generally, an iteration scheme is called convergent of *order* α if there exist error bounds $(\varphi(n))_{n \geq 1}$ and a constant $C > 0$ such that

$$\lim_{n \to \infty} \frac{\varphi(n + 1)}{(\varphi(n))^\alpha} = C.$$

For sufficiently large n, one thus has approximately

$$\varphi(n + 1) \approx C(\varphi(n))^\alpha.$$

Linear convergence ($\alpha = 1$) therefore implies

$$\varphi(n + 1) \approx C\varphi(n) \approx C^2\varphi(n - 1) \approx \ldots \approx C^n\,\varphi(1).$$

Plotting the logarithm of $\varphi(n)$ against n (semi-logarithmic representation, as shown for example in Fig. 8.6) results in a straight line:

$$\log \varphi(n + 1) \approx n \log C + \log \varphi(1).$$

If $C < 1$ then the error bound $\varphi(n + 1)$ tends to 0 and the number of correct decimal places increases with each iteration by a constant. Quadratic convergence would mean that the number of correct decimal places approximately doubles with each iteration.

Derivation of Newton's method. The aim of the construction is to obtain a procedure that provides quadratic convergence ($\alpha = 2$), at least if one starts sufficiently close to a simple zero ξ of a differentiable function. The geometric idea behind Newton's method is simple: Once an approximation x_n is chosen, one calculates x_{n+1} as the intersection of the tangent to the graph of f through $(x_n, f(x_n))$ with the x-axis, see Fig. 8.5. The equation of the tangent is given by

$$y = f(x_n) + f'(x_n)(x - x_n).$$

The point of intersection x_{n+1} with the x-axis is obtained from

$$0 = f(x_n) + f'(x_n)(x_{n+1} - x_n),$$

Fig. 8.5 Two steps of
Newton's method

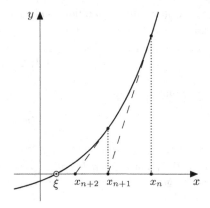

thus

$$x_{n+1} = x_n - \frac{f(x_n)}{f'(x_n)}, \quad n \geq 1.$$

Obviously it has to be assumed that $f'(x_n) \neq 0$. This condition is fulfilled, if f' is
continuous, $f'(\xi) \neq 0$ and x_n is sufficiently close to the zero ξ.

Proposition 8.14 (Convergence of Newton's method) *Let f be a real-valued function, twice differentiable with a continuous second derivative. Further, let $f(\xi) = 0$
and $f'(\xi) \neq 0$. Then there exists a neighbourhood $U_\varepsilon(\xi)$ such that Newton's method
converges quadratically to ξ for every starting value $x_1 \in U_\varepsilon(\xi)$.*

Proof Since $f'(\xi) \neq 0$ and f' is continuous, there exist a neighbourhood $U_\delta(\xi)$ and
a bound $m > 0$ so that $|f'(x)| \geq m$ for all $x \in U_\delta(\xi)$. Applying the mean value
theorem twice gives

$$
\begin{aligned}
|x_{n+1} - \xi| &= \left| x_n - \xi - \frac{f(x_n) - f(\xi)}{f'(x_n)} \right| \\
&\leq |x_n - \xi| \left| 1 - \frac{f'(\eta)}{f'(x_n)} \right| = |x_n - \xi| \frac{|f'(x_n) - f'(\eta)|}{|f'(x_n)|} \\
&\leq |x_n - \xi|^2 \frac{|f''(\zeta)|}{|f'(x_n)|}
\end{aligned}
$$

with η between x_n and ξ and ζ between x_n and η. Let M denote the maximum of
$|f''|$ on $U_\delta(\xi)$. Under the assumption that all iterates x_n lie in the neighbourhood
$U_\delta(\xi)$, we obtain the quadratic error bound

$$\varphi(n+1) = |x_{n+1} - \xi| \leq |x_n - \xi|^2 \frac{M}{m} = (\varphi(n))^2 \frac{M}{m}$$

for the error $\varphi(n) = |x_n - \xi|$. Thus, the assertion of the proposition holds with
the neighbourhood $U_\delta(\xi)$. Otherwise we have to decrease the neighbourhood by

choosing an $\varepsilon < \delta$ which satisfies the inequality $\varepsilon \frac{M}{m} \leq 1$. Then

$$|x_n - \xi| \leq \varepsilon \quad \Rightarrow \quad |x_{n+1} - \xi| \leq \varepsilon^2 \frac{M}{m} \leq \varepsilon.$$

This means that if an approximate value x_n lies in $U_\varepsilon(\xi)$ then so does the subsequent value x_{n+1}. Since $U_\varepsilon(\xi) \subset U_\delta(\xi)$, the quadratic error estimate from above is still valid. Thus the assertion of the proposition is valid with the smaller neighbourhood $U_\varepsilon(\xi)$. $\qquad\qquad \square$

Example 8.15 In computing the root $\xi = \sqrt[3]{2}$ of $x^3 - 2 = 0$, we compare the bisection method with starting interval $[-2, 2]$ and Newton's method with starting value $x_1 = 2$. The interval boundaries $[a_n, b_n]$ and the iterates x_n are listed in Tables 8.1 and 8.2, respectively. Newton's method gives the value

$$\sqrt[3]{2} = 1.25992104989487$$

correct to 14 decimal places after only six iterations.

Table 8.1 Bisection method for calculating the third root of 2

n	a_n	b_n	Error
1	-2.00000000000000	2.00000000000000	4.00000000000000
2	0.00000000000000	2.00000000000000	2.00000000000000
3	1.00000000000000	2.00000000000000	1.00000000000000
4	1.00000000000000	1.50000000000000	0.50000000000000
5	1.25000000000000	1.50000000000000	0.25000000000000
6	1.25000000000000	1.37500000000000	0.12500000000000
7	1.25000000000000	1.31250000000000	0.06250000000000
8	1.25000000000000	1.28125000000000	0.03125000000000
9	1.25000000000000	1.26562500000000	0.01562500000000
10	1.25781250000000	1.26562500000000	0.00781250000000
11	1.25781250000000	1.26171875000000	0.00390625000000
12	1.25976562500000	1.26171875000000	0.00195312500000
13	1.25976562500000	1.26074218750000	0.00097656250000
14	1.25976562500000	1.26025390625000	0.00048828125000
15	1.25976562500000	1.26000976562500	0.00024414062500
16	1.25988769531250	1.26000976562500	0.00012207031250
17	1.25988769531250	1.25994873046875	0.00006103515625
18	1.25991821289063	1.25994873046875	0.00003051757813

Table 8.2 Newton's method
for calculating the third root
of 2

n	x_n	Error
1	2.00000000000000	0.74007895010513
2	1.50000000000000	0.24007895010513
3	1.29629629629630	0.03637524640142
4	1.26093222474175	0.00101117484688
5	1.25992186056593	0.00000081067105
6	1.25992104989539	0.00000000000052
7	1.25992104989487	0.00000000000000

The error curves for the bisection method and Newton's method can be seen in Fig. 8.6. A semi-logarithmic representation (MATLAB command `semilogy`) is used there.

Remark 8.16 The convergence behaviour of Newton's method depends on the conditions of Proposition 8.14. If the starting value x_1 is too far away from the zero ξ, then the method might diverge, oscillate or converge to a different zero. If $f'(\xi) = 0$, which means the zero ξ has a multiplicity > 1, then the order of convergence may be reduced.

Experiment 8.17 Open the applet *Newton's method* and test—using the sine function—how the choice of the starting value influences the result (in the applet the right interval boundary is the initial value). Experiment with the intervals $[-2, x_0]$ for $x_0 = 1, 1.1, 1.2, 1.3, 1.5, 1.57, 1.5707, 1.57079$ and interpret your observations. Also carry out the calculations with the same starting values with the help of the M-file `mat08_2.m`.

Experiment 8.18 With the help of the applet *Newton's method*, study how the order of convergence drops for multiple zeros. For this purpose, use the two polynomial functions given in the applet.

Remark 8.19 Variants of Newton's method can be obtained by evaluating the derivative $f'(x_n)$ numerically. For example, the approximation

$$f'(x_n) \approx \frac{f(x_n) - f(x_{n-1})}{x_n - x_{n-1}}$$

Fig. 8.6 Error of the bisection method and of Newton's method for the calculation of $\sqrt[3]{2}$

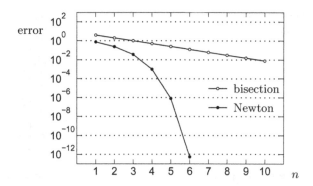

provides the *secant method*

$$x_{n+1} = x_n - \frac{(x_n - x_{n-1})f(x_n)}{f(x_n) - f(x_{n-1})},$$

which computes x_{n+1} as intercept of the secant through $(x_n, f(x_n))$ and $(x_{n-1}, f(x_{n-1}))$ with the x-axis. It has a fractional order less than 2.

8.3 Regression Line Through the Origin

This section is a first digression into data analysis: Given a collection of data points scattered in the plane, find the *line of best fit (regression line)* through the origin. We will discuss this problem as an application of differentiation; it can also be solved by using methods of linear algebra. The general problem of multiple linear regression will be dealt with in Chap. 18.

In the year 2002, the height x [cm] and the weight y [kg] of 70 students in Computer Science at the University of Innsbruck were collected. The data can be obtained from the M-file mat08_3.m.

The measurements $(x_i, y_i), i = 1, \ldots, n$ of height and weight form a scatter plot in the plane as shown in Fig. 8.7. Under the assumption that there is a linear relation of the form $y = kx$ between height and weight, k should be determined such that the straight line $y = kx$ represents the scatter plot *as closely as possible* (Fig. 8.8). The approach that we discuss below goes back to Gauss[2] and understands the data fit in the sense of minimising the sum of squares of the errors.

---

[2]C.F. Gauss, 1777–1855.

Fig. 8.7 Scatter plot
height/weight

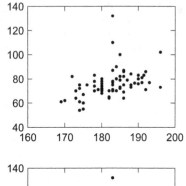

Fig. 8.8 Line of best fit
$y = kx$

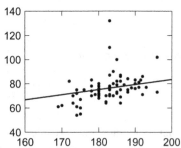

Application 8.20 (Line of best fit through the origin)   A straight line through the
origin

$$y = kx$$

is to be fitted to a scatter plot (x_i, y_i), $i = 1, \ldots, n$. If k is known, one can compute
the square of the deviation of the measurement y_i from the value kx_i given by the
equation of the straight line as

$$(y_i - kx_i)^2$$

(the *square of the error*). We are looking for the specific k which minimises the sum
of squares of the errors; thus

$$F(k) = \sum_{i=1}^{n}(y_i - kx_i)^2 \rightarrow \min$$

Obviously, $F(k)$ is a quadratic function of k. First we compute the derivatives

$$F'(k) = \sum_{i=1}^{n}(-2x_i)(y_i - kx_i), \qquad F''(k) = \sum_{i=1}^{n}2x_i^2.$$

By setting $F'(k) = 0$ we obtain the formula

$$F'(k) = -2\sum_{i=1}^{n}x_i y_i + 2k\sum_{i=1}^{n}x_i^2 = 0.$$

Since evidently $F'' > 0$, its solution

$$k = \frac{\sum x_i y_i}{\sum x_i^2}$$

is the global minimum and gives the slope of the line of best fit.

Example 8.21 To illustrate the regression line through the origin we use the Austrian consumer price index 2010–2016 (data taken from [26]):

year	2010	2011	2012	2013	2014	2015	2016
index	100.0	103.3	105.8	107.9	109.7	110.7	111.7

For the calculation it is useful to introduce new variables x and y, where $x = 0$ corresponds to the year 2010 and $y = 0$ to the index 100. This means that $x =$ (year $- 2010$) and $y =$ (index $- 100$); y describes the relative price increase (in per cent) with respect to the year 2010. The re-scaled data are

x_i	0	1	2	3	4	5	6
y_i	0.0	3.3	5.8	7.9	9.7	10.7	11.7

We are looking for the line of best fit to these data through the origin. For this purpose we have to minimise

$$F(k) = (3.3 - k \cdot 1)^2 + (5.8 - k \cdot 2)^2 + (7.9 - k \cdot 3)^2 + (9.7 - k \cdot 4)^2$$
$$+ (10.7 - k \cdot 5)^2 + (11.7 - k \cdot 6)^2$$

which results in (rounded)

$$k = \frac{1 \cdot 3.3 + 2 \cdot 5.8 + 3 \cdot 7.9 + 4 \cdot 9.7 + 5 \cdot 10.7 + 6 \cdot 11.7}{1 \cdot 1 + 2 \cdot 2 + 3 \cdot 3 + 4 \cdot 4 + 5 \cdot 5 + 6 \cdot 6} = \frac{201.1}{91} = 2.21.$$

The line of best fit is thus

$$y = 2.21x$$

or transformed back

$$\text{index} = 100 + (\text{year} - 2010) \cdot 2.21.$$

The result is shown in Fig. 8.9, in a year/index-scale as well as in the transformed variables. For the year 2017, extrapolation along the regression line would forecast

$$\text{index}(2017) = 100 + 7 \cdot 2.21 = 115.5.$$

Fig. 8.9 Consumer price
index and regression line

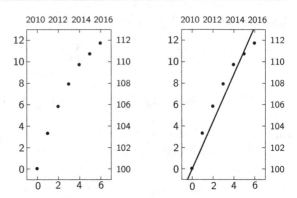

The actual consumer price index in 2017 had the value 114.0. Inspection of Fig. 8.9 shows that the consumer price index stopped growing linearly around 2014; thus the straight line is a bad fit to the data in the period under consideration. How to choose better regression models will be discussed in Chap. 18.

8.4 Exercises

1. Find out which of the following (continuous) functions are differentiable at $x = 0$:

$$y = x|x|; \qquad y = |x|^{1/2}, \qquad y = |x|^{3/2}, \qquad y = x\sin(1/x).$$

2. Find all maxima and minima of the functions

$$f(x) = \frac{x}{x^2 + 1} \quad \text{and} \quad g(x) = x^2 e^{-x^2}.$$

3. Find the maxima of the functions

$$y = \frac{1}{x} e^{-(\log x)^2 / 2}, \ x > 0 \quad \text{and} \quad y = e^{-x} e^{-(e^{-x})}, \ x \in \mathbb{R}.$$

These functions represent the densities of the standard lognormal distribution and of the Gumbel distribution, respectively.

4. Find all maxima and minima of the function

$$f(x) = \frac{x}{\sqrt{x^4 + 1}},$$

determine on what intervals it is increasing or decreasing, analyse its behaviour as $x \to \pm\infty$, and sketch its graph.

Fig. 8.10 Failure wedge
with sliding surface

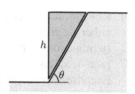

5. Find the proportions of the cylinder which has the smallest surface area F for a given volume V.

   Hint. $F = 2r\pi h + 2r^2\pi \rightarrow$ min. Calculate the height h as a function of the radius r from $V = r^2\pi h$, substitute and minimise $F(r)$.

6. (From mechanics of solids) The moment of inertia with respect to the central axis of a beam with rectangular cross section is $I = \frac{1}{12}bh^3$ (b the width, h the height). Find the proportions of the beam which can be cut from a log with circular cross section of given radius r such that its moment of inertia becomes maximal.

   Hint. Write b as function of h, $I(h) \rightarrow$ max.

7. (From soil mechanics) The mobilised cohesion $c_m(\theta)$ of a failure wedge with sliding surface, inclined by an angle θ, is

$$c_m(\theta) = \frac{\gamma h \sin(\theta - \varphi_m) \cos\theta}{2 \cos\varphi_m}.$$

   Here h is the height of the failure wedge, φ_m the angle of internal friction, γ the specific weight of the soil (see Fig. 8.10). Show that the mobilised cohesion c_m with given h, φ_m, γ is a maximum for the angle of inclination $\theta = \varphi_m/2 + 45°$.

8. This exercise aims at investigating the convergence of Newton's method for solving the equations

$$x^3 - 3x^2 + 3x - 1 = 0,$$
$$x^3 - 3x^2 + 3x - 2 = 0$$

   on the interval $[0, 3]$.

   (a) Open the applet *Newton's method* and carry out Newton's method for both equations with an accuracy of 0.0001. Explain why you need a different number of iterations.

   (b) With the help of the M-file mat08_1.m, generate a list of approximations in each case (starting value x1 = 1.5, tol = 100*eps, maxk = 100) and plot the errors $|x_n - \xi|$ in each case using semilogy. Discuss the results.

9. Apply the MATLAB program mat08_2.m to the functions which are defined by the M-files mat08_f1.m and mat08_f2.m (with respective derivatives mat08_df1.m and mat08_df2.m). Choose x1 = 2, maxk = 250. How do you explain the results?

10. Rewrite the MATLAB program `mat08_2.m` so that termination occurs when either the given number of iterations `maxk` or a given error bound `tol` is reached (termination at the nth iteration, if either $n > $ `maxk` or $|f(x_n)| < $ `tol`). Compute n, x_n and the error $|f(x_n)|$. Test your program using the functions from Exercise 8 and explain the results.

    Hint. Consult the M-file `mat08_ex9.m`.

11. Write a MATLAB program which carries out the secant method for cubic polynomials.

12. (a) By minimising the sum of squares of the errors, derive a formula for the coefficient c of the regression parabola $y = cx^2$ through the data $(x_1, y_1), ..., (x_n, y_n)$.

    (b) A series of measurements of braking distances s [m] (without taking into account the perception-reaction distance) of a certain type of car in dependence on the velocity v [km/h] produced the following values:

v_i	10	20	40	50	60	70	80	100	120
s_i	1	3	8	13	18	23	31	47	63

    Calculate the coefficient c of the regression parabola $s = cv^2$ and plot the result.

13. Show that the best horizontal straight line $y = d$ through the data points (x_i, y_i), $i = 1, \ldots, n$ is given by the arithmetic mean of the y-values:

$$d = \frac{1}{n} \sum_{i=1}^{n} y_i.$$

    Hint. Minimise $G(d) = \sum_{i=1}^{n} (y_i - d)^2$.

14. (From geotechnics) The angle of internal friction of a soil specimen can be obtained by means of a direct shear test, whereby the material is subjected to normal stress σ and the lateral shear stress τ at failure is recorded. In case the cohesion is negligible, the relation between τ and σ can be modelled by a regression line through the origin of the form $\tau = k\sigma$. The slope of the regression line is interpreted as the tangent of the friction angle φ, $k = \tan \varphi$. In a laboratory experiment, the following data have been obtained for a specimen of glacial till (data from [25]):

σ_i [kPa]	100	150	200	300	150	250	300	100	150	250	100	150	200	250
τ_i [kPa]	68	127	135	206	127	148	197	76	78	168	123	97	124	157

    Calculate the angle of internal friction of the specimen.

15. (a) Convince yourself by applying the mean value theorem that the function $f(x) = \cos x$ is a contraction (see Definition C.17) on the interval $[0, 1]$ and compute the *fixed point* $x^* = \cos x^*$ up to two decimal places using the iteration of Proposition C.18.

(b) Write a MATLAB program which carries out the first N iterations for the computation of $x^* = \cos x^*$ for a given initial value $x_1 \in [0, 1]$ and displays $x_1, x_2, \ldots, x_N$ in a column.

Fractals and L-systems

9

In geometry objects are often defined by explicit rules and transformations which can easily be translated into mathematical formulas. For example, a circle is the set of all points which are at a fixed distance r from a centre (a, b):

$$K = \{(x, y) \in \mathbb{R}^2 \; ; \; (x - a)^2 + (y - b)^2 = r^2\}$$

or

$$K = \{(x, y) \in \mathbb{R}^2 \; ; \; x = a + r \cos \varphi, \; y = b + r \sin \varphi, \; 0 \leq \varphi < 2\pi\}.$$

In contrast to that, the objects of *fractal geometry* are usually given by a *recursion*. These fractal sets (*fractals*) have recently found many interesting applications, e.g. in computer graphics (modelling of clouds, plants, trees, landscapes), in image compression and data analysis. Furthermore fractals have a certain importance in modelling growth processes.

Typical properties of fractals are often their *non-integer dimension* and the *self-similarity* of the entire set with its pieces. The latter can frequently be found in nature, e.g. in geology. There it is often difficult to decide from a photograph without a given scale whether the object in question is a grain of sand, a pebble or a large piece of rock. For that reason fractal geometry is often exuberantly called the geometry of nature.

In this chapter we exemplarily have a look at fractals in $\mathbb{R}^2$ and $\mathbb{C}$. Furthermore we give a short introduction to L-systems and discuss, as an application, a simple concept for modelling the growth of plants. For a more in-depth presentation we refer to the textbooks [21, 22].

© Springer Nature Switzerland AG 2018
M. Oberguggenberger and A. Ostermann, *Analysis for Computer Scientists*,
Undergraduate Topics in Computer Science,
https://doi.org/10.1007/978-3-319-91155-7_9

9.1 Fractals

To start with we generalise the notions of *open* and *closed interval* to subsets of $\mathbb{R}^2$. For a fixed $\mathbf{a} = (a, b) \in \mathbb{R}^2$ and $\varepsilon > 0$ the set

$$B(\mathbf{a}, \varepsilon) = \left\{ (x, y) \in \mathbb{R}^2 \; ; \; \sqrt{(x-a)^2 + (y-b)^2} < \varepsilon \right\}$$

is called an ε-*neighbourhood* of $\mathbf{a}$. Note that the set $B(\mathbf{a}, \varepsilon)$ is a circular disc (with centre $\mathbf{a}$ and radius ε) where the *boundary* is missing.

Definition 9.1 Let $A \subseteq \mathbb{R}^2$.

(a) A point $\mathbf{a} \in A$ is called *interior point* of A if there exists an ε-neighbourhood of $\mathbf{a}$ which itself is contained in A.

(b) A is called *open* if each point of A is an interior point.

(c) A point $\mathbf{c} \in \mathbb{R}^2$ is called *boundary point* of A if *every* ε-neighbourhood of $\mathbf{c}$ contains at least one point of A as well as a point of $\mathbb{R}^2 \setminus A$. The set of boundary points of A is denoted by ∂A (*boundary* of A).

(d) A set is called *closed* if it contains all its boundary points.

(e) A is called *bounded* if there is a number $r > 0$ with $A \subseteq B(\mathbf{0}, r)$.

Example 9.2 The square

$$Q = \{(x, y) \in \mathbb{R}^2 \; ; \; 0 < x < 1 \text{ and } 0 < y < 1\}$$

is open since every point of Q has an ε-neighbourhood which is contained in Q, see Fig. 9.1, left picture. The boundary of Q consists of four line segments

$$\{0, 1\} \times [0, 1] \; \cup \; [0, 1] \times \{0, 1\}.$$

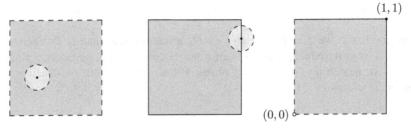

Fig. 9.1 Open (left), closed (middle) and neither open nor closed (right) square with side length 1

Fig. 9.2 Covering a curve using circles

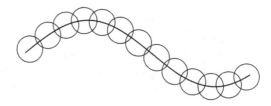

Every ε-neighbourhood of a boundary point also contains points which are outside of Q, see Fig. 9.1, middle picture. The square in Fig. 9.1, right picture,

$$\{(x, y) \in \mathbb{R}^2 \; ; \; 0 < x \leq 1 \text{ and } 0 < y \leq 1\}$$

is neither closed nor open since the boundary point $(x, y) = (0, 0)$ is not an element of the set and the set on the other hand contains the point $(x, y) = (1, 1)$ which is not an inner point. All three sets are bounded since they are, for example, contained in $B(\mathbf{0}, 2)$.

Fractal dimension. Roughly speaking, points have dimension 0, line segments dimension 1 and plane regions dimension 2. The concept of fractal dimension serves to make finer distinctions. If, for example, a curve fills a plane region *densely* one tends to assign to it a higher dimension than 1. Conversely, if a line segment has *many* gaps, its dimension could be between 0 and 1.

Let $A \subseteq \mathbb{R}^2$ be bounded (and not empty) and let $N(A, \varepsilon)$ be the *smallest number* of closed circles with radius ε which are needed to cover A, see Fig. 9.2.

The following intuitive idea stands behind the definition of the fractal dimension d of A: For curve segments the number $N(A, \varepsilon)$ is inverse proportional to ε, for plane regions inverse proportional to ε^2, so

$$N(A, \varepsilon) \approx C \cdot \varepsilon^{-d},$$

where d denotes the dimension. Taking logarithms one obtains

$$\log N(A, \varepsilon) \approx \log C - d \log \varepsilon,$$

and

$$d \approx -\frac{\log N(A, \varepsilon) - \log C}{\log \varepsilon},$$

respectively. This approximation is getting more precise the smaller one chooses $\varepsilon > 0$. Due to

$$\lim_{\varepsilon \to 0^+} \frac{\log C}{\log \varepsilon} = 0$$

this leads to the following definition.

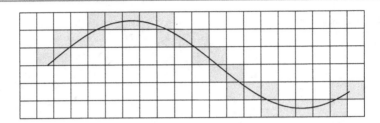

Fig. 9.3 Raster of the plane using squares of side length ε. The boxes that have a non-empty intersection with the fractal are coloured in grey. In the picture we have $N(A, \varepsilon) = 27$

Definition 9.3 Let $A \subseteq \mathbb{R}^2$ be not empty, bounded and $N(A, \varepsilon)$ as above. If the limit

$$d = d(A) = -\lim_{\varepsilon \to 0^+} \frac{\log N(A, \varepsilon)}{\log \varepsilon}$$

exists, then d is called *fractal dimension* of A.

Remark 9.4 In the above definition it is sufficient to choose a zero sequence of the form

$$\varepsilon_n = C \cdot q^n, \qquad 0 < q < 1$$

for ε. Furthermore it is not essential to use circular discs for the covering. One can just as well use squares, see [5, Chap. 5]. Hence the number obtained by Definition 9.3 is also called *box-dimension* of A.

Experimentally the dimension of a fractal can be determined in the following way: For various rasters of the plane with mesh size ε_n one counts the number of boxes which have a non-empty intersection with the fractal, see Fig. 9.3. Let us call this number again $N(A, \varepsilon_n)$. If one plots $\log N(A, \varepsilon_n)$ as a function of $\log \varepsilon_n$ in a double-logarithmic diagram and fits the best line to this graph (Sect. 18.1), then

$$d(A) \approx - \text{ slope of the straight line.}$$

With this procedure one can, for example, determine the fractal dimension of the coastline of Great Britain, see Exercise 1.

Example 9.5 The line segment (Fig. 9.4)

$$A = \{(x, y) \in \mathbb{R}^2 \; ; \; a \le x \le b, \; y = c\}$$

has fractal dimension $d = 1$.

Fig. 9.4 Covering of a straight line segment using circles

Fig. 9.5 A set of points with box-dimension $d = \frac{1}{2}$

We choose

$$\varepsilon_n = (b - a) \cdot 2^{-n}, \qquad q = 1/2.$$

Due to $N(A, \varepsilon_n) = 2^{n-1}$ the following holds

$$-\frac{\log N(A, \varepsilon_n)}{\log \varepsilon_n} = -\frac{(n-1)\log 2}{\log(b-a) - n \log 2} \to 1 \qquad \text{as } n \to \infty.$$

Likewise, it can easily be shown: Every set that consists of finitely many points has fractal dimension 0. Plane regions in $\mathbb{R}^2$ have fractal dimension 2. The fractal dimension is in this way a generalisation of the intuitive notion of dimension. Still, caution is advisable here as can be seen in the following example.

Example 9.6 The set $F = \left\{0, 1, \frac{1}{2}, \frac{1}{3}, \frac{1}{4}, \ldots\right\}$ displayed in Fig. 9.5 has box-dimension $d = 1/2$. We check this claim with the following MATLAB experiment.

Experiment 9.7 To determine the dimension of F approximately with the help of MATLAB we take the following steps. For $j = 1, 2, 3, \ldots$ we split the interval $[0, 1]$ into 4^j equally large subintervals, set $\varepsilon_j = 4^{-j}$ and determine the number $N_j = N(F, \varepsilon_j)$ of subintervals which have a non-empty intersection with F. Then we plot $\log N_j$ as a function of $\log \varepsilon_j$ in a double-logarithmic diagram. The slope of the secant

$$d_j = -\frac{\log N_{j+1} - \log N_j}{\log \varepsilon_{j+1} - \log \varepsilon_j}$$

is an approximation to d which is steadily improving with growing j. The values obtained by using the program mat09_1.m are given in the following table:

4^j	4	16	64	256	1024	4096	16384	65536	262144	1048576
d_j	0.79	0.61	0.55	0.52	0.512	0.5057	0.5028	0.5014	0.5007	0.50035

Verify the given values and determine that the approximations given by Definition 9.3

$$\tilde{d}_j = -\frac{\log N_j}{\log \varepsilon_j}$$

are much worse. Explain this behaviour.

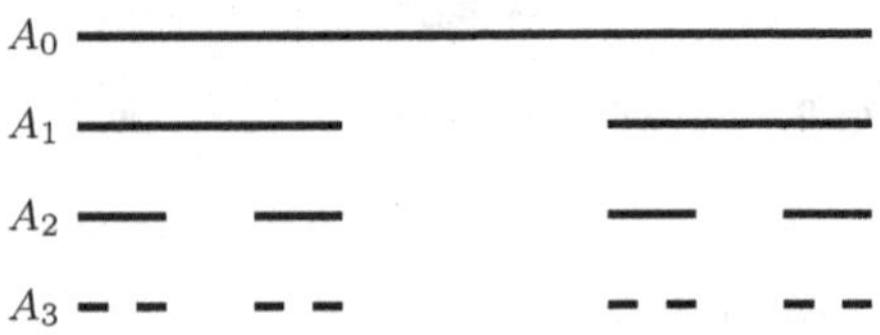

Fig. 9.6 The construction of the Cantor set

Example 9.8 (Cantor set)   We construct this set recursively using

$$A_0 = [0, 1]$$
$$A_1 = \left[0, \tfrac{1}{3}\right] \cup \left[\tfrac{2}{3}, 1\right]$$
$$A_2 = \left[0, \tfrac{1}{9}\right] \cup \left[\tfrac{2}{9}, \tfrac{1}{3}\right] \cup \left[\tfrac{2}{3}, \tfrac{7}{9}\right] \cup \left[\tfrac{8}{9}, 1\right]$$
$$\vdots$$

One obtains A_{n+1} from A_n by removing the middle third of each line segment of A_n, see Fig. 9.6.

The intersection of all these sets

$$A = \bigcap_{n=0}^{\infty} A_n$$

is called Cantor set. Let $|A_n|$ denote the *length* of A_n. Obviously the following holds true: $|A_0| = 1$, $|A_1| = 2/3$, $|A_2| = (2/3)^2$ and $|A_n| = (2/3)^n$. Thus

$$|A| = \lim_{n\to\infty} |A_n| = \lim_{n\to\infty} (2/3)^n = 0,$$

which means that A has length 0. Nevertheless, A does not simply consist of discrete points. More information about the structure of A is given by its fractal dimension d. To determine it we choose

$$\varepsilon_n = \frac{1}{2} \cdot 3^{-n}, \quad \text{i.e. } q = 1/3,$$

and obtain (according to Fig. 9.6) the value $N(A, \varepsilon_n) = 2^n$. Thus

$$d = -\lim_{n\to\infty} \frac{\log 2^n}{\log 3^{-n} - \log 2} = \lim_{n\to\infty} \frac{n \log 2}{n \log 3 + \log 2} = \frac{\log 2}{\log 3} = 0.6309...$$

The Cantor set is thus an object between points and straight lines. The self-similarity of A is also noteworthy. Enlarging certain parts of A results in copies of A. This together with the non-integer dimension is a typical property of fractals.

Fig. 9.7 Snowflakes of depth 0, 1, 2, 3 and 4

Fig. 9.8 Law of formation
of the snowflake

Example 9.9 (Koch's snowflake[1]) This is a figure of finite area whose boundary is a fractal of infinite length. In Fig. 9.7 one can see the first five construction steps of this fractal. In the step from A_n to A_{n+1} we substitute each straight boundary segment by four line segments in the following way: We replace the central third by two sides of an equilateral triangle, see Fig. 9.8.

The perimeter U_n of the figure A_n is computed as

$$U_n = \frac{4}{3} U_{n-1} = \left(\frac{4}{3}\right)^2 U_{n-2} = \cdots = \left(\frac{4}{3}\right)^n U_0 = 3a \left(\frac{4}{3}\right)^n.$$

Hence the perimeter U_∞ of Koch's snowflake A_∞ is

$$U_\infty = \lim_{n \to \infty} U_n = \infty.$$

Next we compute the fractal dimension of ∂A_∞. For that we set

$$\varepsilon_n = \frac{a}{2} \cdot 3^{-n}, \quad \text{i.e. } q = 1/3.$$

Since one can use a circle of radius ε_n to cover each straight boundary piece, we obtain

$$N(\partial A_\infty, \varepsilon_n) \le 3 \cdot 4^n$$

and hence

$$d = d(\partial A_\infty) \le \frac{\log 4}{\log 3} \approx 1.262.$$

A covering using equilateral triangles of side length ε_n shows that $N(\partial A_\infty, \varepsilon_n)$ is proportional to 4^n and thus

$$d = \frac{\log 4}{\log 3}.$$

The boundary of the snowflake ∂A_∞ is hence a geometric object between a curve and a plane region.

---

[1]H. von Koch, 1870–1924.

9.2 Mandelbrot Sets

An interesting class of fractals can be obtained with the help of *iteration methods*.
As an example we consider in $\mathbb{C}$ the iteration

$$z_{n+1} = z_n^2 + c.$$

Setting $z = x + \mathrm{i}y$ and $c = a + \mathrm{i}b$ one obtains, by separating the real and the imaginary part, the equivalent *real form* of the iteration

$$x_{n+1} = x_n^2 - y_n^2 + a,$$
$$y_{n+1} = 2x_n y_n + b.$$

The real representation is important when working with a programming language
that does not support complex arithmetic.

First we investigate for which values of $c \in \mathbb{C}$ the iteration

$$z_{n+1} = z_n^2 + c, \qquad z_0 = 0$$

remains *bounded*. In the present case this is equivalent to $|z_n| \not\to \infty$ for $n \to \infty$.
The set of all c with this property is obviously not empty since it contains $c = 0$. On
the other hand it is bounded since the iteration always diverges for $|c| > 2$ as can
easily be verified with MATLAB.

Definition 9.10 The set

$$M = \{c \in \mathbb{C} \; ; \; |z_n| \not\to \infty \text{ as } n \to \infty\}$$

is called *Mandelbrot set*[2] of the iteration $z_{n+1} = z_n^2 + c$, $z_0 = 0$.

To get an impression of M we carry out a numerical experiment in MATLAB.

Experiment 9.11 To visualise the Mandelbrot set M one first chooses a raster of a
certain region, for example

$$-2 \le \operatorname{Re} c \le 1, \quad -1.15 \le \operatorname{Im} c \le 1.15.$$

Next for each point of the raster one carries out a large number of iterations (e.g. 80)
and decides then whether the iterations remain bounded (for example $|z_n| \le 2$). If
this is the case one colours the point in black. This way one successively obtains
a picture of M. For your experiments use the MATLAB program mat09_2.m and
modify it as required. This way generate in particular the pictures in Fig. 9.9 in high
resolution.

---

[2]B. Mandelbrot, 1924–2010.

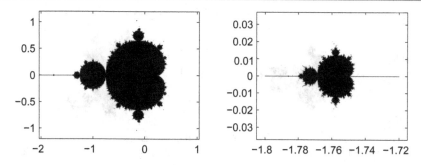

Fig. 9.9 The Mandelbrot set of the iteration $z_{n+1} = z_n^2 + c$, $z_0 = 0$ and enlargement of a section

Figure 9.9 shows as result a *little apple man* which has smaller apple men attached which finally develop into an *antenna*. Here one already recognises the self-similarity. If an enlargement of a certain detail on the antenna ($-1.8 \leq \mathrm{Re}\, c \leq -1.72$, $-0.03 \leq \mathrm{Im}\, c \leq 0.03$) is made, one finds an almost perfect copy of the complete apple man. The Mandelbrot set is one of the most popular fractals and one of the most complex mathematical objects which can be visualised.

9.3 Julia Sets

Again we consider the iteration

$$z_{n+1} = z_n^2 + c.$$

This time, however, we interchange the roles of z_0 and c.

Definition 9.12 For a given $c \in \mathbb{C}$, the set

$$J_c = \{z_0 \in \mathbb{C} \; ; \; |z_n| \nrightarrow \infty \text{ as } n \to \infty\}$$

is called *Julia set*[3] of the iteration $z_{n+1} = z_n^2 + c$.

The Julia set for the parameter value c hence consists of those *initial values* for which the iteration remains bounded. For some values of c the pictures of J_c are displayed in Fig. 9.10. Julia sets have many interesting properties; for example,

$$J_c \text{ is connected } \Leftrightarrow c \in M.$$

---

[3]G. Julia, 1893–1978.

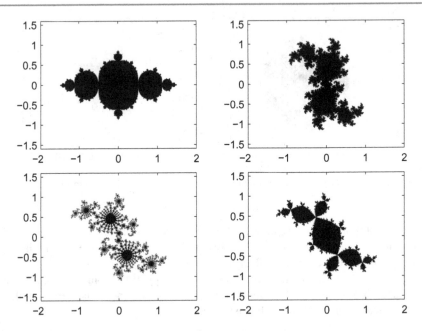

Fig. 9.10 Julia sets of the iteration $z_{n+1} = z_n^2 + c$ for the parameter values $c = -0.75$ (top left), $c = 0.35 + 0.35\,\mathrm{i}$ (top right), $c = -0.03 + 0.655\,\mathrm{i}$ (bottom left) and $-0.12 + 0.74\,\mathrm{i}$ (bottom right)

Thus one can alternatively define the Mandelbrot set M as

$$M = \{c \in \mathbb{C} \ ; \ J_c \text{ is connected}\}.$$

Furthermore the boundary of a Julia set is self-similar and a fractal.

Experiment 9.13 Using the MATLAB program `mat09_3.m` plot the Julia sets J_c in Fig. 9.10 in high definition. Also try other values of c.

9.4 Newton's Method in $\mathbb{C}$

Since the arithmetic in $\mathbb{C}$ is an extension of that in $\mathbb{R}$, many concepts of real analysis can be transferred directly to $\mathbb{C}$. For example, a function $f \colon \mathbb{C} \to \mathbb{C} \colon z \mapsto f(z)$ is called *complex differentiable* if the difference quotient

$$\frac{f(z + \Delta z) - f(z)}{\Delta z}$$

has a limit as $\Delta z \to 0$. This limit is again denoted by

$$f'(z) = \lim_{\Delta z \to 0} \frac{f(z + \Delta z) - f(z)}{\Delta z}$$

and called *complex derivative* of f at the point z. Since differentiation in $\mathbb{C}$ is defined in the same way as differentiation in $\mathbb{R}$, the same differentiation rules hold. In particular any polynomial

$$f(z) = a_n z^n + \cdots + a_1 z + a_0$$

is complex differentiable and has the derivative

$$f'(z) = n a_n z^{n-1} + \cdots + a_1.$$

Like the real derivative (see Sect. 7.3), the complex derivative has an interpretation as a linear approximation

$$f(z) \approx f(z_0) + f'(z_0)(z - z_0)$$

for z close to z_0.

Let $f \colon \mathbb{C} \to \mathbb{C} \colon z \mapsto f(z)$ be a complex differentiable function with $f(\zeta) = 0$ and $f'(\zeta) \neq 0$. In order to compute the zero ζ of the function f, one first computes the linear approximation starting from the initial value z_0, so

$$z_1 = z_0 - \frac{f(z_0)}{f'(z_0)}.$$

Subsequently z_1 is used as the new initial value and the procedure is iterated. In this way one obtains Newton's method in $\mathbb{C}$:

$$z_{n+1} = z_n - \frac{f(z_n)}{f'(z_n)}.$$

For initial values z_0 close to ζ the procedure converges (as in $\mathbb{R}$) quadratically. Otherwise, however, the situation can become very complicated.

In 1983 Eckmann [9] investigated Newton's method for the function

$$f(z) = z^3 - 1 = (z - 1)(z^2 + z + 1).$$

This function has three roots in $\mathbb{C}$

$$\zeta_1 = 1, \quad \zeta_{2,3} = -\frac{1}{2} \pm i \frac{\sqrt{3}}{2}.$$

Naively one could think that the complex plane $\mathbb{C}$ is split into three equally large sectors where the iteration with initial values in sector S_1 converges to ζ_1, the ones in S_2 to ζ_2, etc., see Fig. 9.11.

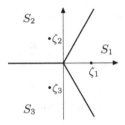

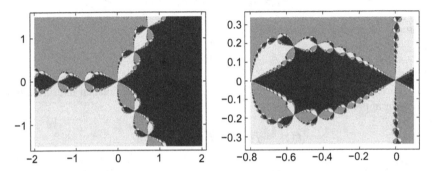

Fig. 9.11 Possible regions of attraction of Newton's iteration for finding the roots of $z^3 - 1$

Fig. 9.12 Actual regions of attraction of Newton's iteration for finding the roots of $z^3 - 1$ and enlargement of a part

A numerical experiment, however, shows that it is not that way. If one colours the initial values according to their convergence, one obtains a very complex picture. One can prove (however, not easily imagine) that at every point where two colours meet, the third colour is also present. The boundaries of the regions of attraction are dominated by pincer-like motifs which reappear again and again when enlarging the scale, see Fig. 9.12. The boundaries of the regions of attraction are Julia sets. Again we have found fractals.

Experiment 9.14 Using the MATLAB program mat09_4.m carry out an experiment. Ascertain yourself of the self-similarity of the appearing Julia sets by producing suitable enlargements of the boundaries of the region of attraction.

9.5 L-systems

The formalism of L-systems was developed by Lindenmayer[4] around 1968 in order to model the growth of plants. It also turned out that many fractals can be created this way. In this section we give a brief introduction to L-systems and discuss a possible implementation in maple.

---

[4]A. Lindenmayer, 1925–1989.

Definition 9.15 An L-system consists of the following five components:

(a) A finite set B of symbols, the so-called alphabet. The elements of B are called *letters*, and any string of letters is called a *word*.

(b) Certain substitution rules. These rules determine how the letters of the current word are to be replaced in each iteration step.

(c) The initial word $w \in W$. The initial word is also called *axiom* or *seed*.

(d) The number of iteration steps which one wants to carry out. In each of these steps, every letter of the current word is replaced according to the substitution rules.

(e) A graphical interpretation of the word.

Let W be the set of all words that can be formed in the given L-system. The substitution rules can be interpreted as a mapping from B to W:

$$S : B \to W : \mathrm{b} \mapsto S(\mathrm{b}).$$

Example 9.16 Consider the alphabet $B = \{\mathrm{f}, \mathrm{p}, \mathrm{m}\}$ consisting of the three letters f, p and m. As substitution rules for this alphabet we take

$$S(\mathrm{f}) = \mathrm{fpfmfmffpfpfmf}, \quad S(\mathrm{p}) = \mathrm{p}, \quad S(\mathrm{m}) = \mathrm{m}$$

and consider the axiom $w = \mathrm{fpfpfpf}$. An application of the substitution rules shows that, after one substitution, the word fpf becomes the new word $\mathrm{fpfmfmffpfpfmfpfpfmfmffpfpfmf}$. If one applies the substitution rules on the axiom then one obtains a new word. Applying the substitution rules on that again gives a new word, and so on. Each of these words can be interpreted as a polygon by assigning the following meaning to the individual letters:

f means forward by one unit;
p stands for a rotation of α radians (plus);
m stands for a rotation of $-\alpha$ radians (minus).

Thereby $0 \le \alpha \le \pi$ is a chosen angle. One plots the polygon by choosing an arbitrary initial point and an arbitrary initial direction. Then one sequentially processes the letters of the word to be displayed according to the rules above.

In maple lists and the substitution command subs lend themselves to the implementation of L-systems. In the example above the axiom would hence be defined by

$$a := [\mathrm{f}, \mathrm{p}, \mathrm{f}, \mathrm{p}, \mathrm{f}, \mathrm{p}, \mathrm{f}]$$

the substitution rules would be

$$a-> \mathtt{subs}(f = (f, p, f, m, f, m, f, f, p, f, p, f, m, f), a).$$

The letters p and m do not change in the example, and they are fixed points in the construction. For the purpose of visualisation one can use polygons in maple, given by lists of points (in the plane). These lists can be plotted easily using the command plot.

Construction of fractals. With the graphical interpretation above and $\alpha = \pi/2$, the axiom fpfpfpf is a square which is passed through in a counterclockwise direction. The substitution rule converts a straight line segment into a zigzag line. By an iterative application of the substitution rule the axiom develops into a fractal.

Experiment 9.17 Using the maple worksheet mp09_1.mws create different fractals. Further, try to understand the procedure fractal in detail.

Example 9.18 The substitution rule for Koch's curve is

$$a \rightarrow \mathtt{subs}(f=(f,p,f,m,m,f,p,f),a).$$

Depending on which axiom one uses, one can build fractal curves or snowflakes from that, see the maple worksheet mp09_1.mws.

Simulation of plant growth. As a new element branchings (ramifications) are added here. Mathematically one can describe this using two new symbols:

v stands for a ramification;
e stands for the end of the branch.

Let us look, for example, at the word

$$[f, p, f, v, p, p, f, p, f, e, v, m, f, m, f, e, f, p, f, v, p, f, p, f, e, f, m, f].$$

If one removes all branchings that start with v and end with e from the list then one obtains the stem of the plant

$$\mathtt{stem} := [f, p, f, f, p, f, f, m, f].$$

After the second f in the stem obviously a double branching is taking place and the branches sprout

$$\mathtt{branch1} := [p, p, f, p, f] \quad \text{and} \quad \mathtt{branch2} := [m, f, m, f].$$

Further up the stem branches again with the branch [p,f,p,f].

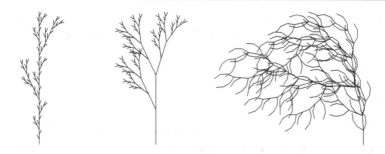

Fig. 9.13 Plants created using the maple worksheet mp09_2.mws

For a more realistic modelling one can introduce additional parameters. For example, asymmetry can be build in by rotating by the positive angle α at p and by the negative angle $-\beta$ at m. In the program mp09_2.mws that was done, see Fig. 9.13.

Experiment 9.19 Using the maple worksheet mp09_2.mws create different artificial plants. Further, try to understand the procedure grow in detail.

To visualise the created plants one can use lists of polygons in maple, i.e. lists of points (in the plane). To implement the branchings one conveniently uses a recursive *stack*. Whenever one comes across the command v for a branching, one saves the current state as the topmost value in the *stack*. A state is described by three numbers (x, y, t) where x and y denote the position in the (x, y)-plane and t the angle enclosed the with the positive x-axis. Conversely one removes the topmost state from the *stack* if one comes across the end of a branch e and returns to this state in order to continue the plot. At the beginning the *stack* is empty (at the end it should be as well).

Extensions. In the context of L-systems many generalisations are possible which can make the emerging structures more realistic. For example one could:

(a) Represent the letter f by shorter segments as one moves further away from the root of the plant. For that, one has to save the distance from the root as a further state parameter in the stack.

(b) Introduce randomness by using different substitution rules for one and the same letter and in each step choosing one at random. For example, the substitution rules for random weeds could be as such:

```
f -> (f,v,p,f,e,f,v,m,f,e,f)    with probability 1/3;
f -> (f,v,p,f,e,f)              with probability 1/3;
f -> (f,v,m,f,e,f)              with probability 1/3.
```

Using random numbers one selects the according rule in each step.

Experiment 9.20 Using the maple worksheet mp09_3.mws create *random* plants. Further, try to understand the implemented substitution rule in detail.

9.6 Exercises

1. Determine experimentally the fractal dimension of the coastline of Great Britain. In order to do that, take a map of Great Britain (e.g. a copy from an atlas) and raster the map using different mesh sizes (e.g. with 1/64th, 1/32th, 1/16th, 1/8th and 1/4th of the North–South expansion). Count the boxes which contain parts of the coastline and display this number as a function of the mesh size in a double-logarithmic diagram. Fit the best line through these points and determine the fractal dimension in question from the slope of the straight line.
2. Using the program mat09_3.m visualise the Julia sets of $z_{n+1} = z_n^2 + c$ for $c = -1.25$ and $c = 0.365 - 0.3\,i$. Search for interesting details.
3. Let $f(z) = z^3 - 1$ with $z = x + iy$. Use Newton's method to solve $f(z) = 0$ and separate the real part and the imaginary part, i.e. find the functions g_1 and g_2 with

$$x_{n+1} = g_1(x_n, y_n),$$
$$y_{n+1} = g_2(x_n, y_n).$$

4. Modify the procedure grow in the program mp09_2.mws by representing the letter f by shorter segments depending on how far it is away from the root. With that plot the *umbel* from Experiment 9.19 again.
5. Modify the program mp09_3.mws by attributing new probabilities to the existing substitution rules (or invent new substitution rules). Use your modified program to plot some plants.
6. Modify the program mat09_3.m to visualise the Julia sets of $z_{n+1} = z_n^{2k} - c$ for $c = -1$ and integer values of k. Observe how varying k affects the shape of the Julia set. Try other values of c as well.
7. Modify the program mat09_3.m to visualise the Julia sets of

$$z_{n+1} = z_n^3 + (c - 1)z_n - c.$$

Study especially the behaviour of the Julia sets when c ranges between 0.60 and 0.65.

Antiderivatives

10

The derivative of a function $y = F(x)$ describes its *local rate of change*, i.e. the change Δy of the y-value with respect to the change Δx of the x-value in the limit $\Delta x \to 0$; more precisely

$$f(x) = F'(x) = \lim_{\Delta x \to 0} \frac{\Delta y}{\Delta x} = \lim_{\Delta x \to 0} \frac{F(x + \Delta x) - F(x)}{\Delta x}.$$

Conversely, the question about the reconstruction of a function F from its local rate of change f leads to the notion of *indefinite integrals* which comprises the totality of all functions that have f as their derivative, the *antiderivatives* of f. Chapter 10 addresses this notion, its properties, some basic examples and applications.

By multiplying the rate of change $f(x)$ with the change Δx one obtains an approximation to the change of the values of the function of the antiderivative F in the segment of length Δx:

$$\Delta y = F(x + \Delta x) - F(x) \approx f(x)\Delta x.$$

Adding up these local changes in an interval, for instance between $x = a$ and $x = b$ in steps of length Δx, gives an approximation to the total change $F(b) - F(a)$. The limit $\Delta x \to 0$ (with an appropriate increase of the number of summands) leads to the notion of the *definite integral* of f in the interval $[a, b]$, which is the subject of Chap. 11.

10.1 Indefinite Integrals

In Sect. 7.2 it was shown that the derivative of a constant is zero. The following proposition shows that the converse is also true.

© Springer Nature Switzerland AG 2018
M. Oberguggenberger and A. Ostermann, *Analysis for Computer Scientists*,
Undergraduate Topics in Computer Science,
https://doi.org/10.1007/978-3-319-91155-7_10

Proposition 10.1 *If the function F is differentiable on (a, b) and $F'(x) = 0$ for all $x \in (a, b)$ then F is constant. This means that $F(x) = c$ for a certain $c \in \mathbb{R}$ and all $x \in (a, b)$.*

Proof We choose an arbitrary $x_0 \in (a, b)$ and set $c = F(x_0)$. If now $x \in (a, b)$ then, according to the mean value theorem (Proposition 8.4),

$$F(x) - F(x_0) = F'(\xi)(x - x_0)$$

for a point ξ between x and x_0. Since $F'(\xi) = 0$ it follows that $F(x) = F(x_0) = c$. This holds for all $x \in (a, b)$, consequently F has to be equal to the constant function with value c.   □

Definition 10.2 *(Antiderivatives)* Let f be a real-valued function on an interval (a, b). An *antiderivative* of f is a differentiable function $F: (a, b) \to \mathbb{R}$ whose derivative F' equals f.

Example 10.3 The function $F(x) = \frac{x^3}{3}$ is an antiderivative of $f(x) = x^2$, as is $G(x) = \frac{x^3}{3} + 5$.

Proposition 10.1 implies that antiderivatives are unique up to an additive constant.

Proposition 10.4 *Let F and G be antiderivatives of f in (a, b). Then $F(x) = G(x) + c$ for a certain $c \in \mathbb{R}$ and all $x \in (a, b)$.*

Proof Since $F'(x) - G'(x) = f(x) - f(x) = 0$ for all $x \in (a, b)$, an application of Proposition 10.1 gives the desired result.   □

Definition 10.5 *(Indefinite integrals)* The *indefinite integral*

$$\int f(x) \, dx$$

denotes the totality of all antiderivatives of f.

Once a particular antiderivative F has been found, one writes accordingly

$$\int f(x) \, dx = F(x) + c.$$

Example 10.6 The indefinite integral of the quadratic function (Example 10.3) is $\int x^2 \, dx = \frac{x^3}{3} + c$.

Example 10.7 (a) An application of indefinite integration to the differential equation of the vertical throw: Let $w(t)$ denote the height (in metres [m]) at time t (in seconds [s]) of an object above ground level ($w = 0$). Then

$$w'(t) = v(t)$$

is the velocity of the object (positive in upward direction) and

$$v'(t) = a(t)$$

the acceleration (positive in upward direction). In this coordinate system the gravitational acceleration

$$g = 9.81 \, [\mathrm{m/s^2}]$$

acts downwards, consequently

$$a(t) = -g.$$

Velocity and distance are obtained by inverting the differentiation process

$$v(t) = \int a(t) \, \mathrm{d}t + c_1 = -gt + c_1,$$

$$w(t) = \int v(t) \, \mathrm{d}t + c_2 = \int (-gt + c_1) \, \mathrm{d}t + c_2 = -\frac{g}{2}t^2 + c_1 t + c_2,$$

where the constants c_1, c_2 are determined by the initial conditions:

$$c_1 = v(0) \quad \ldots \quad \text{initial velocity,}$$
$$c_2 = w(0) \quad \ldots \quad \text{initial height.}$$

(b) A concrete example—the free fall from a height of $100 \, \mathrm{m}$. Here

$$w(0) = 100, \quad v(0) = 0$$

and thus

$$w(t) = -\frac{1}{2} 9.81 t^2 + 100.$$

The travelled distance as a function of time (Fig. 10.1) is given by a parabola.
    The time of impact t_0 is obtained from the condition $w(t_0) = 0$, i.e.

$$0 = -\frac{1}{2} 9.81 t_0^2 + 100, \quad t_0 = \sqrt{200/9.81} \approx 4.5 \, [\mathrm{s}],$$

the velocity at impact is

$$v(t_0) = -gt_0 \approx 44.3 \, [\mathrm{m/s}] \approx 160 \, [\mathrm{km/h}].$$

Fig. 10.1 Free fall: travelled
distance as function of time

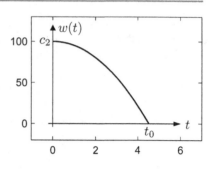

10.2 Integration Formulas

It follows immediately from Definition 10.5 that indefinite integration can be seen
as the inversion of differentiation. It is, however, only unique up to a constant:

$$\left(\int f(x)\,dx\right)' = f(x),$$

$$\int g'(x)\,dx = g(x) + c.$$

With this consideration and the formulas from Sect. 7.4 one easily obtains the *basic
integration formulas* stated in the following table. The formulas are valid in the
according domains.

The formulas in Table 10.1 are a direct consequence of those in Table 7.1.

Experiment 10.8 Antiderivatives can be calculated in maple using the command
int. Explanations and further integration commands can be found in the maple

Table 10.1 Integrals of some elementary functions

$f(x)$	$x^\alpha,\ \alpha \neq -1$	$\dfrac{1}{x}$	e^x	a^x
$\int f(x)\,dx$	$\dfrac{x^{\alpha+1}}{\alpha+1}+c$	$\log|x|+c$	e^x+c	$\dfrac{1}{\log a}a^x+c$
$f(x)$	$\sin x$	$\cos x$	$\dfrac{1}{\sqrt{1-x^2}}$	$\dfrac{1}{1+x^2}$
$\int f(x)\,dx$	$-\cos x+c$	$\sin x+c$	$\arcsin x+c$	$\arctan x+c$
$f(x)$	$\sinh x$	$\cosh x$	$\dfrac{1}{\sqrt{1+x^2}}$	$\dfrac{1}{\sqrt{x^2-1}}$
$\int f(x)\,dx$	$\cosh x+c$	$\sinh x+c$	$\operatorname{arsinh} x+c$	$\operatorname{arcosh} x+c$

worksheet `mp10_1.mws`. Experiment with these maple commands by applying them to the examples of Table 10.1 and other functions of your choice.

Experiment 10.9 Integrate the following expressions

$$xe^{-x^2}, \quad e^{-x^2}, \quad \sin(x^2)$$

with maple.

Functions that are obtained by combining power functions, exponential functions and trigonometric functions, as well as their inverses, are called *elementary* functions. The derivative of an elementary function is again an elementary function and can be obtained using the rules from Chap. 7. In contrast to differentiation there is no general procedure for computing indefinite integrals. Not only does the calculation of an integral often turn out to be a difficult task, but there are also many elementary functions whose antiderivatives are not elementary. An algorithm to decide whether a functions has an elementary indefinite integral was first deduced by Liouville[1] around 1835. This was the starting point for the field of *symbolic integration*. For details, we refer to [7].

Example 10.10 (Higher transcendental functions) Antiderivatives of functions that do not possess elementary integrals are frequently called higher transcendental functions. We give the following examples:

$$\frac{2}{\sqrt{\pi}} \int e^{-x^2}\, dx = \mathrm{Erf}(x) + c \quad \ldots \quad \text{Gaussian error function;}$$

$$\int \frac{e^x}{x}\, dx = \mathcal{E}i(x) + c \quad \ldots \quad \text{exponential integral;}$$

$$\int \frac{1}{\log x}\, dx = \ell i(x) + c \quad \ldots \quad \text{logarithmic integral;}$$

$$\int \frac{\sin x}{x}\, dx = \mathcal{S}i(x) + c \quad \ldots \quad \text{sine integral;}$$

$$\int \sin\left(\frac{\pi}{2}x^2\right)\, dx = \mathcal{S}(x) + c \quad \ldots \quad \text{Fresnel integral.[2]}$$

Proposition 10.11 (Rules for indefinite integration) *For indefinite integration the following rules hold:*

(a) Sum: $\int \big(f(x) + g(x)\big)\, dx = \int f(x)\, dx + \int g(x)\, dx$

---

[1] J. Liouville, 1809–1882.
[2] A.J. Fresnel, 1788–1827.

(b) Constant factor: $\int \lambda f(x)\,dx = \lambda \int f(x)\,dx \quad (\lambda \in \mathbb{R})$

(c) Integration by parts:

$$\int f(x)g'(x)\,dx = f(x)g(x) - \int f'(x)g(x)\,dx$$

(d) Substitution:

$$\int f(g(x))g'(x)\,dx = \int f(y)\,dy\Big|_{y=g(x)}.$$

Proof (a) and (b) are clear; (c) follows from the product rule for the derivative (Sect. 7.4)

$$\int f(x)g'(x)\,dx + \int f'(x)g(x)\,dx = \int \big(f(x)g'(x) + f'(x)g(x)\big)\,dx$$

$$= \int \big(f(x)g(x)\big)'\,dx = f(x)g(x) + c,$$

which can be rewritten as

$$\int f(x)g'(x)\,dx = f(x)g(x) - \int f'(x)g(x)\,dx.$$

In this formula we can drop the integration constant c since it is already contained in the notion of indefinite integrals, which appear on both sides. Point (d) is an immediate consequence of the chain rule according to which an antiderivative of $f(g(x))g'(x)$ is given by the antiderivative of $f(y)$ evaluated at $y = g(x)$. $\qquad\square$

Example 10.12 The following five examples show how the rules of Table 10.1 and Proposition 10.11 can be applied.

(a) $\displaystyle\int \frac{dx}{\sqrt[3]{x}} = \int x^{-1/3}\,dx = \frac{x^{-\frac{1}{3}+1}}{-\frac{1}{3}+1} + c = \frac{3}{2}x^{2/3} + c.$

(b) $\displaystyle\int x\cos x\,dx = x\sin x - \int \sin x\,dx = x\sin x + \cos x + c,$

  which follows via integration by parts:

$$f(x) = x, \quad g'(x) = \cos x,$$
$$f'(x) = 1, \quad g(x) = \sin x.$$

(c) $\displaystyle\int \log x\,dx = \int 1 \cdot \log x\,dx = x\log x - \int \frac{x}{x}\,dx = x\log x - x + c,$

via integration by parts:

$$f(x) = \log x, \quad g'(x) = 1,$$
$$f'(x) = \tfrac{1}{x}, \quad\quad g(x) = x.$$

(d) $\displaystyle\int x \sin(x^2)\,dx = \int \frac{1}{2}\sin y\,dy\Big|_{y=x^2} = -\frac{1}{2}\cos y\Big|_{y=x^2} + c = -\frac{1}{2}\cos(x^2) + c,$

which follows from the substitution rule with $y = g(x) = x^2$, $g'(x) = 2x$, $f(y) = \frac{1}{2}\sin y$.

(e) $\displaystyle\int \tan x\,dx = \int \frac{\sin x}{\cos x}\,dx = -\log|y|\,\Big|_{y=\cos x} + c = -\log|\cos x| + c,$

again after substitution with $y = g(x) = \cos x$, $g'(x) = -\sin x$ and $f(y) = -1/y$.

Example 10.13 (A simple expansion into partial fractions) In order to find the indefinite integral of $f(x) = 1/(x^2 - 1)$, we decompose the quadratic denominator in its linear factors $x^2 - 1 = (x - 1)(x + 1)$ and expand $f(x)$ into partial fractions of the form

$$\frac{1}{x^2 - 1} = \frac{A}{x - 1} + \frac{B}{x + 1}.$$

Resolving the fractions leads to the equation $1 = A(x + 1) + B(x - 1)$. Equating coefficients results in

$$(A + B)x = 0, \quad A - B = 1$$

with the obvious solution $A = 1/2$, $B = -1/2$. Thus

$$\int \frac{1}{x^2 - 1}\,dx = \frac{1}{2}\left(\int \frac{dx}{x - 1} - \int \frac{dx}{x + 1}\right)$$
$$= \frac{1}{2}\Big(\log|x - 1| - \log|x + 1|\Big) + C = \frac{1}{2}\log\left|\frac{x - 1}{x + 1}\right| + C.$$

In view of Example 7.30, another antiderivative of $f(x) = 1/(x^2 - 1)$ is $F(x) = -\operatorname{artanh} x$. Thus, by Proposition 10.4,

$$\operatorname{artanh} x = -\frac{1}{2}\log\left|\frac{x - 1}{x + 1}\right| + C = \frac{1}{2}\log\left|\frac{x + 1}{x - 1}\right| + C.$$

Inserting $x = 0$ on both sides shows that $C = 0$ and yields an expression of the inverse hyperbolic tangent in terms of the logarithm.

10.3 Exercises

1. An object is thrown vertically upwards from the ground with a velocity of
   $10 \,[\text{m/s}]$. Find its height $w(t)$ as a function of time t, the maximum height
   as well as the time of impact on the ground.
   Hint. Integrate $w''(t) = -g \approx 9.81 \,[\text{m/s}^2]$ twice indefinitely and determine the
   integration constants from the initial conditions $w(0) = 0$, $w'(0) = 10$.

2. Compute the following indefinite integrals by hand and with maple:

   (a) $\displaystyle\int (x + 3x^2 + 5x^4 + 7x^6)\,dx,$ (b) $\displaystyle\int \frac{dx}{\sqrt{x}},$

   (c) $\displaystyle\int xe^{-x^2}\,dx$ (substitution),   (d) $\displaystyle\int xe^x\,dx$ (integration by parts).

3. Compute the indefinite integrals

   $$(a) \int \cos^2 x\,dx, \qquad (b) \int \sqrt{1 - x^2}\,dx$$

   by hand and check the results using maple.
   Hints. For (a) use the identity

   $$\cos^2 x = \frac{1}{2}(1 + \cos 2x);$$

   for (b) use the substitution $y = g(x) = \arcsin x$, $f(y) = 1 - \sin^2 y$.

4. Compute the indefinite integrals

   $$(a) \int \frac{dx}{x^2 + 2x + 5}\,dx, \qquad (b) \int \frac{dx}{x^2 + 2x - 3}$$

   by hand and check the results using maple.
   Hints. Write the denominator in (a) in the form $(x + 1)^2 + 4$ and reduce it to
   $y^2 + 1$ by means of a suitable substitution. Factorize the denominator in (b) and
   follow the procedure of Example 10.13.

5. Compute the indefinite integrals

   $$(a) \int \frac{dx}{x^2 + 2x}\,dx, \qquad (b) \int \frac{dx}{x^2 + 2x + 1}$$

   by hand and check the results using maple.

6. Compute the indefinite integrals

   $$(a) \int x^2 \sin x\,dx, \qquad (b) \int x^2 e^{-3x}\,dx.$$

   Hint. Repeated integration by parts.

7. Compute the indefinite integrals

$$\text{(a)} \int \frac{e^x}{e^x + 1} \, dx, \qquad \text{(b)} \int \sqrt{1 + x^2} \, dx.$$

Hint. Substitution $y = e^x$ in case (a), substitution $y = \sinh x$ in case (b), invoking the formula $\cosh^2 y - \sinh^2 y = 1$ and repeated integration by parts or recourse to the definition of the hyperbolic functions.

8. Show that the functions

$$f(x) = \arctan x \quad \text{and} \quad g(x) = \arctan \frac{1 + x}{1 - x}$$

differ in the interval $(-\infty, 1)$ by a constant. Compute this constant. Answer the same question for the interval $(1, \infty)$.

9. Prove the identity $\operatorname{arsinh} x = \log\left(x + \sqrt{1 + x^2}\right)$.

Hint. Recall from Chap. 7 that the functions $f(x) = \operatorname{arsinh} x$ and $g(x) = \log\left(x + \sqrt{1 + x^2}\right)$ have the same derivative. (Compare with the algebraic derivation of the formula in Exercise 15 of Sect. 2.3.)

Definite Integrals

11

In the introduction to Chap. 10 the notion of the *definite integral* of a function f on an interval $[a, b]$ was already mentioned. It arises from summing up expressions of the form $f(x)\Delta x$ and taking limits. Such sums appear in many applications including the calculation of areas, surface areas and volumes as well as the calculation of lengths of curves. This chapter employs the notion of Riemann integrals as the basic concept of definite integration. Riemann's approach provides an intuitive concept in many applications, as will be elaborated in examples at the end of the chapter.

The main part of this chapter is dedicated to the properties of the integral. In particular, the two fundamental theorems of calculus are proven. The first theorem allows one to calculate a definite integral from the knowledge of an antiderivative. The second fundamental theorem states that the definite integral of a function f on an interval $[a, x]$ with variable upper bound provides an antiderivative of f. Since the definite integral can be approximated, for example by Riemann sums, the second fundamental theorem offers a possibility to approximate the antiderivative numerically. This is of importance, for example, for the calculation of distribution functions in statistics.

11.1 The Riemann Integral

Example 11.1 (From velocity to distance) How can one calculate the distance w which a vehicle travels between time a and time b if one only knows its velocity $v(t)$ for all times $a \le t \le b$? If $v(t) \equiv v$ is constant, one simply gets

$$w = v \cdot (b - a).$$

© Springer Nature Switzerland AG 2018

149

M. Oberguggenberger and A. Ostermann, *Analysis for Computer Scientists*,
Undergraduate Topics in Computer Science,
https://doi.org/10.1007/978-3-319-91155-7_11

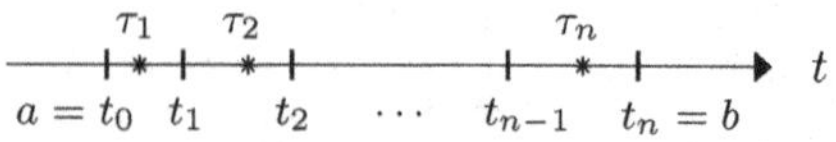

Fig. 11.1 Subdivision of the
time axis

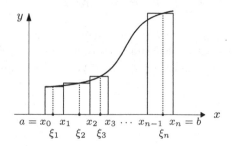

If the velocity $v(t)$ is time-dependent, one divides the time axis into smaller subintervals (Fig. 11.1): $a = t_0 < t_1 < t_2 < \cdots < t_n = b$.

Choosing intermediate points $\tau_j \in [t_{j-1}, t_j]$ one obtains approximately

$$v(t) \approx v(\tau_j) \quad \text{for} \quad t \in [t_{j-1}, t_j],$$

if v is a continuous function of time. The approximation is the more precise, the shorter the intervals $[t_{j-1}, t_j]$ are chosen. The distance travelled in this interval is approximately equal to

$$w_j \approx v(\tau_j)(t_j - t_{j-1}).$$

The total distance covered between time a and time b is then

$$w = \sum_{j=1}^{n} w_j \approx \sum_{j=1}^{n} v(\tau_j)(t_j - t_{j-1}).$$

Letting the length of the subintervals $[t_{j-1}, t_j]$ tend to zero, one expects to obtain the actual value of the distance in the limit.

Example 11.2 (Area under the graph of a nonnegative function) In a similar way one can try to approximate the area under the graph of a function $y = f(x)$ by using rectangles which are successively refined (Fig. 11.2).

The sum of the areas of the rectangles

$$F \approx \sum_{j=1}^{n} f(\xi_j)(x_j - x_{j-1})$$

form an approximation to the actual area under the graph.

Fig. 11.2 Sums of
rectangles as approximation
to the area

The two examples are based on the same concept, the *Riemann integral*,[1] which we will now introduce. Let an interval $[a, b]$ and a function $f = [a, b] \to \mathbb{R}$ be given. Choosing points

$$a = x_0 < x_1 < x_2 < \cdots < x_{n-1} < x_n = b,$$

the intervals $[x_0, x_1], [x_1, x_2], \ldots, [x_{n-1}, x_n]$ form a *partition* Z of the interval $[a, b]$. We denote the length of the largest subinterval by $\Phi(Z)$, i.e.

$$\Phi(Z) = \max_{j=1,\ldots,n} |x_j - x_{j-1}|.$$

For arbitrarily chosen intermediate points $\xi_j \in [x_{j-1}, x_j]$ one calls the expression

$$S = \sum_{j=1}^{n} f(\xi_j)(x_j - x_{j-1})$$

a *Riemann sum*. In order to further specify the idea of the limiting process above, we take a sequence $Z_1, Z_2, Z_3, \ldots$ of partitions such that $\Phi(Z_N) \to 0$ as $N \to \infty$ and corresponding Riemann sums S_N.

Definition 11.3 A function f is called *Riemann integrable* in $[a, b]$ if, for arbitrary sequences of partitions $(Z_N)_{N \geq 1}$ with $\Phi(Z_N) \to 0$, the corresponding Riemann sums $(S_N)_{N \geq 1}$ tend to the same limit $I(f)$, independently of the choice of the intermediate points. This limit

$$I(f) = \int_a^b f(x) \, dx$$

is called the *definite integral* of f on $[a, b]$.

The intuitive approach in the introductory Examples 11.1 and 11.2 can now be made precise. If the respective functions f and v are Riemann integrable, then the integral

$$F = \int_a^b f(x) \, dx$$

represents the area between the x-axis and the graph, and

$$w = \int_a^b v(t) \, dt$$

gives the total distance covered.

---

[1]B. Riemann, 1826–1866.

Experiment 11.4 Open the M-file `mat11_1.m`, study the given explanations and experiment with randomly chosen Riemann sums for the function $f(x) = 3x^2$ in the interval $[0, 1]$. What happens if you take more and more partition points n?

Experiment 11.5 Open the applet *Riemann sums* and study the effects of changing the partition. In particular, vary the maximum length of the subintervals and the choice of intermediate points. How does the sign of the function influence the result?

The following examples illustrate the notion of Riemann integrability.

Example 11.6 (a) Let $f(x) = c = $ constant. Then the area under the graph of the function is the area of the rectangle $c(b - a)$. On the other hand, any Riemann sum is of the form

$$f(\xi_1)(x_1 - x_0) + f(\xi_2)(x_2 - x_1) + \cdots + f(\xi_n)(x_n - x_{n-1})$$
$$= c(x_1 - x_0 + x_2 - x_1 + \cdots + x_n - x_{n-1})$$
$$= c(x_n - x_0) = c(b - a).$$

All Riemann sums are equal and thus, as expected,

$$\int_a^b c \, dx = c(b - a).$$

(b) Let $f(x) = \frac{1}{x}$ for $x \in (0, 1]$, $f(0) = 0$. This function is not integrable in $[0, 1]$. The corresponding Riemann sums are of the form

$$\frac{1}{\xi_1}(x_1 - 0) + \frac{1}{\xi_2}(x_2 - x_1) + \cdots + \frac{1}{\xi_n}(x_n - x_{n-1}).$$

By choosing ξ_1 close to 0 every such Riemann sum can be made arbitrarily large; thus the limit of the Riemann sums does not exist.

(c) *Dirichlet's function*[2]

$$f(x) = \begin{cases} 1, & x \in \mathbb{Q} \\ 0, & x \notin \mathbb{Q} \end{cases}$$

is not integrable in $[0, 1]$. The Riemann sums are of the form

$$S_N = f(\xi_1)(x_1 - x_0) + \cdots + f(\xi_n)(x_n - x_{n-1}).$$

If all $\xi_j \in \mathbb{Q}$ then $S_N = 1$. If one takes all $\xi_j \notin \mathbb{Q}$ then $S_N = 0$; thus the limit depends on the choice of intermediate points ξ_j.

---

[2] P.G.L. Dirichlet, 1805–1859.

Fig. 11.3 A piecewise
continuous function

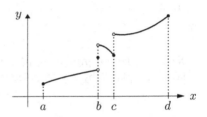

Remark 11.7 Riemann integrable functions $f : [a, b] \to \mathbb{R}$ are necessarily bounded. This fact can easily be shown by generalising the argument in Example 11.6(b).

The most important criteria for Riemann integrability are outlined in the following proposition. Its proof is simple, however, it requires a few technical considerations about refining partitions. For details, we refer to the literature, for instance [4, Chap. 5.1].

Proposition 11.8 *(a) Every function which is bounded and monotonically increasing (monotonically decreasing) on an interval $[a, b]$ is Riemann integrable.*

(b) Every piecewise continuous function on an interval $[a, b]$ is Riemann integrable.   □

A function is called *piecewise continuous* if it is continuous except for a finite number of points. At these points, the graph may have jumps but is required to have left- and right-hand limits (Fig. 11.3).

Remark 11.9 By taking equidistant grid points $a = x_0 < x_1 < \cdots < x_{n-1} < x_n = b$ for the partition, i.e.

$$x_j - x_{j-1} =: \Delta x = \frac{b - a}{n},$$

the Riemann sums can be written as

$$S_N = \sum_{j=1}^{n} f(\xi_j) \Delta x.$$

The transition $\Delta x \to 0$ with simultaneous increase of the number of summands suggests the notation

$$\int_a^b f(x)\,dx.$$

Originally it was introduced by Leibniz[3] with the interpretation as an infinite sum of infinitely small rectangles of width dx. After centuries of dispute, this interpretation

---

[3]G. Leibniz, 1646–1716.

can be rigorously justified today within the framework of *nonstandard analysis* (see, for instance, [27]).

Note that the *integration variable* x in the definite integral is a *bound variable* and can be replaced by any other letter:

$$\int_a^b f(x)\,dx = \int_a^b f(t)\,dt = \int_a^b f(\xi)\,d\xi = \cdots$$

This can be used with advantage in order to avoid possible confusion with other bound variables.

Proposition 11.10 (Properties of the definite integral) *In the following let $a < b$ and f, g be Riemann integrable on $[a, b]$.*

(a) Positivity:

$$f \geq 0 \text{ in } [a, b] \quad \Rightarrow \quad \int_a^b f(x)\,dx \geq 0,$$

$$f \leq 0 \text{ in } [a, b] \quad \Rightarrow \quad \int_a^b f(x)\,dx \leq 0.$$

(b) Monotonicity:

$$f \leq g \text{ in } [a, b] \quad \Rightarrow \quad \int_a^b f(x)\,dx \leq \int_a^b g(x)\,dx.$$

In particular, with

$$m = \inf_{x \in [a,b]} f(x), \quad M = \sup_{x \in [a,b]} f(x),$$

the following inequality holds

$$m(b-a) \leq \int_a^b f(x)\,dx \leq M(b-a).$$

(c) Sum and constant factor (linearity):

$$\int_a^b \big(f(x) + g(x)\big)\,dx = \int_a^b f(x)\,dx + \int_a^b g(x)\,dx$$

$$\int_a^b \lambda f(x)\,dx = \lambda \int_a^b f(x)\,dx \quad (\lambda \in \mathbb{R}).$$

(d) Partition of the integration domain: Let $a < b < c$ and f be integrable in $[a, c]$, then

$$\int_a^b f(x)\,dx + \int_b^c f(x)\,dx = \int_a^c f(x)\,dx.$$

If one defines

$$\int_a^a f(x)\,dx = 0, \qquad \int_b^a f(x)\,dx = -\int_a^b f(x)\,dx,$$

then one obtains the validity of the sum formula even for arbitrary $a, b, c \in \mathbb{R}$ if f is integrable on the respective intervals.

Proof All justifications are easily obtained by considering the corresponding Riemann sums. □

Item (a) from Proposition 11.10 shows that the interpretation of the integral as the area under the graph is only appropriate if $f \geq 0$. On the other hand, the interpretation of the integral of a velocity as travelled distance is also meaningful for negative velocities (change of direction). Item (d) is especially important for the integration of piecewise continuous functions (see Fig. 11.3): the integral is obtained as the sum of the single integrals.

11.2 Fundamental Theorems of Calculus

For a Riemann integrable function f we define a new function

$$F(x) = \int_a^x f(t)\,dt.$$

It is obtained by considering the upper boundary of the integration domain as variable.

Remark 11.11 For positive f, the value $F(x)$ is the area under the graph of the function in the interval $[a, x]$; see Fig. 11.4.

Fig. 11.4 The interpretation of $F(x)$ as area

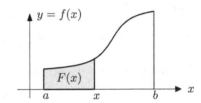

Proposition 11.12 (Fundamental theorems of calculus) *Let f be continuous in* [a, b]. *Then the following assertions hold:*

(a) First fundamental theorem: *If G is an antiderivative of f then*

$$\int_a^b f(x)\, dx = G(b) - G(a).$$

(b) Second fundamental theorem: *The function*

$$F(x) = \int_a^x f(t)\, dt$$

is an antiderivative of f, that is, F is differentiable and $F'(x) = f(x)$.

Proof In the first step we prove the second fundamental theorem. For that let $x \in (a, b)$, $h > 0$ and $x + h \in (a, b)$. According to Proposition 6.15 the function f has a minimum and a maximum in the interval $[x, x + h]$:

$$m(h) = \min_{t \in [x, x+h]} f(t), \qquad M(h) = \max_{t \in [x, x+h]} f(t).$$

The continuity of f implies the convergence $m(h) \to f(x)$ and $M(h) \to f(x)$ as $h \to 0$. According to item (b) in Proposition 11.10 we have that

$$m(h) \cdot h \leq F(x + h) - F(x) = \int_x^{x+h} f(t)\, dt \leq M(h) \cdot h.$$

This shows that F is differentiable at x and

$$F'(x) = \lim_{h \to 0} \frac{F(x + h) - F(x)}{h} = f(x).$$

The first fundamental theorem follows from the second fundamental theorem

$$\int_a^b f(t)\, dt = F(b) = F(b) - F(a),$$

since $F(a) = 0$. If G is another antiderivative then $G = F + c$ according to Proposition 10.1; hence

$$G(b) - G(a) = F(b) + c - (F(a) + c) = F(b) - F(a).$$

Thus $G(b) - G(a) = \int_a^b f(x)\, dx$ as well. $\square$

Remark 11.13 For positive f, the second fundamental theorem of calculus has an intuitive interpretation. The value $F(x + h) - F(x)$ is the area under the graph of the function $y = f(x)$ in the interval $[x, x + h]$, while $hf(x)$ is the area of the approximating rectangle of height $f(x)$. The resulting approximation

$$\frac{F(x + h) - F(x)}{h} \approx f(x)$$

suggests that in the limit as $h \to 0$, $F'(x) = f(x)$. The given proof makes the argument rigorous.

Applications of the first fundamental theorem. The most important application consists in evaluating definite integrals $\int_a^b f(x)\,dx$. For that, one determines an antiderivative $F(x)$, for instance as indefinite integral, and substitutes:

$$\int_a^b f(x)\,dx = F(x)\Big|_{x=a}^{x=b} = F(b) - F(a).$$

Example 11.14 As an application we compute the following integrals.

(a) $\displaystyle\int_1^3 x^2\,dx = \frac{x^3}{3}\Big|_{x=1}^{x=3} = \frac{27}{3} - \frac{1}{3} = \frac{26}{3}.$

(b) $\displaystyle\int_0^{\pi/2} \cos x\,dx = \sin x\Big|_{x=0}^{x=\pi/2} = \sin\frac{\pi}{2} - \sin 0 = 1.$

(c) $\displaystyle\int_0^1 x\sin(x^2)\,dx = -\frac{1}{2}\cos(x^2)\Big|_{x=0}^{x=1} = -\frac{1}{2}\cos 1 - \left(-\frac{1}{2}\cos 0\right)$

$\displaystyle = -\frac{1}{2}\cos 1 + \frac{1}{2}$ (see Example 10.12).

Remark 11.15 In maple the integration of expressions and functions is carried out using the command `int`, which requires the analytic expression and the domain as arguments, for instance

```
int(x^2, x = 1..3);
```

Applications of the second fundamental theorem. Usually, such applications are of theoretical nature, like the description of the relation between travelled distance and velocity,

$$w(t) = w(0) + \int_0^t v(s)\,ds, \quad w'(t) = v(t),$$

where $w(t)$ denotes the travelled distance from 0 to time t and $v(t)$ is the instantaneous velocity. Other applications arise in numerical analysis, for instance

$$\int_0^x e^{-y^2} \, dy \quad \text{is an antiderivative of } e^{-x^2}.$$

The value of such an integral can be approximately calculated using Taylor polynomials (see Application 12.18) or numerical integration methods (see Sect. 13.1). This is of particular interest if the antiderivative is not an elementary function, as it is the case for the Gaussian error function from Example 10.10.

11.3 Applications of the Definite Integral

We now turn to further applications of the definite integral, which confirm the modelling power of the notion of the Riemann integral.

The volume of a solid of revolution. Assume first that for a three-dimensional solid (possibly after choosing an appropriate Cartesian coordinate system) the cross-sectional area $A = A(x)$ is known for every $x \in [a, b]$; see Fig. 11.5. The volume of a thin slice of thickness Δx is approximately equal to $A(x)\Delta x$. Writing down the Riemann sums and taking limits one obtains for the volume V of the solid

$$V = \int_a^b A(x) \, dx.$$

A solid of revolution is obtained by rotating the plane curve $y = f(x)$, $a \leq x \leq b$ around the x-axis. In this case, we have $A(x) = \pi f(x)^2$, and the volume is given by

$$V = \pi \int_a^b f(x)^2 \, dx.$$

Fig. 11.5 Solid of revolution, volume

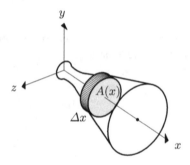

Fig. 11.6 A cone

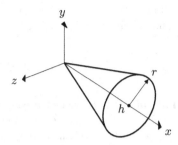

Example 11.16 (Volume of a cone) The rotation of the straight line $y = \frac{r}{h}x$ around the x-axis produces a cone of radius r and height h (Fig. 11.6). Its volume is given by

$$V = \pi \frac{r^2}{h^2} \int_0^h x^2 \, dx = \pi \frac{r^2}{h^2} \cdot \frac{x^3}{3} \bigg|_{x=0}^{x=h} = \pi r^2 \frac{h}{3}.$$

Arc length of the graph of a function. To determine the arc length of the graph of a differentiable function with continuous derivative, we first partition the interval $[a, b]$,

$$a = x_0 < x_1 < x_2 < \cdots < x_n = b,$$

and replace the graph $y = f(x)$ on $[a, b]$ by line segments passing through the points $(x_0, f(x_0)), (x_1, f(x_1)), \ldots, (x_n, f(x_n))$. The total length of the line segments is

$$s_n = \sum_{j=1}^n \sqrt{(x_j - x_{j-1})^2 + (f(x_j) - f(x_{j-1}))^2}.$$

It is simply given by the sum of the lengths of the individual segments (Fig. 11.7). According to the mean value theorem (Proposition 8.4) we have

$$s_n = \sum_{j=1}^n \sqrt{(x_j - x_{j-1})^2 + f'(\xi_j)^2 (x_j - x_{j-1})^2}$$

$$= \sum_{j=1}^n \sqrt{1 + f'(\xi_j)^2} \, (x_j - x_{j-1})$$

Fig. 11.7 The arc length of a graph

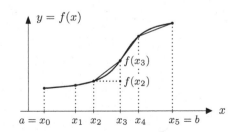

with certain points $\xi_j \in [x_{j-1}, x_j]$. The sums s_n are easily identified as Riemann sums. Their limit is thus given by

$$s = \int_a^b \sqrt{1 + f'(x)^2}\, dx.$$

Lateral surface area of a solid of revolution. The lateral surface of a solid of revolution is obtained by rotating the curve $y = f(x)$, $a \le x \le b$ around the x-axis.

In order to determine its area, we split the solid into small slices of thickness Δx. Each of these slices is approximately a truncated cone with generator of length Δs and mean radius $f(x)$; see Fig. 11.8. According to Exercise 11 of Chap. 3 the lateral surface area of this truncated cone is equal to $2\pi f(x)\Delta s$. According to what has been said previously, $\Delta s \approx \sqrt{1 + f'(x)^2}\, \Delta x$ and thus the lateral surface area of a small slice is approximately equal to

$$2\pi f(x)\sqrt{1 + f'(x)^2}\, \Delta x.$$

Writing down the Riemann sums and taking limits one obtains

$$M = 2\pi \int_a^b f(x)\sqrt{1 + f'(x)^2}\, dx$$

for the lateral surface area.

Example 11.17 (Surface area of a sphere)   The surface of a sphere of radius r is generated by rotation of the graph $f(x) = \sqrt{r^2 - x^2}$, $-r \le x \le r$. One obtains

$$M = 2\pi \int_{-r}^r \sqrt{r^2 - x^2}\, \frac{r}{\sqrt{r^2 - x^2}}\, dx = 4\pi r^2.$$

Fig. 11.8 Solid of rotation, curved surface area

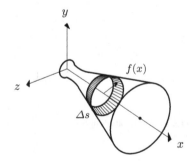

11.4 Exercises

1. Modify the MATLAB program `mat11_1.m` so that it evaluates Riemann sums of given lengths n for polynomials of degree k on arbitrary intervals $[a, b]$ (MATLAB command `polyval`).

2. Prove that every function which is piecewise constant in an interval $[a, b]$ is Riemann integrable (use Definition 11.3).

3. Compute the area between the graphs of $y = \sin x$ and $y = \sqrt{x}$ on the interval $[0, 2\pi]$.

4. (From engineering mechanics; Fig. 11.9) The shear force $Q(x)$ and the bending moment $M(x)$ of a beam of length L under a distributed load $p(x)$ obey the relationships $M'(x) = Q(x)$, $Q'(x) = -p(x)$, $0 \le x \le L$. Compute $Q(x)$ and $M(x)$ and sketch their graphs for

   (a) a simply supported beam with uniformly distributed load: $p(x) = p_0$, $Q(0) = p_0 L/2$, $M(0) = 0$;

   (b) a cantilever beam with triangular load: $p(x) = q_0(1 - x/L)$, $Q(L) = 0$, $M(L) = 0$.

5. Write a MATLAB program which provides a numerical approximation to the integral

$$\int_0^1 e^{-x^2}\, dx.$$

For this purpose, use Riemann sums of the form

$$L = \sum_{j=1}^{n} e^{-x_j^2}\, \Delta x, \quad U = \sum_{j=1}^{n} e^{-x_{j-1}^2}\, \Delta x$$

with $x_j = j\Delta x$, $\Delta x = 1/n$ and try to determine Δx and n, respectively, so that $U - L \le 0.01$; i.e. the result should be correct up to two digits. Compare your result with the value obtained by means of the MATLAB command `sqrt(pi)/2*erf(1)`.

Additional task: Extend your program such that it allows one to compute $\int_0^a e^{-x^2}\, dx$ for arbitrary $a > 0$.

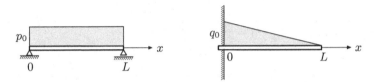

Fig. 11.9 Simply supported beam with uniformly distributed load, cantilever beam with triangular load

6. Show that the error of approximating the integral in Exercise 5 either by L or U is at most $U - L$. Use the applet *Integration* to visualise this fact.
   Hint. Verify the inequality

$$L \le \int_0^1 e^{-x^2} \, dx \le U.$$

Thus, L and U are *lower* and *upper sums*, respectively.

7. Rotation of the parabola $y = 2\sqrt{x}$, $0 \le x \le 1$ around the x-axis produces a paraboloid. Sketch it and compute its volume and its lateral surface area.

8. Compute the arc length of the graph of the following functions:
   (a) the parabola $f(x) = x^2/2$ for $0 \le x \le 2$;
   (b) the catenary $g(x) = \cosh x$ for $-1 \le x \le 3$.
   Hint. See Exercise 7 in Sect. 10.3.

9. The surface of a cooling tower can be described qualitatively by rotating the hyperbola $y = \sqrt{1 + x^2}$ around the x-axis in the bounds $-1 \le x \le 2$.
   (a) Compute the volume of the corresponding solid of revolution.
   (b) Show that the lateral surface area is given by $M = 2\pi \int_{-1}^2 \sqrt{1 + 2x^2} \, dx$. Evaluate the integral directly and by means of maple.
   Hint. Reduce the integral to the one considered in Exercise 7 of Sect. 10.3 by a suitable substitution.

10. A lens-shaped body is obtained by rotating the graph of the sine function $y = \sin x$ around the x-axis in the bounds $0 \le x \le \pi$.
   (a) Compute the volume of the body.
   (b) Compute its lateral surface area.
   Hint. For (a) use the identity $\sin^2 x = \frac{1}{2}(1 - \cos 2x)$; for (b) use the substitution $g(x) = \cos x$.

11. (From probability theory) Let X be a random variable with values in an interval $[a, b]$ which possesses a probability density $f(x)$, that is, $f(x) \ge 0$ and $\int_a^b f(x) \, dx = 1$. Its expectation value $\mu = E(X)$, its second moment $E(X^2)$ and its variance $V(X)$ are defined by

$$E(X) = \int_a^b x f(x) \, dx, \quad E(X^2) = \int_a^b x^2 f(x) \, dx,$$

$$V(X) = \int_a^b (x - \mu)^2 f(x) \, dx.$$

Show that $V(X) = E(X^2) - \mu^2$.

12. Compute the expectation value and the variance of a random variable which has
(a) a uniform distribution on $[a, b]$, i.e. $f(x) = 1/(b - a)$ for $a \le x \le b$;
(b) a (special) beta distribution on $[a, b]$ with density $f(x) = 6(x - a)(b - x)/$
$(b - a)^3$.

13. Compute the expectation value and the variance of a random variable which has
a triangular distribution on $[a, b]$ with modal value m, i.e.

$$f(x) = \begin{cases} \dfrac{2(x - a)}{(b - a)(m - a)} & \text{for } a \le x \le m, \\[3mm] \dfrac{2(b - x)}{(b - a)(b - m)} & \text{for } m \le x \le b. \end{cases}$$

Taylor Series

12

Approximations of complicated functions by simpler functions play a vital part in applied mathematics. Starting with the concept of linear approximation we discuss the approximation of a function by Taylor polynomials and by Taylor series in this chapter. As important applications we will use Taylor series to compute limits of functions and to analyse various approximation formulas.

12.1 Taylor's Formula

In this section we consider the approximation of sufficiently smooth functions by polynomials as well as applications of these approximations. We have already seen an approximation formula in Chap. 7: Let f be a function that is differentiable at a. Then

$$f(x) \approx g(x) = f(a) + f'(a) \cdot (x - a),$$

for all x close to a. The *linear approximation* g is a polynomial of degree 1 in x, and its graph is just the tangent to f at a. We now want to generalise this approximation result.

Proposition 12.1 (Taylor's formula[1]) *Let $I \subseteq \mathbb{R}$ be an open interval and $f: I \to \mathbb{R}$ an $(n + 1)$-times continuously differentiable function (i.e., the derivative of order $(n + 1)$ of f exists and is continuous). Then, for all $x, a \in I$,*

[1]B. Taylor, 1685–1731.
© Springer Nature Switzerland AG 2018
M. Oberguggenberger and A. Ostermann, *Analysis for Computer Scientists*,
Undergraduate Topics in Computer Science,
https://doi.org/10.1007/978-3-319-91155-7_12

$$f(x) = f(a) + f'(a) \cdot (x-a) + \frac{f''(a)}{2!}(x-a)^2 + \cdots + \frac{f^{(n)}(a)}{n!}(x-a)^n$$
$$+ R_{n+1}(x, a)$$

with the remainder term (in integral form)

$$R_{n+1}(x, a) = \frac{1}{n!}\int_a^x (x-t)^n f^{(n+1)}(t)\, dt.$$

Alternatively the remainder term can be expressed by

$$R_{n+1}(x, a) = \frac{f^{(n+1)}(\xi)}{(n+1)!}(x-a)^{n+1},$$

where ξ is a point between a and x (Lagrange's[2] form of the remainder term).

Proof According to the fundamental theorem of calculus, we have

$$\int_a^x f'(t)\, dt = f(x) - f(a),$$

and thus

$$f(x) = f(a) + \int_a^x f'(t)\, dt.$$

We apply integration by parts to this formula. Due to

$$\int_a^x u'(t)v(t)\, dt = u(t)v(t)\Big|_a^x - \int_a^x u(t)v'(t)\, dt$$

with $u(t) = t - x$ and $v(t) = f'(t)$ we get

$$f(x) = f(a) + (t-x)f'(t)\Big|_a^x - \int_a^x (t-x)f''(t)\, dt$$
$$= f(a) + f'(a) \cdot (x-a) + \int_a^x (x-t)f''(t)\, dt.$$

A further integration by parts yields

$$\int_a^x (x-t)f''(t)\, dt = -\frac{(x-t)^2}{2}f''(t)\Big|_a^x + \int_a^x \frac{(x-t)^2}{2}f'''(t)\, dt$$
$$= \frac{f''(a)}{2}(x-a)^2 + \frac{1}{2}\int_a^x (x-t)^2 f'''(t)\, dt,$$

---

[2] J.L. Lagrange, 1736–1813.

and one recognises that repeated integration by parts leads to the desired formula (with the remainder term in integral form). The other representation of the remainder term follows from the mean value theorem for integrals [4, Chap. 5, Theorem 5.4]. □

Example 12.2 (Important special case) If one sets $x = a + h$ and replaces a by x in Taylor's formula, then one obtains

$$f(x + h) = f(x) + h\,f'(x) + \frac{h^2}{2}\,f''(x) + \cdots + \frac{h^n}{n!}\,f^{(n)}(x) + \frac{h^{n+1}}{(n+1)!}\,f^{(n+1)}(\xi)$$

with a point ξ between x and $x + h$. For small h this formula describes how the function f behaves near x.

Remark 12.3 Often one does not know the remainder term

$$R_{n+1}(x, a) = \frac{f^{(n+1)}(\xi)}{(n+1)!}\,(x - a)^{n+1}$$

explicitly since ξ is unknown in general. Let M be the supremum of $\left|f^{(n+1)}\right|$ in the considered interval around a. For x in this interval we obtain the bound

$$\left|R_{n+1}(x, a)\right| \leq \frac{M}{(n+1)!}(x - a)^{n+1}.$$

The remainder term is thus bounded by a constant times h^{n+1}, where $h = x - a$. In this situation, one writes for short

$$R_{n+1}(a + h, a) = \mathcal{O}(h^{n+1})$$

as $h \to 0$ and calls the remainder a term of *order* $n + 1$. This notation is also used by maple.

Definition 12.4 The polynomial

$$T_n(x, a) = f(a) + f'(a) \cdot (x - a) + \cdots + \frac{f^{(n)}(a)}{n!}(x - a)^n$$

is called nth *Taylor polynomial* of f around the *point of expansion* a.

The graphs of the functions $y = T_n(x, a)$ and $y = f(x)$ both pass through the point $(a, f(a))$. Their tangents in this point have the same slope $T_n'(x, a) = f'(a)$ and the graphs have the same curvature (due to $T_n''(x, a) = f''(a)$, see Chap. 14). It depends on the size of the remainder term how well the Taylor polynomial approximates the function.

Example 12.5 (Taylor polynomial of the exponential function)   Let $f(x) = e^x$ and $a = 0$. Due to $(e^x)' = e^x$ we have $f^{(k)}(0) = e^0 = 1$ for all $k \geq 0$ and hence

$$e^x = 1 + x + \frac{x^2}{2} + \cdots + \frac{x^n}{n!} + \frac{e^\xi}{(n+1)!} x^{n+1},$$

where ξ denotes a point between 0 and x. We want to determine the minimal degree of the Taylor polynomial which approximates the function in the interval $[0, 1]$, correct to 5 digits. For that we require the following bound on the remainder term

$$\left| e^x - 1 - x - \cdots - \frac{x^n}{n!} \right| = \frac{e^\xi}{(n+1)!} x^{n+1} \leq 10^{-5}.$$

Note that $x \in [0, 1]$ as well as e^ξ are non-negative. The above remainder will be maximal for $x = \xi = 1$. Thus we determine n from the inequality $e/(n+1)! \leq 10^{-5}$. Due to $e \approx 3$ this inequality is certainly fulfilled from $n = 8$ onwards; in particular,

$$e = 1 + 1 + \frac{1}{2} + \cdots + \frac{1}{8!} \pm 10^{-5}.$$

One has to choose $n \geq 8$ in order to determine the first 5 digits of e.

Experiment 12.6 Repeat the above calculations with the help of the maple worksheet mp12_1.mws. In this worksheet the required maple commands for Taylor's formula are explained.

Example 12.7 (Taylor polynomial of the sine function)   Let $f(x) = \sin x$ and $a = 0$. Recall that $(\sin x)' = \cos x$ and $(\cos x)' = -\sin x$ as well as $\sin 0 = 0$ and $\cos 0 = 1$. Therefore,

$$\sin x = \sum_{k=0}^{2n+1} \frac{\sin^{(k)}(0)}{k!} x^k + R_{2n+2}(x, 0) =$$

$$= x - \frac{x^3}{3!} + \frac{x^5}{5!} - \frac{x^7}{7!} + \cdots + (-1)^n \frac{x^{2n+1}}{(2n+1)!} + R_{2n+2}(x, 0).$$

Note that the Taylor polynomial consists of odd powers of x only. According to Taylor's formula, the remainder has the form

$$R_{2n+2}(x, 0) = \frac{\sin^{(2n+2)}(\xi)}{(2n+2)!} x^{2n+2}.$$

Since all derivatives of the sine function are bounded by 1, we obtain

$$|R_{2n+2}(x, 0)| \leq \frac{x^{2n+2}}{(2n+2)!}.$$

For *fixed* x the remainder term tends to zero as $n \to \infty$, since the expression $x^{2n+2}/(2n+2)!$ is a summand of the exponential series, which converges for all $x \in \mathbb{R}$. The above estimate can be interpreted as follows: For every $x \in \mathbb{R}$ and $\varepsilon > 0$, there exists an integer $N \in \mathbb{N}$ such that the difference of the sine function and its nth Taylor polynomial is small; more precisely,

$$|\sin t - T_n(t, 0)| \leq \varepsilon$$

for all $n \geq N$ and $t \in [-x, x]$.

Experiment 12.8 Using the maple worksheet mp12_2.mws compute the Taylor polynomials of $\sin x$ around the point 0 and determine the accuracy of the approximation (by plotting the difference to $\sin x$). In order to achieve high accuracy for large x, the degree of the polynomials has to be chosen sufficiently high. Due to rounding errors, however, this procedure quickly reaches its limits (unless one increases the number of significant digits).

Example 12.9 The 4th degree Taylor polynomial $T_4(x, 0)$ of the function

$$f(x) = \begin{cases} \dfrac{x}{e^x - 1} & x \neq 0, \\ 1 & x = 0, \end{cases}$$

is given by

$$T_4(x, 0) = 1 - \frac{x}{2} + \frac{1}{12}x^2 - \frac{1}{720}x^4.$$

Experiment 12.10 The maple worksheet mp12_3.mws shows that, for sufficiently large n, the Taylor polynomial of degree n gives a good approximation to the function from Example 12.9 on closed subintervals of $(-2\pi, 2\pi)$. For $x \geq 2\pi$ (as well as for $x \leq -2\pi$) the Taylor polynomial is, however, useless.

12.2 Taylor's Theorem

The last example gives rise to the question for which points the Taylor polynomial converges to the function as $n \to \infty$.

Definition 12.11 Let $I \subseteq \mathbb{R}$ be an open interval and let $f: I \to \mathbb{R}$ have arbitrarily many derivatives. Given $a \in I$, the series

$$T(x, a, f) = \sum_{k=0}^{\infty} \frac{f^{(k)}(a)}{k!}(x - a)^k$$

is called *Taylor series* of f around the point a.

Proposition 12.12 (Taylor's theorem) *Let* $f: I \to \mathbb{R}$ *be a function with arbitrarily many derivatives and let* $T(x, a, f)$ *be its Taylor series around the point a. Then the function and its Taylor series coincide at* $x \in I$, *i.e.,*

$$f(x) = \sum_{k=0}^{\infty} \frac{f^{(k)}(a)}{k!}(x-a)^k,$$

if and only if the remainder term

$$R_n(x, a) = \frac{f^{(n)}(\xi)}{n!}(x-a)^n$$

tends to 0 as $n \to \infty$.

Proof According to Taylor's formula (Proposition 12.1),

$$f(x) - T_n(x, a) = R_{n+1}(x, a)$$

and hence

$$f(x) = \lim_{n \to \infty} T_n(x, a) = T(x, a, f) \quad \Leftrightarrow \quad \lim_{n \to \infty} R_n(x, a) = 0,$$

which was to be shown. $\square$

Example 12.13 Let $f(x) = \sin x$ and $a = 0$. Due to $R_n(x, 0) = \frac{\sin^{(n)}(\xi)}{n!} x^n$ we have

$$|R_n(x, 0)| \le \frac{|x|^n}{n!} \to 0$$

for x fixed and $n \to \infty$. Hence for all $x \in \mathbb{R}$

$$\sin x = \sum_{k=0}^{\infty} (-1)^k \frac{x^{2k+1}}{(2k+1)!} = x - \frac{x^3}{3!} + \frac{x^5}{5!} - \frac{x^7}{7!} + \frac{x^9}{9!} \mp \cdots$$

12.3 Applications of Taylor's Formula

To complete this chapter we discuss a few important applications of Taylor's formula.

Application 12.14 (Extremum test) Let the function $f: I \to \mathbb{R}$ be n-times continuously differentiable in the interval I and assume that

$$f'(a) = f''(a) = \cdots = f^{(n-1)}(a) = 0 \text{ and } f^{(n)}(a) \neq 0.$$

Then the following assertions hold:

(a) The function f has an extremum at a if and only if n is even;

(b) if n is even and $f^{(n)}(a) > 0$, then a is a local minimum of f;
    if n is even and $f^{(n)}(a) < 0$, then a is a local maximum of f.

Proof Due to Taylor's formula, we have

$$f(x) - f(a) = \frac{f^{(n)}(\xi)}{n!}(x-a)^n, \qquad x \in I.$$

If x is close to a, $f^{(n)}(\xi)$ and $f^{(n)}(a)$ have the same sign (since $f^{(n)}$ is continuous). For n *odd* the right-hand side changes its sign at $x = a$ because of the term $(x-a)^n$. Hence an extremum can only occur for n *even*. If now n is even and $f^{(n)}(a) > 0$ then $f(x) > f(a)$ for all x close to a with $x \neq a$. Thus a is a local minimum.   □

Example 12.15 The polynomial $f(x) = 6 + 4x + 6x^2 + 4x^3 + x^4$ has the derivatives

$$f'(-1) = f''(-1) = f'''(-1) = 0, \ f^{(4)}(-1) = 24$$

at the point $x = -1$. Hence $x = -1$ is a local minimum of f.

Application 12.16 (Computation of limits of functions) As an example, we investigate the function

$$g(x) = \frac{x^2 \log(1+x)}{(1 - \cos x) \sin x}$$

in the neighbourhood of $x = 0$. For $x = 0$ we obtain the undefined expression $\frac{0}{0}$. In order to determine the limit when x tends to 0, we expand all appearing functions in Taylor polynomials around the point $a = 0$. Exercise 1 yields that $\cos x = 1 - \frac{x^2}{2} + \mathcal{O}(x^4)$. Taylor's formula for $\log(1+x)$ around the point $a = 0$ reads

$$\log(1 + x) = x + \mathcal{O}(x^2)$$

because of $\log 1 = 0$ and $\log(1 + x)'|_{x=0} = 1$. We thus obtain

$$g(x) = \frac{x^2(x + \mathcal{O}(x^2))}{(1 - 1 + \frac{x^2}{2} + \mathcal{O}(x^4))(x + \mathcal{O}(x^3))} = \frac{x^3 + \mathcal{O}(x^4)}{\frac{x^3}{2} + \mathcal{O}(x^5)} = \frac{1 + \mathcal{O}(x)}{\frac{1}{2} + \mathcal{O}(x^2)}$$

and consequently $\lim_{x \to 0} g(x) = 2$.

Application 12.17 (Analysis of approximation formulas) When differentiating numerically in Chap. 7, we considered the symmetric difference quotient

$$f''(x) \approx \frac{f(x+h) - 2f(x) + f(x-h)}{h^2}$$

as an approximation to the second derivative $f''(x)$. We are now in the position to investigate the accuracy of this formula. From

$$f(x+h) = f(x) + hf'(x) + \frac{h^2}{2}f''(x) + \frac{h^3}{6}f'''(x) + \mathcal{O}(h^4),$$

$$f(x-h) = f(x) - hf'(x) + \frac{h^2}{2}f''(x) - \frac{h^3}{6}f'''(x) + \mathcal{O}(h^4)$$

we infer that

$$f(x+h) + f(x-h) = 2f(x) + h^2 f''(x) + \mathcal{O}(h^4)$$

and hence

$$\frac{f(x+h) - 2f(x) + f(x-h)}{h^2} = f''(x) + \mathcal{O}(h^2).$$

One calls this formula second-order accurate. If one reduces h by the factor λ, then the error reduces by the factor λ^2, as long as rounding errors do not play a decisive role.

Application 12.18 (Integration of functions that do not possess elementary integrals) As already mentioned in Sect. 10.2 there are functions whose antiderivatives cannot be expressed as combinations of elementary functions. For example, the function $f(x) = e^{-x^2}$ does not have an elementary integral. In order to compute the definite integral

$$\int_0^1 e^{-x^2}\, dx,$$

we approximate e^{-x^2} by the Taylor polynomial of degree 8

$$e^{-x^2} \approx 1 - x^2 + \frac{x^4}{2} - \frac{x^6}{6} + \frac{x^8}{24}$$

and approximate the integral sought after by

$$\int_0^1 \left(1 - x^2 + \frac{x^4}{2} - \frac{x^6}{6} + \frac{x^8}{24}\right) dx = \frac{5651}{7560}.$$

The error of this approximation is $6.63 \cdot 10^{-4}$. For more precise results one takes a Taylor polynomial of a higher degree.

Experiment 12.19 Using the maple worksheet mp12_4.mws repeat the calculations from Application 12.18. Subsequently modify the program such that you can integrate $g(x) = \cos(x^2)$ with it.

12.4 Exercises

1. Compute the Taylor polynomials of degree 0, 1, 2, 3 and 4 of the function $g(x) = \cos x$ around the point of expansion $a = 0$. For which $x \in \mathbb{R}$ does the Taylor series of $\cos x$ converge?

2. Compute the Taylor polynomials of degree 1, 3 and 5 of the function $\sin x$ around the point of expansion $a = 9\pi$. Further, compute the Taylor polynomial of degree 39 with maple and plot the graph together with the graph of the function in the interval $[0, 18\pi]$. In order to be able to better distinguish the two graphs you should plot them in different colours.

3. Compute the Taylor polynomials of degree 1, 2 and 3 of the function $f(t) = \sqrt{1 + t}$ around the point of expansion $a = 0$. Further compute the Taylor polynomial of degree 10 with maple.

4. Compute the following limits using Taylor series expansion:

$$\lim_{x \to 0} \frac{x \sin x - x^2}{2 \cos x - 2 + x^2}, \qquad \lim_{x \to 0} \frac{e^{2x} - 1 - 2x}{\sin^2 x},$$

$$\lim_{x \to 0} \frac{e^{-x^2} - 1}{\sin^2(3x)}, \qquad \lim_{x \to 0} \frac{x^2 (\log(1 - 2x))^2}{1 - \cos(x^2)}.$$

Verify your results with maple.

5. For the approximate evaluation of the integral

$$\int_0^1 \frac{\sin(t^2)}{t} \, dt$$

replace the integrand by its Taylor polynomial of degree 9 and integrate this polynomial. Verify your result with maple.

6. Prove the formula

$$e^{i\varphi} = \cos \varphi + i \sin \varphi$$

by substituting the value $i\varphi$ for x into the series of the exponential function

$$e^x = \sum_{k=0}^{\infty} \frac{x^k}{k!}$$

and separating real and imaginary parts.

7. Compute the Taylor series of the hyperbolic functions $f(x) = \sinh x$ and $g(x) = \cosh x$ around the point of expansion $a = 0$ and verify the convergence of the series.

   Hint. Compute the Taylor polynomials of degree $n - 1$ and show that the remainder terms $R_n(x, 0)$ can be estimated by $(\cosh M)M^n/n!$ whenever $|x| \le M$.

8. Show that the Taylor series of $f(x) = \log(1 + x)$ around $a = 0$ is given by

$$\log(1 + x) = \sum_{k=1}^{\infty} (-1)^{k-1} \frac{x^k}{k} = x - \frac{x^2}{2} + \frac{x^3}{3} - \frac{x^4}{4} \pm \ldots$$

for $|x| < 1$.

Hint. A formal calculation, namely an integration of the geometric series expansion

$$\frac{1}{1+t} = \frac{1}{1-(-t)} = \sum_{j=0}^{\infty} (-1)^j t^j$$

from $t = 0$ to $t = x$, suggests the result. For a rigorous proof of convergence, the remainder term has to be estimated. This can be done by integrating the remainder term in the geometric series

$$\frac{1}{1+t} - \sum_{j=0}^{n-1} (-1)^j t^j = \frac{1}{1+t} - \frac{1-(-1)^n t^n}{1+t} = \frac{(-1)^n t^n}{1+t},$$

observing that $1 + t \geq \delta > 0$ for some positive constant δ as long as $|t| \leq |x| < 1$.

Numerical Integration

13

The fundamental theorem of calculus suggests the following approach to the calculation of definite integrals: one determines an antiderivative F of the integrand f and computes from that the value of the integral

$$\int_a^b f(x)\,dx = F(b) - F(a).$$

In *practice*, however, it is difficult and often even impossible to find an antiderivative F as a combination of *elementary* functions. Apart from that, antiderivatives can also be fairly complex, as the example $\int x^{100} \sin x\, dx$ shows. Finally, in concrete applications the integrand is often given numerically and *not* by an explicit formula. In all these cases one reverts to numerical methods. In this chapter the basic concepts of numerical integration (quadrature formulas and their order) are introduced and explained. By means of instructive examples we analyse the achievable accuracy for the Gaussian quadrature formulas and the required computational effort.

13.1 Quadrature Formulas

For the numerical computation of $\int_a^b f(x)\,dx$ we first split the interval of integration $[a, b]$ into subintervals with *grid points* $a = x_0 < x_1 < x_2 < \ldots < x_{N-1} < x_N = b$, see Fig. 13.1. From the additivity of the integral (Proposition 11.10 (d)) we get

$$\int_a^b f(x)\,dx = \sum_{j=0}^{N-1} \int_{x_j}^{x_{j+1}} f(x)\,dx.$$

© Springer Nature Switzerland AG 2018
M. Oberguggenberger and A. Ostermann, *Analysis for Computer Scientists*,
Undergraduate Topics in Computer Science,
https://doi.org/10.1007/978-3-319-91155-7_13

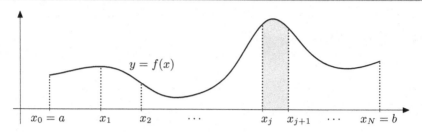

Fig. 13.1 Partition of the interval of integration into subintervals

Hence it is sufficient to find an approximation formula for a (small) subinterval of length $h_j = x_{j+1} - x_j$. One example of such a formula is the *trapezoidal rule* through which the area under the graph of a function is approximated by the area of the corresponding trapezoid (Fig. 13.2)

$$\int_{x_j}^{x_{j+1}} f(x)\, dx \approx h_j \frac{1}{2}\Big(f(x_j) + f(x_{j+1})\Big).$$

For the derivation and analysis of such approximation formulas it is useful to carry out a transformation onto the interval $[0, 1]$. By setting $x = x_j + \tau h_j$ one obtains from $dx = h_j\, d\tau$ that

$$\int_{x_j}^{x_{j+1}} f(x)\, dx = \int_0^1 f(x_j + \tau h_j) h_j\, d\tau = h_j \int_0^1 g(\tau)\, d\tau$$

with $g(\tau) = f(x_j + \tau h_j)$. Thus it is sufficient to find approximation formulas for $\int_0^1 g(\tau)\, d\tau$. The trapezoidal rule in this case is

$$\int_0^1 g(\tau)\, d\tau \approx \frac{1}{2}\Big(g(0) + g(1)\Big).$$

Obviously, it is *exact* if $g(\tau)$ is a polynomial of degree 0 or 1.

Fig. 13.2 Trapezoidal rule

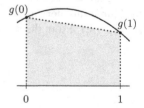

In order to obtain a more accurate formula, we demand that quadratic polynomials are integrated exactly as well. For the moment let

$$g(\tau) = \alpha + \beta\tau + \gamma\tau^2$$

be a general polynomial of degree 2. Due to $g(0) = \alpha$, $g(\frac{1}{2}) = \alpha + \frac{1}{2}\beta + \frac{1}{4}\gamma$ and $g(1) = \alpha + \beta + \gamma$ we get by a short calculation

$$\int_0^1 (\alpha + \beta\tau + \gamma\tau^2)\, d\tau = \alpha + \frac{1}{2}\beta + \frac{1}{3}\gamma = \frac{1}{6}\Big(g(0) + 4g(\tfrac{1}{2}) + g(1)\Big).$$

The corresponding approximation formula for general g reads

$$\int_0^1 g(\tau)\, d\tau \approx \frac{1}{6}\Big(g(0) + 4g(\tfrac{1}{2}) + g(1)\Big).$$

By construction, it is exact for polynomials of degree less than or equal to 2; it is called *Simpson's rule*.[1]

The special forms of the trapezoidal and of Simpson's rule motivate the following definition.

Definition 13.1 The approximation formula

$$\int_0^1 g(\tau)\, d\tau \approx \sum_{i=1}^{s} b_i\, g(c_i)$$

is called a *quadrature formula*. The numbers $b_1, \ldots, b_s$ are called *weights*, and the numbers $c_1, \ldots, c_s$ are called *nodes* of the quadrature formula; the integer s is called the number of *stages*.

A quadrature formula is determined by the specification of the weights and nodes. Thus we denote a quadrature formula by $\{(b_i, c_i),\ i = 1, \ldots, s\}$ for short. Without loss of generality the weights b_i are not zero, and the nodes are pairwise different ($c_i \neq c_k$ for $i \neq k$).

Example 13.2 (a) The trapezoidal rule has $s = 2$ stages and is given by

$$b_1 = b_2 = \frac{1}{2}, \quad c_1 = 0, \quad c_2 = 1.$$

---

[1]T. Simpson, 1710–1761.

(b) Simpson's rule has $s = 3$ stages and is given by

$$b_1 = \frac{1}{6}, \quad b_2 = \frac{2}{3}, \quad b_3 = \frac{1}{6}, \quad c_1 = 0, \quad c_2 = \frac{1}{2}, \quad c_3 = 1.$$

In order to compute the original integral $\int_a^b f(x)\,dx$ by quadrature formulas, one has to reverse the transformation from f to g. Due to $g(\tau) = f(x_j + \tau h_j)$ one obtains

$$\int_{x_j}^{x_{j+1}} f(x)\,dx = h_j \int_0^1 g(\tau)\,dt \approx h_j \sum_{i=1}^{s} b_i g(c_i) = h_j \sum_{i=1}^{s} b_i f(x_j + c_i h_j),$$

and thus the approximation formula

$$\int_a^b f(x)\,dx = \sum_{j=0}^{N-1} \int_{x_j}^{x_{j+1}} f(x)\,dx \approx \sum_{j=0}^{N-1} h_j \sum_{i=1}^{s} b_i f(x_j + c_i h_j).$$

We now look for quadrature formulas that are as accurate as possible. Since the integrand is typically well approximated by Taylor polynomials on small intervals, a *good* quadrature formula is characterised by the property that it integrates *exactly* as many polynomials as possible. This idea motivates the following definition.

Definition 13.3 (Order) The quadrature formula $\{(b_i, c_i),\ i = 1, \ldots, s\}$ has *order* p if all polynomials g of degree less or equal to $p - 1$ are integrated exactly by the quadrature formula; i.e.,

$$\int_0^1 g(\tau)\,d\tau = \sum_{i=1}^{s} b_i\, g(c_i)$$

for all polynomials g of degree smaller than or equal to $p - 1$.

Example 13.4 (a) The trapezoidal rule has order 2.

(b) Simpson's rule has (by construction) at least order 3.

The following proposition yields an algebraic characterisation of the order of quadrature formulas.

Proposition 13.5 *A quadrature formula* $\{(b_i, c_i),\ i = 1, \ldots, s\}$ *has order p if and only if*

$$\sum_{i=1}^{s} b_i\, c_i^{q-1} = \frac{1}{q} \quad \textit{for}\ \ 1 \le q \le p.$$

Proof One uses the fact that a polynomial g of degree $p - 1$

$$g(\tau) = \alpha_0 + \alpha_1 \tau + \ldots + \alpha_{p-1} \tau^{p-1}$$

is a linear combination of monomials, and that both integration and application of a quadrature formula are *linear* processes. Thus it is sufficient to prove the result for the monomials

$$g(\tau) = \tau^{q-1}, \qquad 1 \le q \le p.$$

The proposition now follows directly from the identity

$$\frac{1}{q} = \int_0^1 \tau^{q-1} \, d\tau = \sum_{i=1}^s b_i \, g(c_i) = \sum_{i=1}^s b_i c_i^{q-1}. \qquad \square$$

The conditions of the proposition

$$b_1 + b_2 + \ldots + b_s = 1$$
$$b_1 c_1 + b_2 c_2 + \ldots + b_s c_s = \tfrac{1}{2}$$
$$b_1 c_1^2 + b_2 c_2^2 + \ldots + b_s c_s^2 = \tfrac{1}{3}$$
$$\vdots$$
$$b_1 c_1^{p-1} + b_2 c_2^{p-1} + \ldots + b_s c_s^{p-1} = \tfrac{1}{p}$$

are called *order conditions* of order p. If s nodes $c_1, \ldots, c_s$ are given then the order conditions form a *linear* system of equations for the unknown weights b_i. If the nodes are pairwise different then the weights can be determined uniquely from that. This shows that for s *different* nodes there always exists a *unique* quadrature formula of order $p \ge s$.

Example 13.6 We determine once more the order of Simpson's rule. Due to

$$b_1 + b_2 + b_3 = \tfrac{1}{6} + \tfrac{2}{3} + \tfrac{1}{6} = 1$$
$$b_1 c_1 + b_2 c_2 + b_3 c_3 = \tfrac{2}{3} \cdot \tfrac{1}{2} + \tfrac{1}{6} = \tfrac{1}{2}$$
$$b_1 c_1^2 + b_2 c_2^2 + b_3 c_3^2 = \tfrac{2}{3} \cdot \tfrac{1}{4} + \tfrac{1}{6} = \tfrac{1}{3}$$

its order is at least 3 (as we already know from the construction). However, additionally

$$b_1 c_1^3 + b_2 c_2^3 + b_3 c_3^3 = \tfrac{4}{6} \cdot \tfrac{1}{8} + \tfrac{1}{6} = \tfrac{3}{12} = \tfrac{1}{4},$$

i.e., Simpson's rule even has order 4.

The best quadrature formulas (high accuracy with little computational effort) are the Gaussian quadrature formulas. For that we state the following result whose proof can be found in [23, Chap. 10, Corollary 10.1].

Proposition 13.7 *There is no quadrature formula with s stages of order $p > 2s$. On the other hand, for every $s \in \mathbb{N}$ there exists a (unique) quadrature formula of order $p = 2s$. This formula is called s-stage* Gaussian quadrature formula.

The Gaussian quadrature formulas for $s \leq 3$ are

$$s = 1: \quad c_1 = \frac{1}{2}, \quad b_1 = 1, \quad \text{order 2 (midpoint rule)};$$

$$s = 2: \quad c_1 = \frac{1}{2} - \frac{\sqrt{3}}{6}, \quad c_2 = \frac{1}{2} + \frac{\sqrt{3}}{6}, \quad b_1 = b_2 = \frac{1}{2}, \quad \text{order 4};$$

$$s = 3: \quad c_1 = \frac{1}{2} - \frac{\sqrt{15}}{10}, \quad c_2 = \frac{1}{2}, \quad c_3 = \frac{1}{2} + \frac{\sqrt{15}}{10},$$

$$b_1 = \frac{5}{18}, \quad b_2 = \frac{8}{18}, \quad b_3 = \frac{5}{18}, \quad \text{order 6}.$$

13.2 Accuracy and Efficiency

In the following numerical experiment the accuracy of quadrature formulas will be illustrated. With the help of the Gaussian quadrature formulas of order 2, 4 and 6 we compute the two integrals

$$\int_0^3 \cos x \, \mathrm{d}x = \sin 3 \quad \text{and} \quad \int_0^1 x^{5/2} \, \mathrm{d}x = \frac{2}{7}.$$

For that we choose equidistant grid points

$$x_j = a + jh, \quad j = 0, \ldots, N$$

with $h = (b - a)/N$ and $N = 1, 2, 4, 8, 16, \ldots, 512$. Finally, we plot the costs of the calculation as a function of the achieved accuracy in a double-logarithmic diagram.

A measure for the computational cost of a quadrature formula is the number of required *function evaluations*, abbreviated by `fe`. For an s-stage quadrature formula, it is the number

$$\mathtt{fe} = s \cdot N.$$

The achieved accuracy `err` is the absolute value of the error. The according results are presented in Fig. 13.3. One makes the following observations:

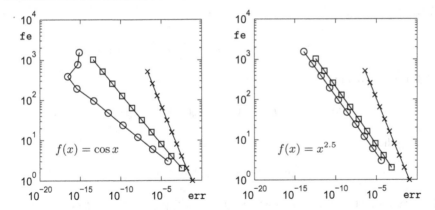

Fig. 13.3 Accuracy-cost-diagram of the Gaussian quadrature formulas. The crosses are the results of the one-stage Gaussian method of order 2, the squares the ones of the two-stage method of order 4 and the circles the ones of the three-stage method of order 6

(a) The curves are straight lines (as long as one does not get into the range of rounding errors, like with the three-stage method in the left picture).

(b) In the left picture the straight lines have slope $-1/p$, where p is the order of the quadrature formula. In the right picture this is only true for the method of order 2, and the other two methods result in straight lines with slope $-2/7$.

(c) For given costs the formulas of higher order are more accurate.

In order to understand this behaviour, we expand the integrand into a Taylor series. On the subinterval $[\alpha, \alpha + h]$ of length h we obtain

$$f(\alpha + \tau h) = \sum_{q=0}^{p-1} \frac{h^q}{q!} f^{(q)}(\alpha)\tau^q + \mathcal{O}(h^p).$$

Since a quadrature formula of order p integrates polynomials of degree less than or equal to $p - 1$ exactly, the Taylor polynomial of f of degree $p - 1$ is being integrated exactly. The error of the quadrature formula on this subinterval is proportional to the length of the interval times the size of the remainder term of the integrand, so

$$h \cdot \mathcal{O}(h^p) = \mathcal{O}(h^{p+1}).$$

In total we have N subintervals; hence, the total error of the quadrature formula is

$$N \cdot \mathcal{O}(h^{p+1}) = Nh \cdot \mathcal{O}(h^p) = (b - a) \cdot \mathcal{O}(h^p) = \mathcal{O}(h^p).$$

Thus we have shown that (for small h) the error err behaves like

$$\text{err} \approx c_1 \cdot h^p.$$

Since furthermore

$$\mathtt{fe} = sN = s \cdot Nh \cdot h^{-1} = s \cdot (b - a) \cdot h^{-1} = c_2 \cdot h^{-1}$$

holds true, we obtain

$$\log(\mathtt{fe}) = \log c_2 - \log h \quad \text{and} \quad \log(\mathtt{err}) \approx \log c_1 + p \cdot \log h,$$

so altogether

$$\log(\mathtt{fe}) \approx c_3 - \frac{1}{p} \cdot \log(\mathtt{err}).$$

This explains why straight lines with slope $-1/p$ appear in the left picture.

In the right picture it has to be noted that the second derivative of the integrand is *discontinuous* at 0. Hence the above considerations with the Taylor series are not valid anymore. The quadrature formula also detects this discontinuity of the high derivatives and reacts with a so-called *order reduction*; i.e., the methods show a lower order (in our case $p = 7/2$).

Experiment 13.8 Compute the integrals

$$\int_0^3 \sqrt{x}\, dx \quad \text{and} \quad \int_1^2 \frac{dx}{x}$$

using the Gaussian quadrature formulas and generate an accuracy-cost-diagram. For that purpose modify the programs `mat13_1.m`, `mat13_2.m`, `mat13_3.m`, `mat13_4.m` and `mat13_5.m` with which Fig. 13.3 was produced.

Commercial programs for numerical integration determine the grid points adaptively based on automatic error estimates. The user can usually specify the desired accuracy. In MATLAB the routines `quad.m` and `quadl.m` serve this purpose.

13.3 Exercises

1. For the calculation of $\int_0^1 x^{100} \sin x\, dx$ first determine an antiderivative F of the integrand f using maple. Then evaluate $F(1) - F(0)$ to 10, 50, 100, 200 and 400 digits and explain the surprising results.

2. Determine the order of the quadrature formula given by

$$b_1 = b_4 = \frac{1}{8}, \quad b_2 = b_3 = \frac{3}{8}, \quad c_1 = 0, \quad c_2 = \frac{1}{3}, \quad c_3 = \frac{2}{3}, \quad c_4 = 1.$$

3. Determine the unique quadrature formula of order 3 with the nodes

$$c_1 = \frac{1}{3}, \quad c_2 = \frac{2}{3}, \quad c_3 = 1.$$

4. Determine the unique quadrature formula with the nodes

$$c_1 = \frac{1}{4}, \quad c_2 = \frac{1}{2}, \quad c_3 = \frac{3}{4}.$$

Which order does it have?

5. Familiarise yourself with the MATLAB programs `quad.m` and `quadl.m` for the computation of definite integrals and test the programs for

$$\int_0^1 e^{-x^2}\, dx \quad \text{and} \quad \int_0^1 \sqrt[3]{x}\, dx.$$

6. Justify the formulas

$$\pi = 4 \int_0^1 \frac{dx}{1+x^2} \quad \text{and} \quad \pi = 4 \int_0^1 \sqrt{1-x^2}\, dx$$

and use them to calculate π by numerical integration. To do so divide the interval $[0, 1]$ into N equally large parts ($N = 10, 100, \ldots$) and use Simpson's rule on those subintervals. Why are the results obtained with the first formula always more accurate?

7. Write a MATLAB program that allows you to evaluate the integral of any given (continuous) function on a given interval $[a, b]$, both by the trapezoidal rule and by Simpson's rule. Use your program to numerically answering the questions of Exercises 7–9 from Sect. 11.4 and Exercise 5 from Sect. 12.4.

8. Use your program from Exercise 7 to produce tables (for $x = 0$ to $x = 10$ in steps of 0.5) of some higher transcendental functions:

(a) the Gaussian error function

$$\text{Erf}(x) = \frac{2}{\sqrt{\pi}} \int_0^x e^{-y^2}\, dy,$$

(b) the sine integral

$$Si(x) = \int_0^x \frac{\sin y}{y}\, dy,$$

(c) the Fresnel integral

$$S(x) = \int_0^x \sin\left(\frac{\pi}{2} y^2\right) dy.$$

9. (Experimental determination of expectation values) The family of standard beta distributions on the interval $[0, 1]$ is defined through the probability densities

$$f(x; r, s) = \frac{1}{B(r, s)} x^{r-1}(1 - x)^{s-1}, \quad 0 \le x \le 1,$$

where $r, s > 0$. Here $B(r, s) = \int_0^1 y^{r-1}(1 - y)^{s-1} \, dy$ is the beta function, which is a higher transcendental function for non-integer values of r, s. For integer values of $r, s \ge 1$ it is given by

$$B(r, s) = \frac{(r - 1)!(s - 1)!}{(r + s - 1)!}.$$

With the help of the MATLAB program quad.m, compute the expectation values $\mu(r, s) = \int_0^1 x f(x; r, s) \, dx$ for various integer values of r and s and guess a general formula for $\mu(r, s)$ from your experimental results.

Curves

14

The graph of a function $y = f(x)$ represents a curve in the plane. This concept, however, is too tight to represent more intricate curves, like loops, self-intersections, or even curves of fractal dimension. The aim of this chapter is to introduce the concept of parametrised curves and to study, in particular, the case of differentiable curves. For the visualisation of the trajectory of a curve, the notions of velocity vector, moving frame, and curvature are important. The chapter contains a collection of geometrically interesting examples of curves and several of their construction principles. Further, the computation of the arc length of differentiable curves is discussed, and an example of a continuous, bounded curve of infinite length is given. The chapter ends with a short outlook on spatial curves. For the vector algebra used in this chapter, we refer to Appendix A.

14.1 Parametrised Curves in the Plane

Definition 14.1 A *parametrised plane curve* is a continuous mapping

$$t \mapsto \mathbf{x}(t) = \begin{bmatrix} x(t) \\ y(t) \end{bmatrix}$$

of an interval $[a, b]$ to $\mathbb{R}^2$; i.e., both the components $t \mapsto x(t)$ and $t \mapsto y(t)$ are continuous functions.[1] The variable $t \in [a, b]$ is called *parameter of the curve*.

---

[1] Concerning the vector notation we remark that $x(t)$, $y(t)$ actually represent the coordinates of a point in $\mathbb{R}^2$. It is, however, common practise and useful to write this point as a position vector, thus the column notation.

© Springer Nature Switzerland AG 2018

M. Oberguggenberger and A. Ostermann, *Analysis for Computer Scientists*,
Undergraduate Topics in Computer Science,
https://doi.org/10.1007/978-3-319-91155-7_14

Example 14.2 An object that is thrown at height h with horizontal velocity v_H and vertical velocity v_V has the trajectory

$$
\begin{aligned}
x(t) &= v_H t, \\
y(t) &= h + v_V t - \tfrac{g}{2} t^2,
\end{aligned} \qquad 0 \le t \le t_0,
$$

where t_0 is the positive solution of the equation $h + v_V t_0 - \tfrac{g}{2} t_0^2 = 0$ (time of impact, see Fig. 14.1). In this example, we can eliminate t and represent the trajectory as the graph of a function (ballistic curve). We have $t = x/v_H$, and thus

$$
y = h + \frac{v_V}{v_H} x - \frac{g}{2 v_H^2} x^2.
$$

Example 14.3 A circle of radius R with centre at the origin has the parametric representation

$$
\begin{aligned}
x(t) &= R \cos t, \\
y(t) &= R \sin t,
\end{aligned} \qquad 0 \le t \le 2\pi.
$$

In this case, t can be interpreted as the angle between the position vector and the positive x-axis (Fig. 14.1). The components $x = x(t)$, $y = y(t)$ satisfy the quadratic equation

$$
x^2 + y^2 = R^2;
$$

however, one cannot represent the circle in its entirety as the graph of a function.

Experiment 14.4 Open the M-file `mat14_1.m` and discuss which curve is being represented. Compare with the M-files `mat14_2.m` to `mat14_4.m`. Are these the same curves?

Experiment 14.4 suggests that one can view curves statically as a set of points in the plane or dynamically as the trajectory of a moving point. Both perspectives are of importance in applications.

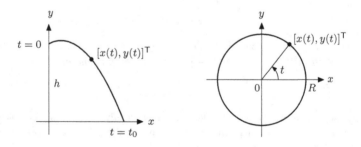

Fig. 14.1 Parabolic trajectory and circle

The kinematic point of view. In the kinematic interpretation, one considers the parameter t of the curve as time and the curve as path. Different parametrisations of the same geometric object are viewed as different curves.

The geometric point of view. In the geometric interpretation, the location, the moving sense and the number of cycles are considered as the defining properties of a curve. The particular parametrisation, however, is irrelevant.

   A strictly monotonically increasing, continuous mapping of an interval $[\alpha, \beta]$ to $[a, b]$,

$$\varphi : [\alpha, \beta] \to [a, b]$$

is called a *change of parameter*. The curve

$$\tau \mapsto \boldsymbol{\xi}(\tau), \quad \alpha \le \tau \le \beta$$

is called a *reparametrisation* of the curve

$$t \mapsto \mathbf{x}(t), \quad a \le t \le b,$$

if it is obtained through a change of parameter $t = \varphi(\tau)$; i.e.,

$$\boldsymbol{\xi}(\tau) = \mathbf{x}(\varphi(\tau)).$$

From the geometric point of view, the parametrised curves $\tau \mapsto \boldsymbol{\xi}(\tau)$ and $t \mapsto \mathbf{x}(t)$ are identified. A *plane curve* Γ is an *equivalence class of parametrised curves* which can be transformed to one another by reparametrisation.

Example 14.5 We consider the segment of a parabola, parametrised by

$$\Gamma : \mathbf{x}(t) = \begin{bmatrix} t \\ t^2 \end{bmatrix}, \quad -1 \le t \le 1.$$

Reparametrisations are for instance

$$\varphi : \left[-\tfrac{1}{2}, \tfrac{1}{2} \right] \to [-1, 1], \quad \varphi(\tau) = 2\tau,$$
$$\widetilde{\varphi} : [-1, 1] \to [-1, 1], \quad \widetilde{\varphi}(t) = \tau^3.$$

Consequently,

$$\boldsymbol{\xi}(\tau) = \begin{bmatrix} 2\tau \\ 4\tau^2 \end{bmatrix}, \quad -\tfrac{1}{2} \le \tau \le \tfrac{1}{2}$$

and

$$\eta(\tau) = \begin{bmatrix} \tau^3 \\ \tau^6 \end{bmatrix}, \quad -1 \le \tau \le 1$$

geometrically represent the same curve. However,

$$\psi : [-1, 1] \to [-1, 1], \quad \psi(\tau) = -\tau,$$
$$\widetilde{\psi} : [0, 1] \to [-1, 1], \quad \widetilde{\psi}(\tau) = -1 + 8\tau(1 - \tau)$$

are not reparametrisations and yield other curves, namely

$$\mathbf{y}(\tau) = \begin{bmatrix} -\tau \\ \tau^2 \end{bmatrix}, \quad -1 \le \tau \le 1,$$

$$\mathbf{z}(\tau) = \begin{bmatrix} -1 + 8\tau(1 - \tau) \\ (-1 + 8\tau(1 - \tau))^2 \end{bmatrix}, \quad 0 \le \tau \le 1.$$

In the first case, the moving sense of $\varGamma$ is reversed, and in the second case, the curve is traversed twice.

Experiment 14.6 Modify the M-files from Experiment 14.4 so that the curves from Example 14.5 are represented.

Algebraic curves. These are obtained as the set of zeros of polynomials in two variables. As examples we had already parabola and circle

$$y - x^2 = 0, \quad x^2 + y^2 - R^2 = 0.$$

One can also create cusps and loops in this way.

Example 14.7 Neil's [2] parabola

$$y^2 - x^3 = 0$$

has a cusp at $x = y = 0$ (Fig. 14.2). Generally, one obtains algebraic curves from

$$y^2 - (x + p)x^2 = 0, \quad p \in \mathbb{R}.$$

For $p > 0$ they have a loop. A parametric representation of this curve is, for instance,

$$x(t) = t^2 - p, \quad -\infty < t < \infty.$$
$$y(t) = t(t^2 - p),$$

---

[2]W. Neil, 1637–1670.

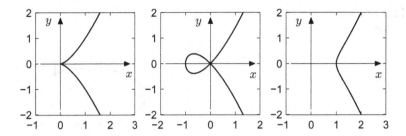

Fig. 14.2 *Neil's parabola*, the α *-curve* and an *elliptic curve*

In the following we will primarily deal with curves which are given by differentiable parametrisations.

Definition 14.8 If a plane curve $\Gamma : t \mapsto \mathbf{x}(t)$ has a parametrisation whose components $t \mapsto x(t)$, $t \mapsto y(t)$ are differentiable, then Γ is called a *differentiable curve*. If the components are k-times differentiable, then Γ is called a k-times differentiable curve.

The graphical representation of a differentiable curve does not have to be smooth but may have cusps and corners, as Example 14.7 shows.

Example 14.9 (Straight line and half ray) The parametric representation

$$t \mapsto \mathbf{x}(t) = \begin{bmatrix} x_0 \\ y_0 \end{bmatrix} + t \begin{bmatrix} r_1 \\ r_2 \end{bmatrix}, \quad -\infty < t < \infty$$

describes a straight line through the point $\mathbf{x}_0 = [x_0, y_0]^\mathsf{T}$ in the direction $\mathbf{r} = [r_1, r_2]^\mathsf{T}$. If one restricts the parameter t to $0 \le t < \infty$ one obtains a half ray. The parametrisation

$$\mathbf{x}_H(t) = \begin{bmatrix} x_0 \\ y_0 \end{bmatrix} + t^2 \begin{bmatrix} r_1 \\ r_2 \end{bmatrix}, \quad -\infty < t < \infty$$

leads to a double passage through the half ray.

Example 14.10 (Parametric representation of an ellipse) The equation of an ellipse is

$$\frac{x^2}{a^2} + \frac{y^2}{b^2} = 1.$$

Fig. 14.3 Parametric
representation of the ellipse

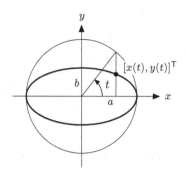

A parametric representation (single passage in counterclockwise sense) is obtained
by

$$x(t) = a \cos t, \qquad 0 \le t \le 2\pi.$$
$$y(t) = b \sin t,$$

This can be seen by substituting these expressions into the equation of the ellipse.
The meaning of the parameter t can be seen from Fig. 14.3.

Example 14.11 (Parametric representation of a hyperbola) The hyperbolic sine and
the hyperbolic cosine have been introduced in Sect. 2.2. The important identity

$$\cosh^2 t - \sinh^2 t = 1$$

has been noted there. It shows that

$$x(t) = a \cosh t, \qquad -\infty < t < \infty$$
$$y(t) = b \sinh t,$$

is a parametric representation of the right branch of the hyperbola

$$\frac{x^2}{a^2} - \frac{y^2}{b^2} = 1,$$

which is highlighted in Fig. 14.4.

Example 14.12 (Cycloids) A circle with radius R rolls (without sliding) along the
x-axis. If the starting position of the centre M is initially $M = (0, R)$, its position will be $M_t = (Rt, R)$ after a turn of angle t. A point P with starting position
$P = (0, R - A)$ thus moves to $P_t = M_t - (A \sin t, A \cos t)$.

The trajectory of the point P is called a *cycloid*. It is parametrised by

$$\begin{aligned} x(t) &= Rt - A\sin t, \\ y(t) &= R - A\cos t, \end{aligned} \qquad -\infty < t < \infty.$$

Compare Fig. 14.5 for the derivation and Fig. 14.6 for some possible shapes of cycloids.

Definition 14.13 Let $\Gamma : t \mapsto \mathbf{x}(t)$ be a differentiable curve. The rate of change of the position vector with regard to the parameter of the curve

$$\dot{\mathbf{x}}(t) = \lim_{h \to 0} \frac{1}{h}\Big(\mathbf{x}(t+h) - \mathbf{x}(t)\Big) = \begin{bmatrix} \dot{x}(t) \\ \dot{y}(t) \end{bmatrix}$$

is called the *velocity vector* at the point $\mathbf{x}(t)$ of the curve. If $\dot{\mathbf{x}}(t) \neq \mathbf{0}$ one defines the *tangent vector*

$$\mathbf{T}(t) = \frac{\dot{\mathbf{x}}(t)}{\|\dot{\mathbf{x}}(t)\|} = \frac{1}{\sqrt{\dot{x}(t)^2 + \dot{y}(t)^2}} \begin{bmatrix} \dot{x}(t) \\ \dot{y}(t) \end{bmatrix}$$

and the *normal vector*

$$\mathbf{N}(t) = \frac{1}{\sqrt{\dot{x}(t)^2 + \dot{y}(t)^2}} \begin{bmatrix} -\dot{y}(t) \\ \dot{x}(t) \end{bmatrix}$$

Fig. 14.4 Parametric representation of the right branch of a hyperbola

Fig. 14.5 Parametrisation of a cycloid

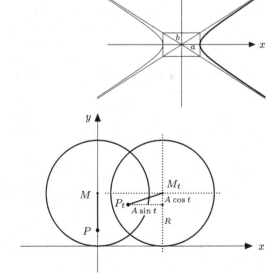

Fig. 14.6 Cycloids for
$A = R/2, R, 3R/2$

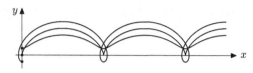

Fig. 14.7 Velocity vector,
acceleration vector, tangent
vector, normal vector

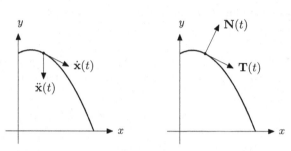

of the curve. The pair $(\mathbf{T}(t), \mathbf{N}(t))$ is called *moving frame*. If the curve Γ is twice
differentiable then the *acceleration vector* is given by

$$\ddot{\mathbf{x}}(t) = \begin{bmatrix} \ddot{x}(t) \\ \ddot{y}(t) \end{bmatrix}.$$

In the kinematic interpretation the parameter t is the time and $\dot{\mathbf{x}}(t)$ the velocity
vector in the physical sense. If it is different from zero, it points in the direction of
the tangent (as limit of secant vectors). The tangent vector is just the unit vector of
the same direction. By rotation of $90°$ in the counterclockwise sense we obtain the
normal vector of the curve, see Fig. 14.7.

Experiment 14.14 Open the Java applet *Parametric curves in the plane*. Plot the
curves from Example 14.5 and the corresponding velocity and acceleration vectors.
Use the moving frame to visualise the kinematic curve progression.

Example 14.15 For the parabola from Example 14.2 we get

$$\dot{x}(t) = v_H, \qquad \ddot{x}(t) = 0,$$
$$\dot{y}(t) = v_V - gt, \qquad \ddot{y}(t) = -g,$$

$$\mathbf{T}(t) = \frac{1}{\sqrt{v_H^2 + (v_V - gt)^2}} \begin{bmatrix} v_H \\ v_V - gt \end{bmatrix},$$

$$\mathbf{N}(t) = \frac{1}{\sqrt{v_H^2 + (v_V - gt)^2}} \begin{bmatrix} gt - v_V \\ v_H \end{bmatrix}.$$

14.2   Arc Length and Curvature

We start with the question whether and how a length can be assigned to a curve segment. Let a continuous curve

$$\Gamma : t \mapsto \mathbf{x}(t) = \begin{bmatrix} x(t) \\ y(t) \end{bmatrix}, \quad a \le t \le b$$

be given. For a partition $Z: a = t_0 < t_1 < \cdots < t_n = b$ of the parameter interval we consider the (inscribed) polygonal chain through the points

$$\mathbf{x}(t_0), \mathbf{x}(t_1), \ldots, \mathbf{x}(t_n).$$

The length of the largest subinterval is again denoted by $\Phi(Z)$. The length of the polygonal chain is

$$L_n = \sum_{i=1}^{n} \sqrt{(x(t_i) - x(t_{i-1}))^2 + (y(t_i) - y(t_{i-1}))^2}.$$

Definition 14.16 (Curves of finite length) A plane curve Γ is called *rectifiable* or *of finite length* if the lengths L_n of all inscribed polygonal chains Z_n converge towards one (and the same) limit provided that $\Phi(Z_n) \to 0$.

Example 14.17 (Koch's snowflake) Koch's snowflake was introduced in Sect. 9.1 as an example of a finite region whose boundary has the fractal dimension $d = \log 4 / \log 3$ and infinite length. This was proven by the fact that the boundary can be constructed as the limit of polygonal chains whose lengths tend to infinity. It remains to verify that the boundary of Koch's snowflake is indeed a continuous, parametrised curve. This can be seen as follows. The snowflake of depth 0 is an equilateral triangle, for instance with the vertices $\mathbf{p}_1, \mathbf{p}_2, \mathbf{p}_3 \in \mathbb{R}^2$. Using the unit interval $[0, 1]$ we obtain a continuous parametrisation

$$\mathbf{x}_0(t) = \begin{cases} \mathbf{p}_1 + 3t(\mathbf{p}_2 - \mathbf{p}_1), & 0 \le t \le \frac{1}{3}, \\ \mathbf{p}_2 + (3t - 1)(\mathbf{p}_3 - \mathbf{p}_2), & \frac{1}{3} \le t \le \frac{2}{3}, \\ \mathbf{p}_3 + (3t - 2)(\mathbf{p}_1 - \mathbf{p}_3), & \frac{2}{3} \le t \le 1. \end{cases}$$

We parametrise the snowflake of depth 1 by splitting the three intervals $[0, \frac{1}{3}]$, $[\frac{1}{3}, \frac{2}{3}]$, $[\frac{2}{3}, 1]$ into three parts each and using the middle parts for the parametrisation of the inserted next smaller angle (Fig. 14.8). Continuing in this way we obtain a sequence of parametrisations

$$t \mapsto \mathbf{x}_0(t), \ t \mapsto \mathbf{x}_1(t), \ \ldots, \ t \mapsto \mathbf{x}_n(t), \ \ldots$$

Fig. 14.8 Parametrisation of
the boundary of Koch's
snowflake

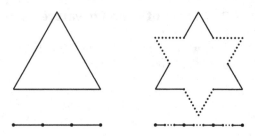

This is a sequence of continuous functions $[0, 1] \to \mathbb{R}^2$ which, due to its construction, converges uniformly (see Definition C.5). According to Proposition C.6 the limit function

$$\mathbf{x}(t) = \lim_{n \to \infty} \mathbf{x}_n(t), \quad t \in [0, 1]$$

is continuous (and obviously parametrises the boundary of Koch's snowflake).

This example shows that continuous curves can be infinitely long even if the parameter of the curve only varies in a bounded interval $[a, b]$. That such a behaviour does not appear for differentiable curves is shown by the next proposition.

Proposition 14.18 (Length of differentiable curves) *Every continuously differentiable curve $t \mapsto \mathbf{x}(t), t \in [a, b]$ is rectifiable. Its length is*

$$L = \int_a^b \|\dot{\mathbf{x}}(t)\| \, dt = \int_a^b \sqrt{\dot{x}(t)^2 + \dot{y}(t)^2} \, dt.$$

Proof We only give the proof for the somewhat simpler case that the components of the velocity vector $\dot{\mathbf{x}}(t)$ are Lipschitz continuous (see Appendix C.4), for instance with a Lipschitz constant C. We start with a partition $Z: a = t_0 < t_1 < \cdots < t_n = b$ of $[a, b]$ with corresponding $\Phi(Z)$. The integral defining L is the limit of Riemann sums

$$\int_a^b \sqrt{\dot{x}(t)^2 + \dot{y}(t)^2} \, dt = \lim_{n \to \infty, \Phi(Z) \to 0} \sum_{i=1}^n \sqrt{\dot{x}(\tau_i)^2 + \dot{y}(\tau_i)^2} \, (t_i - t_{i-1}),$$

where $\tau_i \in [t_{i-1}, t_i]$. On the other hand, according to the mean value theorem, Proposition 8.4, the length of the inscribed polygonal chain through $\mathbf{x}(t_0), \mathbf{x}(t_1), \ldots, \mathbf{x}(t_n)$ is equal to

$$\sum_{i=1}^n \sqrt{(x(t_i) - x(t_{i-1}))^2 + (y(t_i) - y(t_{i-1}))^2}$$

$$= \sum_{i=1}^n \sqrt{\dot{x}(\rho_i)^2 + \dot{y}(\sigma_i)^2} \, (t_i - t_{i-1})$$

for certain $\rho_i, \sigma_i \in [t_{i-1}, t_i]$. In order to be able to estimate the difference between the Riemann sums and the lengths of the inscribed polygonal chains, we use the inequality (triangle inequality for vectors in the plane)

$$\left| \sqrt{a^2 + b^2} - \sqrt{c^2 + d^2} \right| \leq \sqrt{(a - c)^2 + (b - d)^2},$$

which can be checked directly by squaring. Applying this inequality shows that

$$\left| \sqrt{\dot{x}(\tau_i)^2 + \dot{y}(\tau_i)^2} - \sqrt{\dot{x}(\rho_i)^2 + \dot{y}(\sigma_i)^2} \right|$$

$$\leq \sqrt{(\dot{x}(\tau_i) - \dot{x}(\rho_i))^2 + (\dot{y}(\tau_i) - \dot{y}(\sigma_i))^2}$$

$$\leq \sqrt{C^2(\tau_i - \rho_i)^2 + C^2(\tau_i - \sigma_i)^2}$$

$$\leq \sqrt{2} C \Phi(Z).$$

For the difference between the Riemann sums and the lengths of the polygonal chains one obtains the estimate

$$\left| \sum_{i=1}^{n} \left(\sqrt{\dot{x}(\tau_i)^2 + \dot{y}(\tau_i)^2} - \sqrt{\dot{x}(\rho_i)^2 + \dot{y}(\sigma_i)^2} \right) (t_i - t_{i-1}) \right|$$

$$\leq \sqrt{2} C \Phi(Z) \sum_{i=1}^{n} (t_i - t_{i-1}) = \sqrt{2} C \Phi(Z)(b - a).$$

For $\Phi(Z) \to 0$, this difference tends to zero. Thus the Riemann sums and the lengths of the inscribed polygonal chains have the same limit, namely L.

The proof of the general case, where the components of the velocity vector are not Lipschitz continuous, is similar. However, one additionally needs the fact that continuous functions on bounded, closed intervals are uniformly continuous. This is briefly addressed near the end of Appendix C.4. $\square$

Example 14.19 (Length of a circular arc)   The parametric representation of a circle of radius R and its derivative is

$$x(t) = R \cos t, \quad \dot{x}(t) = -R \sin t, \qquad 0 \leq t \leq 2\pi.$$
$$y(t) = R \sin t, \quad \dot{y}(t) = R \cos t,$$

The circumference of the circle is thus

$$L = \int_0^{2\pi} \sqrt{(-R \sin t)^2 + (R \cos t)^2} \, dt = \int_0^{2\pi} R \, dt = 2R\pi.$$

Experiment 14.20 Use the MATLAB program `mat14_5.m` to approximate the circumference of the unit circle using inscribed polygonal chains. Modify the program so that it approximates the lengths of arbitrary differentiable curves.

Definition 14.21 (Arc length) Let $t \mapsto \mathbf{x}(t)$ be a differentiable curve. The length of the curve segment from the initial parameter value a to the current parameter value t is called the *arc length*,

$$s = L(t) = \int_a^t \sqrt{\dot{x}(\tau)^2 + \dot{y}(\tau)^2}\, d\tau.$$

The arc length s is a strictly monotonically increasing, continuous (even continuously differentiable) function. It is thus suitable for a reparametrisation $t = L^{-1}(s)$. The curve

$$s \mapsto \boldsymbol{\xi}(s) = \mathbf{x}(L^{-1}(s))$$

is called *parametrised by arc length*.

In the following let $t \mapsto \mathbf{x}(t)$ be a differentiable curve (in the plane). The angle of the tangent vector with the positive x-axis is denoted by $\varphi(t)$; that is,

$$\tan \varphi(t) = \frac{\dot{y}(t)}{\dot{x}(t)}.$$

Definition 14.22 (Curvature of a plane curve) The *curvature* of a differentiable curve in the plane is the rate of change of the angle φ with respect to the arc length,

$$\kappa = \frac{d\varphi}{ds} = \frac{d}{ds}\varphi(L^{-1}(s)).$$

Figure 14.9 illustrates this definition. If φ is the angle at the length s of the arc and $\varphi + \Delta\varphi$ the angle at the length $s + \Delta s$, then $\kappa = \lim_{\Delta s \to 0} \frac{\Delta\varphi}{\Delta s}$. This shows that the value of κ actually corresponds to the intuitive meaning of curvature. Note that the curvature of a plane curve comes with a sign; when reversing the moving sense, the sign changes.

Proposition 14.23 *The curvature of a twice continuously differentiable curve at the point $(x(t), y(t))$ of the curve is*

$$\kappa(t) = \frac{\dot{x}(t)\ddot{y}(t) - \dot{y}(t)\ddot{x}(t)}{\left(\dot{x}(t)^2 + \dot{y}(t)^2\right)^{3/2}}.$$

Fig. 14.9 Curvature

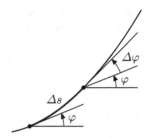

Proof According to the chain rule and the inverse function rule, one gets

$$\kappa = \frac{d}{ds}\,\varphi(L^{-1}(s)) = \dot{\varphi}(L^{-1}(s)) \cdot \frac{d}{ds} L^{-1}(s) = \dot{\varphi}(L^{-1}(s)) \cdot \frac{1}{\dot{L}(L^{-1}(s))}.$$

Differentiating the arc length

$$s = L(t) = \int_a^t \sqrt{\dot{x}(\tau)^2 + \dot{y}(\tau)^2}\, d\tau$$

with respect to t gives

$$\frac{ds}{dt} = \dot{L}(t) = \sqrt{\dot{x}(t)^2 + \dot{y}(t)^2}.$$

Differentiating the relationship $\tan \varphi(t) = \dot{y}(t)/\dot{x}(t)$ leads to

$$\dot{\varphi}(t)\bigl(1 + \tan^2 \varphi(t)\bigr) = \frac{\dot{x}(t)\ddot{y}(t) - \dot{y}(t)\ddot{x}(t)}{\dot{x}(t)^2},$$

which gives, after substituting the above expression for $\tan \varphi(t)$ and simplifying,

$$\dot{\varphi}(t) = \frac{\dot{x}(t)\ddot{y}(t) - \dot{y}(t)\ddot{x}(t)}{\dot{x}(t)^2 + \dot{y}(t)^2}.$$

If one takes into account the relation $t = L^{-1}(s)$ and substitutes the derived expressions for $\dot{\varphi}(t)$ and $\dot{L}(t)$ into the formula for κ at the beginning of the proof, one obtains

$$\kappa(t) = \frac{\dot{\varphi}(t)}{\dot{L}(t)} = \frac{\dot{x}(t)\ddot{y}(t) - \dot{y}(t)\ddot{x}(t)}{\bigl(\dot{x}(t)^2 + \dot{y}(t)^2\bigr)^{3/2}},$$

which is the desired assertion.   □

Remark 14.24 As a special case, the curvature of the graph of a twice differentiable function $y = f(x)$ can be obtained as

$$\kappa(x) = \frac{f''(x)}{\left(1 + f'(x)^2\right)^{3/2}}.$$

This follows easily from the above proposition by using the parametrisation $x = t$, $y = f(t)$.

Example 14.25 The curvature of a circle of radius R, traversed in the positive direction, is constant and equal to $\kappa = \frac{1}{R}$. Indeed

$$x(t) = R\cos t, \quad \dot{x}(t) = -R\sin t, \quad \ddot{x}(t) = -R\cos t,$$
$$y(t) = R\sin t, \quad \dot{y}(t) = R\cos t, \quad \ddot{y}(t) = -R\sin t,$$

and thus

$$\kappa = \frac{R^2 \sin^2 t + R^2 \cos^2 t}{(R^2 \sin^2 t + R^2 \cos^2 t)^{3/2}} = \frac{1}{R}.$$

One obtains the same result from the following geometric consideration. At the point $(x, y) = (R\cos t, R\sin t)$ the angle φ of the tangent vector with the positive x-axis is equal to $t + \pi/2$, and the arc length is $s = Rt$. Therefore $\varphi = s/R + \pi/2$ which differentiated with respect to s gives $\kappa = 1/R$.

Definition 14.26 The *osculating circle* at a point of a differentiable curve is the circle which has the same tangent and the same curvature as the curve.

According to Example 14.25 it follows that the osculating circle has the radius $\frac{1}{|\kappa(t)|}$ and its centre $\mathbf{x}_c(t)$ lies on the normal of the curve. It is given by

$$\mathbf{x}_c(t) = \mathbf{x}(t) + \frac{1}{\kappa(t)}\mathbf{N}(t).$$

Example 14.27 (Clothoid) The *clothoid* is a curve whose curvature is proportional to its arc length. In applications it serves as a connecting link from a straight line (with curvature 0) to a circular arc (with curvature $\frac{1}{R}$). It is used in railway engineering and road design. Its defining property is

$$\kappa(s) = \frac{d\varphi}{ds} = c \cdot s$$

for a certain $c \in \mathbb{R}$. If one starts with curvature 0 at $s = 0$ then the angle is equal to

$$\varphi(s) = \frac{c}{2}s^2.$$

Fig. 14.10 Clothoid

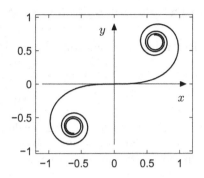

We use s as the curve parameter.

Differentiating the relation

$$s = \int_0^s \sqrt{\dot{x}(\sigma)^2 + \dot{y}(\sigma)^2} \, d\sigma$$

shows that

$$1 = \sqrt{\dot{x}(s)^2 + \dot{y}(s)^2};$$

thus, the velocity vector of a curve parametrised by arc length has length one. This implies in particular

$$\frac{dx}{ds} = \cos \varphi(s), \qquad \frac{dy}{ds} = \sin \varphi(s).$$

From there we can compute the parametrisation of the curve:

$$x(s) = \int_0^s \frac{dx}{ds}(\sigma) \, d\sigma = \int_0^s \cos \varphi(\sigma) \, d\sigma = \int_0^s \cos \left(\frac{c}{2} \sigma^2 \right) d\sigma,$$

$$y(s) = \int_0^s \frac{dy}{ds}(\sigma) \, d\sigma = \int_0^s \sin \varphi(\sigma) \, d\sigma = \int_0^s \sin \left(\frac{c}{2} \sigma^2 \right) d\sigma.$$

The components of the curve are thus given by Fresnel's integrals. The shape of the curve is displayed in Fig. 14.10, its numerical calculation can be seen in the MATLAB program `mat14_6.m`.

14.3 Plane Curves in Polar Coordinates

By writing the parametric representation in the form

$$x(t) = r(t) \cos t,$$
$$y(t) = r(t) \sin t$$

in polar coordinates with t as angle and $r(t)$ as radius, one obtains a simple way of representing many curves. By convention negative radii are plotted in opposite direction of the ray with angle t.

Example 14.28 (Spirals) The *Archimedean*[3] *spiral* is defined by

$$r(t) = t, \qquad 0 \le t < \infty,$$

the *logarithmic spiral* by

$$r(t) = e^t, \qquad -\infty < t < \infty,$$

the *hyperbolic spiral* by

$$r(t) = \frac{1}{t}, \qquad 0 < t < \infty.$$

Typical parts of these spirals are displayed in Fig. 14.11.

Experiment 14.29 Study the behaviour of the logarithmic spiral near the origin using the zoom tool (use the M-file `mat14_7.m`).

Example 14.30 (Loops) Loops are obtained by choosing $r(t) = \cos nt$, $n \in \mathbb{N}$. In Cartesian coordinates the parametric representation thus reads

$$x(t) = \cos nt \, \cos t,$$
$$y(t) = \cos nt \, \sin t.$$

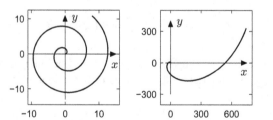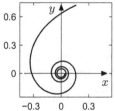

Fig. 14.11 Archimedean, logarithmic and hyperbolic spirals

---

[3] Archimedes of Syracuse, 287–212 B.C.

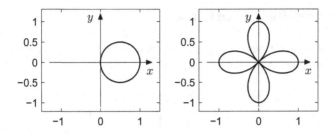

Fig. 14.12 Loops with $r = \cos t$ and $r = \cos 2t$

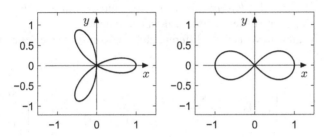

Fig. 14.13 Loops with $r = \cos 3t$ and $r = \pm\sqrt{\cos 2t}$

The choice $n = 1$ results in a circle of radius $\frac{1}{2}$ about $(\frac{1}{2}, 0)$, for odd n one obtains n leaves, for even n one obtains $2n$ leaves, see Figs. 14.12 and 14.13.

The *figure eight* from Fig. 14.13 is obtained by $r(t) = \sqrt{\cos 2t}$ and $r(t) = -\sqrt{\cos 2t}$, respectively, for $-\frac{\pi}{4} < t < \frac{\pi}{4}$, where the positive root gives the right leave and the negative root the left leave. This curve is called *lemniscate*.

Example 14.31 (Cardioid) The *cardioid* is a special epicycloid, where one circle is rolling around another circle with the same radius A. Its parametric representation is

$$x(t) = 2A \cos t + A \cos 2t,$$
$$y(t) = 2A \sin t + A \sin 2t$$

for $0 \le t \le 2\pi$. The cardioid with radius $A = 1$ is shown in Fig. 14.14.

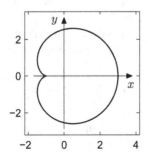

Fig. 14.14 Cardioid with $A = 1$

14.4 Parametrised Space Curves

In the same way as for plane curves, a *parametrised curve in space* is defined as a continuous mapping of an interval $[a, b]$ to $\mathbb{R}^3$,

$$t \mapsto \mathbf{x}(t) = \begin{bmatrix} x(t) \\ y(t) \\ z(t) \end{bmatrix}, \quad a \leq t \leq b.$$

The curve is called *differentiable*, if all three components $t \mapsto x(t)$, $t \mapsto y(t)$, $t \mapsto z(t)$ are differentiable real-valued functions.

Velocity and tangent vector of a differentiable curve in space are defined as in the planar case by

$$\dot{\mathbf{x}}(t) = \begin{bmatrix} \dot{x}(t) \\ \dot{y}(t) \\ \dot{z}(t) \end{bmatrix}, \quad \mathbf{T}(t) = \frac{\dot{\mathbf{x}}(t)}{\|\dot{\mathbf{x}}(t)\|} = \frac{1}{\sqrt{\dot{x}(t)^2 + \dot{y}(t)^2 + \dot{z}(t)^2}} \begin{bmatrix} \dot{x}(t) \\ \dot{y}(t) \\ \dot{z}(t) \end{bmatrix}.$$

The second derivative $\ddot{\mathbf{x}}(t)$ is the acceleration vector. In the spatial case there is a *normal plane* to the curve which is spanned by the *normal vector*

$$\mathbf{N}(t) = \frac{1}{\|\dot{\mathbf{T}}(t)\|} \dot{\mathbf{T}}(t)$$

and the *binormal vector*

$$\mathbf{B}(t) = \mathbf{T}(t) \times \mathbf{N}(t),$$

provided that $\dot{\mathbf{x}}(t) \neq \mathbf{0}$, $\dot{\mathbf{T}}(t) \neq \mathbf{0}$. The formula

$$0 = \frac{d}{dt} 1 = \frac{d}{dt} \|\mathbf{T}(t)\|^2 = 2\langle \mathbf{T}(t), \dot{\mathbf{T}}(t) \rangle$$

(which is verified by a straightforward computation) implies that $\dot{\mathbf{T}}(t)$ is perpendicular to $\mathbf{T}(t)$. Therefore, the three vectors $(\mathbf{T}(t), \mathbf{N}(t), \mathbf{B}(t))$ form an orthogonal basis in $\mathbb{R}^3$, called the *moving frame* of the curve.

Rectifiability of a curve in space is defined in analogy to Definition 14.16 for plane curves. The *length* of a differentiable curve in space can be computed by

$$L = \int_a^b \|\dot{\mathbf{x}}(t)\| \, dt = \int_a^b \sqrt{\dot{x}(t)^2 + \dot{y}(t)^2 + \dot{z}(t)^2} \, dt.$$

Also, the *arc length* can be defined similarly to the planar case (Definition 14.21).

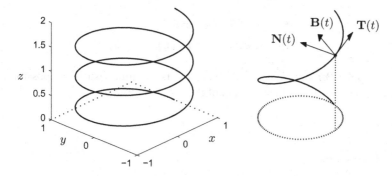

Fig. 14.15 Helix with tangent, normal and binormal vector

Example 14.32 (Helix) The parametric representation of the helix is

$$\mathbf{x}(t) = \begin{bmatrix} \cos t \\ \sin t \\ t \end{bmatrix}, \quad -\infty < t < \infty.$$

We obtain

$$\dot{\mathbf{x}}(t) = \begin{bmatrix} -\sin t \\ \cos t \\ 1 \end{bmatrix}, \quad \mathbf{T}(t) = \frac{1}{\sqrt{2}} \begin{bmatrix} -\sin t \\ \cos t \\ 1 \end{bmatrix},$$

$$\dot{\mathbf{T}}(t) = \frac{1}{\sqrt{2}} \begin{bmatrix} -\cos t \\ -\sin t \\ 0 \end{bmatrix}, \quad \mathbf{N}(t) = \begin{bmatrix} -\cos t \\ -\sin t \\ 0 \end{bmatrix}$$

with binormal vector

$$\mathbf{B}(t) = \frac{1}{\sqrt{2}} \begin{bmatrix} -\sin t \\ \cos t \\ 1 \end{bmatrix} \times \begin{bmatrix} -\cos t \\ -\sin t \\ 0 \end{bmatrix} = \frac{1}{\sqrt{2}} \begin{bmatrix} \sin t \\ -\cos t \\ 1 \end{bmatrix}.$$

The formula for the arc length of the helix, counting from the origin, is particularly simple:

$$L(t) = \int_0^t \|\dot{\mathbf{x}}(\tau)\| \, d\tau = \int_0^t \sqrt{2} \, d\tau = \sqrt{2}\, t.$$

Figure 14.15 was drawn using the MATLAB commands

```
t=0 : pi/100 : 6*pi;
plot3(cos(t), sin(t), t/10).
```

The Java applet *Parametric curves in space* offers dynamic visualising possibilities of those and other curves in space and of their moving frames.

14.5 Exercises

1. Find out which geometric formation is represented by the set of zeros of the polynomial $y^2 - x(x^2 - 1) = 0$. Visualise the curve in maple using the command `implicitplot`. Can you parametrise it as a continuous curve?

2. Verify that the algebraic curves $y^2 - (x + p)x^2 = 0$, $p \in \mathbb{R}$ (Example 14.7) can be parametrised by

$$\begin{aligned} x(t) &= t^2 - p, \\ y(t) &= t(t^2 - p), \end{aligned} \quad -\infty < t < \infty.$$

Visualise the curves for $p = -1, 0, 1$ in maple using the command `implicitplot`.

3. Using MATLAB or maple, investigate the shape of Lissajous figures[4]

$$x(t) = \sin(w_1 t), \quad y(t) = \cos(w_2 t)$$

and

$$x(t) = \sin(w_1 t), \quad y(t) = \cos\left(w_2 t + \frac{\pi}{2}\right).$$

Consider the cases $w_2 = w_1$, $w_2 = 2w_1$, $w_2 = \frac{3}{2}w_1$ and explain the results.

The following exercises use the Java applets *Parametric curves in the plane* and *Parametric curves in space*.

4. (a) Using the Java applet analyse where the cycloid

$$\begin{aligned} x(t) &= t - 2\sin t, \\ y(t) &= 1 - 2\cos t, \end{aligned} \quad -2\pi \le t \le 2\pi.$$

has its maximal speed ($\|\dot{\mathbf{x}}(t)\| \to$ max) and check your result by hand.

---

[4]J.A. Lissajous, 1822–1880.

(b) Discuss and explain the shape of the loops

$$x(t) = \cos nt \cos t,$$
$$y(t) = \cos nt \sin t,$$
$$0 \le t \le 2\pi.$$

for $n = 1, 2, 3, 4, 5$ using the Java applets (plot the moving frame).

5. Study the velocity and the acceleration of the following curves by using the Java applet. Verify your results by computing the points where the curve has either a horizontal tangent ($\dot{x}(t) \ne 0$, $\dot{y}(t) = 0$) or a vertical tangent ($\dot{x}(t) = 0$, $\dot{y}(t) \ne 0$), or is singular ($\dot{x}(t) = 0$, $\dot{y}(t) = 0$).

(a) Cycloid:

$$x(t) = t - \sin t,$$
$$y(t) = 1 - \cos t,$$
$$-2\pi \le t \le 2\pi.$$

(b) Cardioid:

$$x(t) = 2 \cos t + \cos 2t,$$
$$y(t) = 2 \sin t + \sin 2t,$$
$$0 \le t \le 2\pi.$$

6. Analyse and explain the trajectories of the curves

$$\mathbf{x}(t) = \begin{bmatrix} 1 - 2t^2 \\ (1 - 2t^2)^2 \end{bmatrix}, \quad -1 \le t \le 1,$$

$$\mathbf{y}(t) = \begin{bmatrix} \cos t \\ \cos^2 t \end{bmatrix}, \quad 0 \le t \le 2\pi,$$

$$\mathbf{z}(t) = \begin{bmatrix} t \cos t \\ t^2 \cos^2 t \end{bmatrix}, \quad -2 \le t \le 2.$$

Are these curves (geometrically) equivalent?

7. (a) Compute the curvature $\kappa(t)$ of the branch of the hyperbola

$$x(t) = \cosh t,$$
$$y(t) = \sinh t,$$
$$-\infty < t < \infty.$$

(b) Determine its osculating circle (centre and radius) at $t = 0$.

8. Consider the ellipse

$$\mathbf{x}(t) = \begin{bmatrix} 2 \cos t \\ \sin t \end{bmatrix}, \quad -\pi \le t \le \pi.$$

(a) Compute its velocity vector $\dot{\mathbf{x}}(t)$, its acceleration vector $\ddot{\mathbf{x}}(t)$ as well as the moving frame $(\mathbf{T}(t), \mathbf{N}(t))$.

(b) Compute its curvature $\kappa(t)$ and determine the osculating circle (centre and radius) at $t = 0$.

9. (a) Analyse the trajectory of the astroid

$$\mathbf{x}(t) = \begin{bmatrix} \cos^3 t \\ \sin^3 t \end{bmatrix}, \quad 0 \le t \le 2\pi.$$

   (b) Compute the length of the part of the astroid which lies in the first quadrant.

10. (a) Compute the velocity vector $\dot{\mathbf{x}}(t)$ and the moving frame $(\mathbf{T}(t), \mathbf{N}(t))$ for the segment

$$\mathbf{x}(t) = \begin{bmatrix} e^t \cos t \\ e^t \sin t \end{bmatrix}, \quad 0 \le t \le \pi/2$$

   of the logarithmic spiral. At what point in the interval $[0, \pi/2]$ does it have a vertical tangent?

   (b) Compute the length of the segment. Deduce a formula for its arc length $s = L(t)$.

   (c) Reparametrise the spiral by its arc length, i.e., compute $\boldsymbol{\xi}(s) = \mathbf{x}\big(L^{-1}(s)\big)$ and verify that $\|\dot{\boldsymbol{\xi}}(s)\| = 1$.

11. (Application of the secant and cosecant functions) Analyse what plane curves are determined in polar coordinates by

$$r(t) = \sec t, \; -\pi/2 < t < \pi/2 \quad \text{and} \quad r(t) = \csc t, \; 0 < t < \pi.$$

12. (a) Determine the tangent and the normal to the graph of the function $y = 1/x$ at $(x_0, y_0) = (1, 1)$ and compute its curvature at that point.

   (b) Suppose the graph of the function $y = 1/x$ is to be replaced by a circular arc at x_0, i.e., for $x \ge 1$. Find the centre and the radius of a circle which admits a smooth transition (same tangent, same curvature).

13. (a) Analyse the space curve

$$\mathbf{x}(t) = \begin{bmatrix} \cos t \\ \sin t \\ 2 \sin \frac{t}{2} \end{bmatrix}, \quad 0 \le t \le 4\pi$$

   using the applet.

   (b) Check that the curve is the intersection of the cylinder $x^2 + y^2 = 1$ with the sphere $(x + 1)^2 + y^2 + z^2 = 4$.

   Hint. Use $\sin^2 \frac{t}{2} = \frac{1}{2}(1 - \cos t)$.

14. Using MATLAB, maple or the applet, sketch and discuss the space curves

$$\mathbf{x}(t) = \begin{bmatrix} t \cos t \\ t \sin t \\ 2t \end{bmatrix}, \quad 0 \le t < \infty,$$

and

$$\mathbf{y}(t) = \begin{bmatrix} \cos t \\ \sin t \\ 0 \end{bmatrix}, \quad 0 \leq t \leq 4\pi.$$

15. Sketch and discuss the space curves

$$\mathbf{x}(t) = \begin{bmatrix} t \\ t \\ t^3 \end{bmatrix}, \quad \mathbf{y}(t) = \begin{bmatrix} t \\ t^2 \\ t^3 \end{bmatrix}, \quad 0 \leq t < 1.$$

Compute their velocity vectors $\dot{\mathbf{x}}(t)$, $\dot{\mathbf{y}}(t)$ and their acceleration vectors $\ddot{\mathbf{x}}(t)$, $\ddot{\mathbf{y}}(t)$.

16. Sketch the space curve

$$\mathbf{x}(t) = \begin{bmatrix} \sqrt{2}\, t \\ \cosh t \\ \cosh t \end{bmatrix}, \quad 0 \leq t < 1.$$

Compute its moving frame $(\mathbf{T}(t), \mathbf{N}(t), \mathbf{B}(t))$ as well as its length.

17. Sketch the space curve

$$\mathbf{x}(t) = \begin{bmatrix} \cos t \\ \sin t \\ t^{3/2} \end{bmatrix}, \quad 0 \leq t < 2\pi,$$

and compute its length.

Scalar-Valued Functions of Two Variables

15

This chapter is devoted to differential calculus of functions of two variables. In particular we will study geometrical objects such as tangents and tangent planes, maxima and minima, as well as linear and quadratic approximations. The restriction to two variables has been made for simplicity of presentation. All ideas in this and the next chapter can easily be extended (although with slightly more notational effort) to the case of n variables.

We begin by studying the graph of a function with the help of vertical cuts and level sets. As a further tool we introduce partial derivatives, which describe the rate of change of the function in the direction of the coordinate axes. Finally the notion of the Fréchet derivative allows us to define the tangent plane to the graph. As for functions of one variable the Taylor formula plays a central role. We use it, e.g., to determine extrema of functions of two variables.

In the entire chapter D denotes a subset of $\mathbb{R}^2$, and

$$f : D \subset \mathbb{R}^2 \to \mathbb{R} : (x, y) \mapsto z = f(x, y)$$

denotes a *scalar-valued* function of two variables. Details of vector and matrix algebra used in this chapter can be found in Appendices A and B.

15.1 Graph and Partial Mappings

The *graph*

$$G = \left\{ (x, y, z) \in D \times \mathbb{R} \; ; \; z = f(x, y) \right\} \subset \mathbb{R}^3$$

© Springer Nature Switzerland AG 2018
M. Oberguggenberger and A. Ostermann, *Analysis for Computer Scientists*,
Undergraduate Topics in Computer Science,
https://doi.org/10.1007/978-3-319-91155-7_15

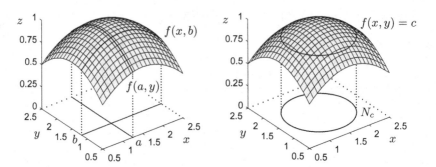

Fig. 15.1 Graph of a function as surface in space with coordinate curves (left) and level curve N_c (right)

of a function of two variables $f : D \to \mathbb{R}$ is a surface in space, if f is sufficiently regular. To describe the properties of this surface we consider particular curves on it.

The *partial mappings*

$$x \mapsto f(x, b), \qquad y \mapsto f(a, y)$$

are obtained by fixing one of the two variables $y = b$ or $x = a$. The partial mappings can be used to introduce the space curves

$$x \mapsto \begin{bmatrix} x \\ b \\ f(x, b) \end{bmatrix}, \qquad y \mapsto \begin{bmatrix} a \\ y \\ f(a, y) \end{bmatrix}.$$

These curves lie on the graph G of the function and are called *coordinate curves*. Geometrically they are obtained as the intersection of G with the vertical planes $y = b$ and $x = a$ respectively, see Fig. 15.1, left.

The *level curves* are the projections of the intersections of the graph G with the horizontal planes $z = c$ to the (x, y)-plane,

$$N_c = \{(x, y) \in D ; \ f(x, y) = c\},$$

see Fig. 15.1, right. The set N_c is called level curve at level c.

Example 15.1 The graph of the quadratic function

$$f : \mathbb{R}^2 \to \mathbb{R} : (x, y) \mapsto z = \frac{x^2}{a^2} - \frac{y^2}{b^2}$$

describes a surface in space which is shaped like a saddle and called *hyperbolic paraboloid*. Figure 15.2 shows the graph of $z = x^2/4 - y^2/5$ with coordinate curves (left) as well as some level curves (right).

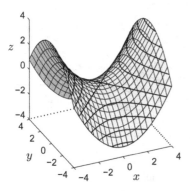 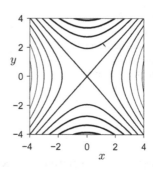

Fig. 15.2 The picture on the left shows the graph of the function $z = x^2/4 - y^2/5$ with coordinate curves. Furthermore, it shows the intersections with the planes $z = c$ for selected values of c. The picture on the right illustrates the level curves of the function for the same values of c (lower levels correspond to thicker lines). The two intersecting straight lines are the level curves at level $c = 0$

Experiment 15.2 With the help of the MATLAB program `mat15_1.m` visualise the elliptic paraboloid $z = x^2 + 2y^2 - 4x + 1$. Choose a suitable domain D and plot the graph and some level curves.

15.2 Continuity

Like for functions in one variable (see Chap. 6) we characterise the continuity of functions of two variables by means of sequences. Thus we need the concept of convergence of vector-valued sequences.

Let $(\mathbf{a}_n)_{n \geq 1} = (\mathbf{a}_1, \mathbf{a}_2, \mathbf{a}_3, \ldots)$ be a sequence of points in D with terms

$$\mathbf{a}_n = (a_n, b_n) \in D \subset \mathbb{R}^2.$$

The sequence $(\mathbf{a}_n)_{n \geq 1}$ is said to *converge* to $\mathbf{a} = (a, b) \in D$ as $n \to \infty$, if and only if both components of the sequence converge, i.e.

$$\lim_{n \to \infty} a_n = a \quad \text{and} \quad \lim_{n \to \infty} b_n = b.$$

This is denoted by

$$(a_n, b_n) = \mathbf{a}_n \to \mathbf{a} = (a, b) \quad \text{as } n \to \infty \quad \text{or} \quad \lim_{n \to \infty} \mathbf{a}_n = \mathbf{a}.$$

Otherwise the sequence is called *divergent*.

Fig. 15.3 A function which
is discontinuous along a
straight line. For every
sequence $(\mathbf{a}_n)$ which
converges to $\mathbf{a}$, the images of
the sequence $(f(\mathbf{a}_n))$
converge to $f(\mathbf{a})$. For the
point $\mathbf{b}$, however, this does
not hold; f is discontinuous
at that point

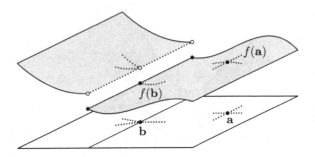

An example of a convergent vector-valued sequence is

$$\lim_{n\to\infty} \left(\frac{1}{n}, \frac{2n}{3n+4} \right) = \left(0, \frac{2}{3} \right).$$

Definition 15.3 A function $f : D \to \mathbb{R}$ is called *continuous* at the point $\mathbf{a} \in D$, if

$$\lim_{n\to\infty} f(\mathbf{a}_n) = f(\mathbf{a})$$

for all sequences $(\mathbf{a}_n)_{n\geq 1}$ which converge to $\mathbf{a}$ in D.

For continuous functions, the limit and the function sign can be interchanged.
Figure 15.3 shows a function which is discontinuous along a straight line but continuous everywhere else.

15.3   Partial Derivatives

The partial derivatives of a function of two variables are the derivatives of the partial mappings.

Definition 15.4 Let $D \subset \mathbb{R}^2$ be open, $f : D \to \mathbb{R}$ and $\mathbf{a} = (a, b) \in D$. The function
f is called *partially differentiable with respect to x* at the point $\mathbf{a}$, if the limit

$$\frac{\partial f}{\partial x}(a, b) = \lim_{x\to a} \frac{f(x, b) - f(a, b)}{x - a}$$

exists. It is called *partially differentiable with respect to y* at the point $\mathbf{a}$, if the limit

$$\frac{\partial f}{\partial y}(a, b) = \lim_{y\to b} \frac{f(a, y) - f(a, b)}{y - b}$$

Fig. 15.4 Geometric
interpretation of partial
derivatives

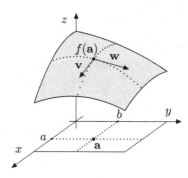

exists. The expressions

$$\frac{\partial f}{\partial x}(a, b) \quad \text{and} \quad \frac{\partial f}{\partial y}(a, b)$$

are called *partial derivatives* of f with respect to x and y, respectively, at the point (a, b). Further f is called *partially differentiable* at **a**, if both partial derivatives exist.

Another notation for partial derivatives at the point (x, y) is

$$\frac{\partial f}{\partial x}(x, y) = \frac{\partial}{\partial x} f(x, y) = \partial_1 f(x, y)$$

and likewise

$$\frac{\partial f}{\partial y}(x, y) = \frac{\partial}{\partial y} f(x, y) = \partial_2 f(x, y).$$

Geometrically, partial derivatives can be interpreted as slopes of the tangents to the coordinate curves $x \mapsto [x, b, f(x, b)]^{\mathsf{T}}$ and $y \mapsto [a, y, f(a, y)]^{\mathsf{T}}$, see. Fig. 15.4.

The two tangent vectors **v** and **w** to the coordinate curves at the point $(a, b, f(a, b))$ can therefore be represented as

$$\mathbf{v} = \begin{bmatrix} 1 \\ 0 \\ \dfrac{\partial f}{\partial x}(a, b) \end{bmatrix}, \quad \mathbf{w} = \begin{bmatrix} 0 \\ 1 \\ \dfrac{\partial f}{\partial y}(a, b) \end{bmatrix}.$$

Since partial differentiation is nothing else but ordinary differentiation with respect to one variable (while fixing the other one), the usual rules of differentiation apply, e.g. the product rule

$$\frac{\partial}{\partial y}\Big(f(x, y) \cdot g(x, y)\Big) = \frac{\partial f}{\partial y}(x, y) \cdot g(x, y) + f(x, y) \cdot \frac{\partial g}{\partial y}(x, y).$$

Fig. 15.5 Partially differentiable, discontinuous function

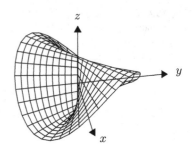

Example 15.5 Let $r : \mathbb{R}^2 \to \mathbb{R} : (x, y) \mapsto \sqrt{x^2 + y^2}$. This function is everywhere partially differentiable with the exception of $(x, y) = (0, 0)$. The partial derivatives are

$$\frac{\partial r}{\partial x}(x, y) = \frac{1}{2}\frac{2x}{\sqrt{x^2 + y^2}} = \frac{x}{r(x, y)}, \qquad \frac{\partial r}{\partial y}(x, y) = \frac{1}{2}\frac{2y}{\sqrt{x^2 + y^2}} = \frac{y}{r(x, y)}.$$

In maple one can use the commands `diff` and `Diff` in order to calculate partial derivatives, e.g. in the above example:

```
r:=sqrt(x^2+y^2);
diff(r,x);
```

Remark 15.6 In contrast to functions in one variable (see Application 7.16), partial differentiability does not imply continuity

$$f \text{ partially differentiable} \nRightarrow f \text{ continuous.}$$

An example is given by the function (see Fig. 15.5)

$$f(x, y) = \begin{cases} \dfrac{xy}{x^2 + y^2}, & (x, y) \neq (0, 0), \\ 0, & (x, y) = (0, 0). \end{cases}$$

This function is everywhere partially differentiable. In particular, at the point $(x, y) = (0, 0)$ one obtains

$$\frac{\partial f}{\partial x}(0, 0) = \lim_{x \to 0}\frac{f(x, 0) - f(0, 0)}{x} = 0 = \lim_{y \to 0}\frac{f(0, y) - f(0, 0)}{y} = \frac{\partial f}{\partial y}(0, 0).$$

However, the function is discontinuous at $(0, 0)$. In order to see this, we choose two sequences which converge to $(0, 0)$:

$$\mathbf{a}_n = \left(\tfrac{1}{n}, \tfrac{1}{n}\right) \quad \text{and} \quad \mathbf{c}_n = \left(\tfrac{1}{n}, -\tfrac{1}{n}\right).$$

We have

$$\lim_{n\to\infty} f(\mathbf{a}_n) = \lim_{n\to\infty} \frac{1/n^2}{2/n^2} = \frac{1}{2},$$

but also

$$\lim_{n\to\infty} f(\mathbf{c}_n) = \lim_{n\to\infty} \frac{-1/n^2}{2/n^2} = -\frac{1}{2}.$$

The limits do not coincide, in particular, they differ from $f(0, 0) = 0$.

Experiment 15.7 Visualise the function given in Remark 15.6 with the help of MATLAB and maple. Using the command

```
plot3d(-x*y/(x^2+y^2), x=-1..1, y=-1..1, shading=zhue);
```

the corresponding plot can be obtained in maple.

Higher-order partial derivatives. Let $D \subset \mathbb{R}^2$ be open and $f : D \to \mathbb{R}$ partially differentiable. The assignments

$$\frac{\partial f}{\partial x} : D \to \mathbb{R} \quad \text{and} \quad \frac{\partial f}{\partial y} : D \to \mathbb{R}$$

define themselves scalar-valued functions of two variables. If these functions are also partially differentiable, then f is called *twice* partially differentiable. The notation in this case is

$$\frac{\partial^2 f}{\partial x^2} = \frac{\partial}{\partial x}\left(\frac{\partial f}{\partial x}\right), \quad \frac{\partial^2 f}{\partial y \partial x} = \frac{\partial}{\partial y}\left(\frac{\partial f}{\partial x}\right), \quad \text{etc.}$$

Note that there are four partial derivatives of second order.

Definition 15.8 A function $f : D \to \mathbb{R}$ is k-times *continuously (partially) differentiable*, denoted $f \in C^k(D)$, if f is k-times partially differentiable and all partial derivatives up to order k are continuous.

Example 15.9 The function $f(x, y) = e^{xy^2}$ is arbitrarily often partially differentiable, $f \in C^\infty(D)$, and the following holds

$$\frac{\partial f}{\partial x}(x, y) = e^{xy^2} y^2,$$

$$\frac{\partial f}{\partial y}(x, y) = e^{xy^2} 2xy,$$

$$\frac{\partial^2 f}{\partial x^2}(x, y) = e^{xy^2} y^4,$$

$$\frac{\partial^2 f}{\partial y^2}(x, y) = e^{xy^2}(4x^2y^2 + 2x),$$

$$\frac{\partial^2 f}{\partial y \partial x}(x, y) = \frac{\partial}{\partial y}\left(\frac{\partial f}{\partial x}(x, y)\right) = e^{xy^2}(2xy^3 + 2y),$$

$$\frac{\partial^2 f}{\partial x \partial y}(x, y) = \frac{\partial}{\partial x}\left(\frac{\partial f}{\partial y}(x, y)\right) = e^{xy^2}(2xy^3 + 2y).$$

The identity

$$\frac{\partial^2 f}{\partial y \partial x}(x, y) = \frac{\partial^2 f}{\partial x \partial y}(x, y)$$

which is evident in this example is generally valid for twice *continuously* differentiable functions f. This observation is also true for higher derivatives: For k-times continuously differentiable functions the order of differentiation of the kth partial derivatives is irrelevant (Theorem of Schwarz[1]), see [3, Chap. 15, Theorem 1.1].

15.4 The Fréchet Derivative

Our next topic is the study of a *simultaneous* variation of both variables of the function. This leads us to the notion of the Fréchet[2] derivative. For functions of one variable, $\varphi : \mathbb{R} \to \mathbb{R}$, the derivative was defined by the limit

$$\varphi'(a) = \lim_{x \to a} \frac{\varphi(x) - \varphi(a)}{x - a}.$$

For functions of two variables this expression does not make sense anymore as one cannot divide by vectors. We therefore will make use of the equivalent definition of the derivative as a linear approximation

$$\varphi(x) = \varphi(a) + A \cdot (x - a) + R(x, a)$$

with $A = \varphi'(a)$ and the remainder term $R(x, a)$ satisfying

$$\lim_{x \to a} \frac{R(x, a)}{|x - a|} = 0.$$

This formula can be generalised to functions of two variables.

---

[1]H.A. Schwarz, 1843–1921.
[2]M. Fréchet, 1878–1973.

Definition 15.10 Let $D \subset \mathbb{R}^2$ be open and $f : D \to \mathbb{R}$. The function f is called *Fréchet differentiable* at the point $(a, b) \in D$, if there exists a *linear* mapping $A : \mathbb{R}^2 \to \mathbb{R}$ such that

$$f(x, y) = f(a, b) + A(x - a, y - b) + R(x, y; a, b)$$

with a remainder $R(x, y; a, b)$ fulfilling the condition

$$\lim_{(x,y) \to (a,b)} \frac{R(x, y; a, b)}{\sqrt{(x - a)^2 + (y - b)^2}} = 0.$$

The linear mapping A is called *derivative* of f at the point (a, b). Instead of A we also write $Df(a, b)$. The (1×2)-matrix of the linear mapping is called *Jacobian*[3] of f. We denote it by $f'(a, b)$.

The question whether the derivative of a function is unique and how it can be calculated, is answered in the following proposition.

Proposition 15.11 *Let $D \subset \mathbb{R}^2$ be open and $f : D \to \mathbb{R}$. If f is Fréchet differentiable at $(x, y) \in D$, then f is also partially differentiable at (x, y) and*

$$f'(x, y) = \left[\frac{\partial f}{\partial x}(x, y), \frac{\partial f}{\partial y}(x, y) \right].$$

The components of the Jacobian are the partial derivatives. In particular, the Jacobian and consequently the Fréchet derivative are unique.

Proof Exemplarily, we compute the second component and show that

$$\left(f'(x, y) \right)_2 = \frac{\partial f}{\partial y}(x, y).$$

Since f is Fréchet differentiable at (x, y), it holds that

$$f(x, y + h) = f(x, y) + f'(x, y) \begin{bmatrix} 0 \\ h \end{bmatrix} + R(x, y + h; x, y).$$

Therefore

$$\frac{f(x, y + h) - f(x, y)}{h} - \left(f'(x, y) \right)_2 = \frac{R(x, y + h; x, y)}{h} \to 0 \quad \text{as } h \to 0.$$

Consequently f is partially differentiable with respect to y, and the second component of the Jacobian is the partial derivative of f with respect to y.

---

[3]C.G.J. Jacobi, 1804–1851.

The next proposition follows immediately from the identity

$$\lim_{(x,y)\to(a,b)} f(x, y) = \lim_{(x,y)\to(a,b)} \Big(f(a, b) + \mathrm{D}f(a, b)(x - a, y - b) + R(x, y; a, b) \Big)$$
$$= f(a, b).$$

Proposition 15.12 *If f is Fréchet differentiable then f is continuous.* $\square$

In particular, the function

$$f(x, y) = \begin{cases} \dfrac{xy}{x^2 + y^2}, & (x, y) \neq (0, 0), \\ 0, & (x, y) = (0, 0) \end{cases}$$

is not Fréchet differentiable at the point $(0, 0)$.

Fréchet differentiability follows from partial differentiability under certain regularity assumptions. In fact, one can show that a *continuously* partially differentiable function is Fréchet differentiable, see [4, Chap. 7, Theorem 7.12].

Example 15.13 The function $f : \mathbb{R}^2 \to \mathbb{R} : (x, y) \mapsto x^2 e^{3y}$ is Fréchet differentiable, its derivative is

$$f'(x, y) = \left[2x e^{3y}, 3x^2 e^{3y} \right] = x e^{3y} [2, 3x].$$

Example 15.14 The affine function $f : \mathbb{R}^2 \to \mathbb{R}$ with

$$f(x, y) = \alpha x + \beta y + \gamma = [\alpha, \beta] \begin{bmatrix} x \\ y \end{bmatrix} + \gamma$$

is Fréchet differentiable and $f'(x, y) = [\alpha, \beta]$.

Example 15.15 The quadratic function $f : \mathbb{R}^2 \to \mathbb{R}$ with

$$f(x, y) = \alpha x^2 + 2\beta xy + \gamma y^2 + \delta x + \varepsilon y + \zeta$$
$$= [x, y] \begin{bmatrix} \alpha & \beta \\ \beta & \gamma \end{bmatrix} \begin{bmatrix} x \\ y \end{bmatrix} + [\delta, \varepsilon] \begin{bmatrix} x \\ y \end{bmatrix} + \zeta$$

is Fréchet differentiable with the Jacobian

$$f'(x, y) = [2\alpha x + 2\beta y + \delta, 2\beta x + 2\gamma y + \varepsilon] = 2[x, y] \begin{bmatrix} \alpha & \beta \\ \beta & \gamma \end{bmatrix} + [\delta, \varepsilon].$$

The chain rule. Now we are in the position to generalise the chain rule to the case of two variables.

Proposition 15.16 *Let $D \subset \mathbb{R}^2$ be open and $f : D \to \mathbb{R} : (x, y) \mapsto f(x, y)$ Fréchet differentiable. Furthermore let $I \subset \mathbb{R}$ be an open interval and $\phi, \psi : I \to \mathbb{R}$ differentiable. Then the composition of functions*

$$F : I \to \mathbb{R} : t \mapsto F(t) = f\big(\phi(t), \psi(t)\big)$$

is also differentiable and

$$\frac{\mathrm{d}F}{\mathrm{d}t}(t) = \frac{\partial f}{\partial x}\big(\phi(t), \psi(t)\big)\frac{\mathrm{d}\phi}{\mathrm{d}t}(t) + \frac{\partial f}{\partial y}\big(\phi(t), \psi(t)\big)\frac{\mathrm{d}\psi}{\mathrm{d}t}(t).$$

Proof From the Fréchet differentiability of f it follows that

$$F(t + h) - F(t) = f\big(\phi(t + h), \psi(t + h)\big) - f\big(\phi(t), \psi(t)\big)$$

$$= f'\big(\phi(t), \psi(t)\big)\begin{bmatrix} \phi(t + h) - \phi(t) \\ \psi(t + h) - \psi(t) \end{bmatrix} + R\big(\phi(t + h), \psi(t + h); \phi(t), \psi(t)\big).$$

We divide this expression by h and subsequently examine the limit as $h \to 0$. Let $g(t, h) = \big(\phi(t + h) - \phi(t)\big)^2 + \big(\psi(t + h) - \psi(t)\big)^2$. Then, due to the differentiability of f, ϕ and ψ, we have

$$\lim_{h \to 0} \frac{R\big(\phi(t + h), \psi(t + h); \phi(t), \psi(t)\big)}{\sqrt{g(t, h)}} \cdot \frac{\sqrt{g(t, h)}}{h} = 0.$$

Therefore, the function F is differentiable and the formula stated in the proposition is valid. $\square$

Example 15.17 Let $D \subset \mathbb{R}^2$ be an open set that contains the circle $x^2 + y^2 = 1$ and let $f : D \to \mathbb{R}$ be a differentiable function. Then the restriction F of f to the circle

$$F : \mathbb{R} \to \mathbb{R} : t \mapsto f(\cos t, \sin t)$$

is differentiable as a function of the angle t and

$$\frac{\mathrm{d}F}{\mathrm{d}t}(t) = -\frac{\partial f}{\partial x}(\cos t, \sin t) \cdot \sin t + \frac{\partial f}{\partial y}(\cos t, \sin t) \cdot \cos t.$$

For instance, for $f(x, y) = x^2 - y^2$ the derivative is $\frac{\mathrm{d}F}{\mathrm{d}t}(t) = -4 \cos t \sin t$.

Interpretation of the Fréchet derivative. Using the Fréchet derivative we obtain, like in the case of one variable, the linear approximation $g(x, y)$ to the graph of the function at (a, b)

$$g(x, y) = f(a, b) + f'(a, b)\begin{bmatrix} x - a \\ y - b \end{bmatrix} \approx f(x, y).$$

Now we want to interpret the plane

$$z = f(a, b) + f'(a, b) \begin{bmatrix} x - a \\ y - b \end{bmatrix}$$

geometrically. For this we use the fact that the components of the Jacobian are the partial derivatives. With that we can write the above equation as

$$z = f(a, b) + \frac{\partial f}{\partial x}(a, b) \cdot (x - a) + \frac{\partial f}{\partial y}(a, b) \cdot (y - b),$$

or alternatively in parametric form $(x - a = \lambda, \ y - b = \mu)$

$$\begin{bmatrix} x \\ y \\ z \end{bmatrix} = \begin{bmatrix} a \\ b \\ f(a, b) \end{bmatrix} + \lambda \begin{bmatrix} 1 \\ 0 \\ \frac{\partial f}{\partial x}(a, b) \end{bmatrix} + \mu \begin{bmatrix} 0 \\ 1 \\ \frac{\partial f}{\partial y}(a, b) \end{bmatrix}.$$

The plane intersects the graph of f at the point $(a, b, f(a, b))$ and is spanned by the tangent vectors to the coordinate curves. The equation

$$z = f(a, b) + \frac{\partial f}{\partial x}(a, b) \cdot (x - a) + \frac{\partial f}{\partial y}(a, b) \cdot (y - b),$$

consequently describes the *tangent plane* to the graph of f at the point (a, b).

The example shows that the graph of a function which is Fréchet differentiable at the point (x, y) possesses a tangent plane at this point. Note that the existence of tangents to the coordinate curves does *not* imply the existence of a tangent plane, see Remark 15.6.

Example 15.18 We calculate the tangent plane at a point on the northern hemisphere (with radius r)

$$f(x, y) = z = \sqrt{r^2 - x^2 - y^2}.$$

Let $c = f(a, b) = \sqrt{r^2 - a^2 - b^2}$. The partial derivatives of f at (a, b) are

$$\frac{\partial f}{\partial x}(a, b) = -\frac{a}{\sqrt{r^2 - a^2 - b^2}} = -\frac{a}{c}, \qquad \frac{\partial f}{\partial y}(a, b) = -\frac{b}{\sqrt{r^2 - a^2 - b^2}} = -\frac{b}{c}.$$

Therefore, the equation of the tangent plane is

$$z = c - \frac{a}{c}(x - a) - \frac{b}{c}(y - b),$$

or alternatively

$$a(x - a) + b(y - b) + c(z - c) = 0.$$

The last formula actually holds for all points on the surface of the sphere.

15.5 Directional Derivative and Gradient

So far functions $f : D \subset \mathbb{R}^2 \to \mathbb{R}$ were defined on $\mathbb{R}^2$ as a point space. For the purpose of directional derivatives it is useful and customary to write the arguments $(x, y) \in \mathbb{R}^2$ as position vectors $\mathbf{x} = [x, y]^\mathsf{T}$. In this way each function $f : D \subset \mathbb{R}^2 \to \mathbb{R}$ can also be considered as a function of column vectors. We identify these two functions and will not distinguish between $f(x, y)$ and $f(\mathbf{x})$ henceforth.

In Sect. 15.3 we have defined partial derivatives along coordinate axes. Now we want to generalise this concept to differentiation in *any* direction.

Definition 15.19 Let $D \subset \mathbb{R}^2$ be open, $\mathbf{x} = [x, y]^\mathsf{T} \in D$ and $f : D \to \mathbb{R}$. Furthermore let $\mathbf{v} \in \mathbb{R}^2$ with $\|\mathbf{v}\| = 1$. The limit

$$\partial_\mathbf{v} f(\mathbf{x}) = \frac{\partial f}{\partial \mathbf{v}}(\mathbf{x}) = \lim_{h \to 0} \frac{f(\mathbf{x} + h\mathbf{v}) - f(\mathbf{x})}{h}$$

$$= \lim_{h \to 0} \frac{f(x + hv_1, y + hv_2) - f(x, y)}{h}$$

(in case it exists) is called *directional derivative* of f at $\mathbf{x}$ *in direction* $\mathbf{v}$.

The partial derivatives are special cases of the directional derivative, namely the derivatives in direction of the coordinate axes.

The directional derivative $\partial_\mathbf{v} f(\mathbf{x})$ describes the rate of change of the function f at the point $\mathbf{x}$ in the direction of $\mathbf{v}$. Indeed, this can been seen from the following. Consider the straight line $\{\mathbf{x} + t\mathbf{v} \mid t \in \mathbb{R}\} \subset \mathbb{R}^2$ and the function

$$g(t) = f(\mathbf{x} + t\mathbf{v}) \qquad (f \text{ restricted to this straight line})$$

with $g(0) = f(\mathbf{x})$. Then

$$g'(0) = \lim_{h \to 0} \frac{g(h) - g(0)}{h} = \lim_{h \to 0} \frac{f(\mathbf{x} + h\mathbf{v}) - f(\mathbf{x})}{h} = \partial_\mathbf{v} f(\mathbf{x}).$$

Next we clarify how the directional derivative can be computed. For that we need the following definition.

Definition 15.20 Let $D \subset \mathbb{R}^2$ be open and $f : D \to \mathbb{R}$ partially differentiable. The vector

$$\nabla f(x, y) = \begin{bmatrix} \dfrac{\partial f}{\partial x}(x, y) \\[2mm] \dfrac{\partial f}{\partial y}(x, y) \end{bmatrix} = f'(x, y)^\mathsf{T}$$

is called *gradient* of f.

Fig. 15.6 Geometric
interpretation of ∇f

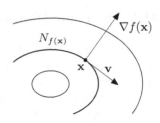

Proposition 15.21 *Let* $D \subset \mathbb{R}^2$ *be open,* $\mathbf{v} = [v_1, v_2]^\mathsf{T} \in \mathbb{R}^2$, $\|\mathbf{v}\| = 1$ *and* $f : D \to$ $\mathbb{R}$ *Fréchet differentiable at* $\mathbf{x} = [x, y]^\mathsf{T}$. *Then*

$$\partial_\mathbf{v} f(\mathbf{x}) = \langle \nabla f(\mathbf{x}), \mathbf{v} \rangle = f'(x, y)\,\mathbf{v} = \frac{\partial f}{\partial x}(x, y)\,v_1 + \frac{\partial f}{\partial y}(x, y)\,v_2.$$

Proof Since f is Fréchet differentiable at $\mathbf{x}$, it holds that

$$f(\mathbf{x} + h\mathbf{v}) = f(\mathbf{x}) + f'(\mathbf{x}) \cdot h\mathbf{v} + R(x + hv_1, y + hv_2; x, y)$$

and hence

$$\frac{f(\mathbf{x} + h\mathbf{v}) - f(\mathbf{x})}{h} = f'(\mathbf{x}) \cdot \mathbf{v} + \frac{R(x + hv_1, y + hv_2; x, y)}{h}.$$

Letting $h \to 0$ proves the desired assertion. $\square$

Proposition 15.22 (Geometric interpretation of ∇) *Let* $D \subset \mathbb{R}^2$ *be open and* $f :$ $D \to \mathbb{R}$ *continuously differentiable at* $\mathbf{x} = (x, y)$ *with* $f'(\mathbf{x}) \neq [0, 0]$. *Then* $\nabla f(\mathbf{x})$ *is* perpendicular *to the level curve* $N_{f(\mathbf{x})} = \{\widetilde{\mathbf{x}} \in \mathbb{R}^2 \,;\, f(\widetilde{\mathbf{x}}) = f(\mathbf{x})\}$ *and points in direction of the* steepest ascent *of* f, *see Fig. 15.6.*

Proof Let $\mathbf{v}$ be a tangent vector to the level curve at the point $\mathbf{x}$. From the implicit function theorem (see [4, Chap. 14.1]) it follows that $N_{f(\mathbf{x})}$ can be parametrised as a differentiable curve $\gamma(t) = [x(t), y(t)]^\mathsf{T}$, with

$$\gamma(0) = \mathbf{x} \quad \text{and} \quad \dot{\gamma}(0) = \mathbf{v}$$

in a neighbourhood of $\mathbf{x}$. Thus, for all t near $t = 0$,

$$f\big(\gamma(t)\big) = f(\mathbf{x}) = \text{const.}$$

Since f and γ are differentiable, it follows from the chain rule (Proposition 15.16) that

$$0 = \frac{d}{dt} f\big(\gamma(t)\big)\Big|_{t=0} = f'\big(\gamma(0)\big)\,\dot{\gamma}(0) = \langle \nabla f(\mathbf{x}), \mathbf{v} \rangle$$

because $\gamma(0) = \mathbf{x}$ and $\dot{\gamma}(0) = \mathbf{v}$. Hence $\nabla f(\mathbf{x})$ is perpendicular to $\mathbf{v}$. Let $\mathbf{w} \in \mathbb{R}^2$ be a further unit vector. Then

$$\partial_{\mathbf{w}} f(\mathbf{x}) = \frac{\partial f}{\partial \mathbf{w}}(\mathbf{x}) = \langle \nabla f(\mathbf{x}), \mathbf{w} \rangle = \|\nabla f(\mathbf{x})\| \cdot \|\mathbf{w}\| \cdot \cos \sphericalangle,$$

where $\sphericalangle$ denotes the angle enclosed by $\nabla f(\mathbf{x})$ and $\mathbf{w}$. From this formula one deduces that $\partial_{\mathbf{w}} f(\mathbf{x})$ is maximal if and only if $\cos \sphericalangle = 1$, which means $\nabla f(\mathbf{x}) = \lambda \mathbf{w}$ for some $\lambda > 0$. $\qquad\square$

Example 15.23 Let $f(x, y) = x^2 + y^2$. Then $\nabla f(x, y) = 2[x, y]^{\mathsf{T}}$.

15.6 The Taylor Formula in Two Variables

Let $f : D \subset \mathbb{R}^2 \to \mathbb{R}$ be a function of two variables. In the following calculation we assume that f is at least three times continuously differentiable. In order to expand $f(x + h, y + k)$ into a Taylor series in a neighbourhood of (x, y), we first fix the second variable and expand with respect to the first:

$$f(x+h, y+k) = f(x, y+k) + \frac{\partial f}{\partial x}(x, y+k) \cdot h + \frac{1}{2}\frac{\partial^2 f}{\partial x^2}(x, y+k) \cdot h^2 + \mathcal{O}(h^3).$$

Then we also expand the terms on the right-hand side with respect to the second variable (while fixing the first one):

$$f(x, y+k) = f(x, y) + \frac{\partial f}{\partial y}(x, y) \cdot k + \frac{1}{2}\frac{\partial^2 f}{\partial y^2}(x, y) \cdot k^2 + \mathcal{O}(k^3),$$

$$\frac{\partial f}{\partial x}(x, y+k) = \frac{\partial f}{\partial x}(x, y) + \frac{\partial^2 f}{\partial y \partial x}(x, y) \cdot k + \mathcal{O}(k^2),$$

$$\frac{\partial^2 f}{\partial x^2}(x, y+k) = \frac{\partial^2 f}{\partial x^2}(x, y) + \mathcal{O}(k).$$

Inserting these expressions into the equation above, we obtain

$$f(x + h, y + k) = f(x, y) + \frac{\partial f}{\partial x}(x, y) \cdot h + \frac{\partial f}{\partial y}(x, y) \cdot k$$

$$+ \frac{1}{2}\frac{\partial^2 f}{\partial x^2}(x, y) \cdot h^2 + \frac{1}{2}\frac{\partial^2 f}{\partial y^2}(x, y) \cdot k^2 + \frac{\partial^2 f}{\partial y \partial x}(x, y) \cdot hk$$

$$+ \mathcal{O}(h^3) + \mathcal{O}(h^2 k) + \mathcal{O}(hk^2) + \mathcal{O}(k^3).$$

In matrix-vector notation we can also write this equation as

$$f(x + h, y + k) = f(x, y) + f'(x, y)\begin{bmatrix} h \\ k \end{bmatrix} + \frac{1}{2}[h, k] \cdot H_f(x, y)\begin{bmatrix} h \\ k \end{bmatrix} + \cdots$$

with the *Hessian matrix*[4]

$$
H_f(x, y) =
\begin{bmatrix}
\dfrac{\partial^2 f}{\partial x^2}(x, y) & \dfrac{\partial^2 f}{\partial y \partial x}(x, y) \\[4mm]
\dfrac{\partial^2 f}{\partial x \partial y}(x, y) & \dfrac{\partial^2 f}{\partial y^2}(x, y)
\end{bmatrix}
$$

collecting the second-order partial derivatives. By the above assumptions, these derivatives are continuous. Thus the Hessian matrix is symmetric due to Schwarz's theorem.

Example 15.24 We compute the second-order approximation to the function $f : \mathbb{R}^2 \to \mathbb{R} : (x, y) \mapsto x^2 \sin y$ at the point $(a, b) = (2, 0)$. The partial derivatives are

	f	$\dfrac{\partial f}{\partial x}$	$\dfrac{\partial f}{\partial y}$	$\dfrac{\partial^2 f}{\partial x^2}$	$\dfrac{\partial^2 f}{\partial y \partial x}$	$\dfrac{\partial^2 f}{\partial y^2}$
General	$x^2 \sin y$	$2x \sin y$	$x^2 \cos y$	$2 \sin y$	$2x \cos y$	$-x^2 \sin y$
At $(2, 0)$	0	0	4	0	4	0

Therefore, the quadratic approximation $g(x, y) \approx f(x, y)$ is given by the formula

$$
\begin{aligned}
g(x, y) &= f(2, 0) + f'(2, 0) \begin{bmatrix} x - 2 \\ y \end{bmatrix} + \frac{1}{2}[x - 2, y] \cdot H_f(2, 0) \begin{bmatrix} x - 2 \\ y \end{bmatrix} \\
&= 0 + [0, 4] \begin{bmatrix} x - 2 \\ y \end{bmatrix} + \frac{1}{2}[x - 2, y] \begin{bmatrix} 0 & 4 \\ 4 & 0 \end{bmatrix} \begin{bmatrix} x - 2 \\ y \end{bmatrix} \\
&= 4y + 4y(x - 2) \ = \ 4y(x - 1).
\end{aligned}
$$

15.7 Local Maxima and Minima

Let $D \subset \mathbb{R}^2$ be open and $f : D \to \mathbb{R}$. In this section we investigate the graph of the function f with respect to maxima and minima.

Definition 15.25 The scalar function f has a *local maximum* (respectively, *local minimum*) at $(a, b) \in D$, if

$$
f(x, y) \leq f(a, b) \quad \text{(respectively, } f(x, y) \geq f(a, b)\text{).}
$$

---

[4]L.O. Hesse, 1811–1874.

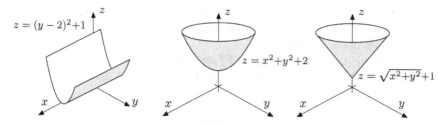

Fig. 15.7 Local and isolated local minima. The function in the picture on the left has local minima along the straight line $y = 2$. The minima are not isolated. The function in the middle picture has an isolated minimum at $(x, y) = (0, 0)$. This minimum is even a global minimum. Finally, the function in the picture on the right-hand side has also an isolated minimum at $(x, y) = (0, 0)$. However, the function is not differentiable at that point

for all (x, y) in a neighbourhood of (a, b). The maximum (minimum) is called *isolated*, if (a, b) is the only point in a neighbourhood with this property.

Figure 15.7 shows a few typical examples. One observes that the existence of a horizontal tangent plane is a necessary condition for *extrema* (i.e. maxima or minima) of *differentiable* functions.

Proposition 15.26 *Let f be partially differentiable. If f has a local maximum or minimum at $(a, b) \in D$, then the partial derivatives vanish at (a, b):*

$$\frac{\partial f}{\partial x}(a, b) = \frac{\partial f}{\partial y}(a, b) = 0.$$

If, in addition, f is Fréchet differentiable, then $f'(a, b) = [0, 0]$, i.e. f has a horizontal tangent plane at (a, b).

Proof Due to the assumptions, the function $g(h) = f(a + h, b)$ has an extremum at $h = 0$. Thus, Proposition 8.2 implies

$$g'(0) = \frac{\partial f}{\partial x}(a, b) = 0.$$

Likewise one can show that $\frac{\partial f}{\partial y}(a, b) = 0$.   □

Definition 15.27 Let f be a Fréchet differentiable function with $f'(a, b) = [0, 0]$. Then (a, b) is called a *stationary point* of f.

Stationary points are consequently candidates for extrema. Conversely, not all stationary points are extrema, they can also be *saddle points*. We call (a, b) a saddle point of f, if there is a vertical cut through the graph which has a local maximum at

(a, b), and a second vertical cut which has a local minimum at (a, b), see, for example, Fig. 15.2. In order to decide what is the case, one resorts to Taylor expansion, similarly as for functions of one variable.

Let $\mathbf{a} = [a, b]^\mathsf{T}$ be a stationary point of f and $\mathbf{v} \in \mathbb{R}^2$ any unit vector. We investigate the behaviour of f, restricted to the straight line $\mathbf{a} + \lambda\mathbf{v}$, $\lambda \in \mathbb{R}$. Taylor expansion shows that

$$f(\mathbf{a} + \lambda\mathbf{v}) = f(\mathbf{a}) + f'(\mathbf{a}) \cdot \lambda\mathbf{v} + \tfrac{1}{2}\lambda^2\mathbf{v}^\mathsf{T} H_f(\mathbf{a})\,\mathbf{v} + \mathcal{O}(\lambda^3).$$

Since $\mathbf{a}$ is a stationary point, it follows that $f'(\mathbf{a}) = [0, 0]$ and consequently

$$\frac{f(\mathbf{a} + \lambda\mathbf{v}) - f(\mathbf{a})}{\lambda^2} = \tfrac{1}{2}\,\mathbf{v}^\mathsf{T} H_f(\mathbf{a})\,\mathbf{v} + \mathcal{O}(\lambda).$$

For small λ the sign on the left-hand side is therefore determined by the sign of $\mathbf{v}^\mathsf{T} H_f(\mathbf{a})\,\mathbf{v}$. We ask how this can be expressed by conditions on $H_f(\mathbf{a})$. Writing

$$H_f(\mathbf{a}) = \begin{bmatrix} \alpha & \beta \\ \beta & \gamma \end{bmatrix} \quad \text{and} \quad \mathbf{v} = \begin{bmatrix} v \\ w \end{bmatrix}$$

we get

$$\mathbf{v}^T H_f(\mathbf{a})\,\mathbf{v} = \alpha v^2 + 2\beta vw + \gamma w^2.$$

For an isolated local minimum this expression has to be positive for all $\mathbf{v} \neq 0$. If $w = 0$ and $v \neq 0$, then $\alpha v^2 > 0$ and therefore necessarily

$$\alpha > 0.$$

If $w \neq 0$, we substitute $v = tw$ with $t \in \mathbb{R}$ and obtain

$$\alpha t^2 w^2 + 2\beta t w^2 + \gamma w^2 > 0,$$

or alternatively (multiplying by $\alpha > 0$ and simplifying by w^2)

$$t^2\alpha^2 + 2t\alpha\beta + \alpha\gamma > 0.$$

Therefore,

$$(t\alpha + \beta)^2 + \alpha\gamma - \beta^2 > 0$$

for all $t \in \mathbb{R}$. The left-hand side is smallest for $t = -\beta/\alpha$. Inserting this we obtain the second condition

$$\det H_f(\mathbf{a}) = \alpha\gamma - \beta^2 > 0$$

in terms of the determinant, see Appendix B.1.

We have thus shown the following result.

Proposition 15.28 *The function f has an isolated local minimum at the stationary point* **a**, *if the conditions*

$$\frac{\partial^2 f}{\partial x^2}(\mathbf{a}) > 0 \quad and \quad \det H_f(\mathbf{a}) > 0$$

are fulfilled.

By replacing f by $-f$ one gets the corresponding result for isolated maxima.

Proposition 15.29 *The function f has an isolated local maximum at the stationary point* **a**, *if the conditions*

$$\frac{\partial^2 f}{\partial x^2}(\mathbf{a}) < 0 \quad and \quad \det H_f(\mathbf{a}) > 0$$

are fulfilled.

In a similar way one can prove the following assertion.

Proposition 15.30 *The stationary point* **a** *of the function f is a saddle point, if* $\det H_f(\mathbf{a}) < 0$.

If the determinant of the Hessian matrix equals zero, the behaviour of the function needs to be investigated along vertical cuts. One example is given in Exercise 12.

Example 15.31 We determine the maxima, minima and saddle points of the function $f(x, y) = x^6 + y^6 - 3x^2 - 3y^2$. The condition

$$f'(x, y) = [6x^5 - 6x, 6y^5 - 6y] = [0, 0]$$

gives the following nine stationary points

$$x_1 = 0, \quad x_{2,3} = \pm 1, \quad y_1 = 0, \quad y_{2,3} = \pm 1.$$

The Hessian matrix of the function is

$$H_f(x, y) = \begin{bmatrix} 30x^4 - 6 & 0 \\ 0 & 30y^4 - 6 \end{bmatrix}.$$

Applying the criteria of Propositions 15.28 through 15.30, we obtain the following results: The point $(0, 0)$ is an isolated local maximum of f, the points $(-1, -1)$, $(-1, 1)$, $(1, -1)$ and $(1, 1)$ are isolated local minima, and the points $(-1, 0)$, $(1, 0)$, $(0, -1)$ and $(0, 1)$ are saddle points. The reader is advised to visualise this function with maple.

15.8 Exercises

1. Compute the partial derivatives of the functions

$$f(x, y) = \arcsin\left(\frac{y}{x}\right), \qquad g(x, y) = \log\frac{1}{\sqrt{x^2 + y^2}}.$$

Verify your results with maple.

2. Show that the function

$$v(x, t) = \frac{1}{\sqrt{t}} \exp\left(\frac{-x^2}{4t}\right)$$

satisfies the *heat equation*

$$\frac{\partial v}{\partial t} = \frac{\partial^2 v}{\partial x^2}$$

for $t > 0$ and $x \in \mathbb{R}$.

3. Show that the function $w(x, t) = g(x - kt)$ satisfies the *transport equation*

$$\frac{\partial w}{\partial t} + k\frac{\partial w}{\partial x} = 0$$

for any differentiable function g.

4. Show that the function $g(x, y) = \log(x^2 + 2y^2)$ satisfies the equation

$$\frac{\partial^2 g}{\partial x^2} + \frac{1}{2}\frac{\partial^2 g}{\partial y^2} = 0$$

for $(x, y) \neq (0, 0)$.

5. Represent the ellipsoid $x^2 + 2y^2 + z^2 = 1$ as graph of a function $(x, y) \mapsto f(x, y)$. Distinguish between positive and negative z-coordinates, respectively. Compute the partial derivatives of f and sketch the level curves of f. Find the direction in which ∇f points.

6. Solve Exercise 5 for the hyperboloid $x^2 + 2y^2 - z^2 = 1$.

7. Compute the directional derivative of the function $f(x, y) = xy$ in the direction $\mathbf{v}$ at the four points $\mathbf{a}_1, \ldots \mathbf{a}_4$, where

$$\mathbf{a}_1 = (1, 2),\ \mathbf{a}_2 = (-1, 2),\ \mathbf{a}_3 = (1, -2),\ \mathbf{a}_4 = (-1, -2)\ \text{ and }\ \mathbf{v} = \frac{1}{\sqrt{5}}\begin{bmatrix} 2 \\ 1 \end{bmatrix}.$$

At the given points $\mathbf{a}_1, \ldots \mathbf{a}_4$, determine the direction for which the directional derivative is maximal.

8. Consider the function $f(x, y) = 4 - x^2 - y^2$.

(a) Determine and sketch the level curves $f(x, y) = c$ for $c = 4, 2, 0, -2$ and the graphs of the coordinate curves

$$x \mapsto \begin{bmatrix} x \\ b \\ f(x, b) \end{bmatrix}, \qquad y \mapsto \begin{bmatrix} a \\ y \\ f(a, y) \end{bmatrix}$$

for $a, b = -1, 0, 1$.

(b) Compute the gradient of f at the point $(1, 1)$ and determine the equation of the tangent plane at $(1, 1, 2)$. Verify that the gradient is perpendicular to the level curve through $(1, 1, 2)$.

(c) Compute the directional derivatives of f at $(1, 1)$ in the directions

$$\mathbf{v}_1 = \frac{1}{\sqrt{2}} \begin{bmatrix} 1 \\ 1 \end{bmatrix}, \ \mathbf{v}_2 = \frac{1}{\sqrt{2}} \begin{bmatrix} -1 \\ 1 \end{bmatrix}, \ \mathbf{v}_3 = \frac{1}{\sqrt{2}} \begin{bmatrix} -1 \\ -1 \end{bmatrix}, \ \mathbf{v}_4 = \frac{1}{\sqrt{2}} \begin{bmatrix} 1 \\ -1 \end{bmatrix}.$$

Sketch the vectors $\mathbf{v}_1, \ldots, \mathbf{v}_4$ in the (x, y)-plane and interpret the value of the directional derivatives.

9. Consider the function $f(x, y) = ye^{2x-y}$, where $x = x(t)$ and $y = y(t)$ are differentiable functions satisfying

$$x(0) = 2, \qquad y(0) = 4, \qquad \dot{x}(0) = -1, \qquad \dot{y}(0) = 4.$$

From this information compute the derivative of $z(t) = f\big(x(t), y(t)\big)$ at the point $t = 0$.

10. Find all stationary points of the function

$$f(x, y) = x^3 - 3xy^2 + 6y.$$

Determine whether they are maxima, minima or saddle points.

11. Find the stationary point of the function

$$f(x, y) = e^x + ye^y - x$$

and determine whether it is a maximum, minimum or a saddle point.

12. Investigate the function

$$f(x, y) = x^4 - 3x^2y + y^3$$

for local extrema and saddle points. Visualise the graph of the function.
Hint. To study the behaviour of the function at $(0, 0)$ consider the partial mappings $f(x, 0)$ and $f(0, y)$.

13. Determine for the function

$$f(x, y) = x^2 e^{y/3}(y - 3) - \tfrac{1}{2}y^2$$

   (a) the gradient and the Hessian matrix;

   (b) the second-order Taylor approximation at $(0, 0)$;

   (c) all stationary points. Find out whether they are maxima, minima or saddle points.

14. Expand the polynomial $f(x, y) = x^2 + xy + 3y^2$ in powers of $x - 1$ and $y - 2$, i.e. in the form

$$f(x, y) = \alpha(x - 1)^2 + \beta(x - 1)(y - 2) + \gamma(y - 2)^2$$
$$+ \delta(x - 1) + \varepsilon(y - 2) + \zeta.$$

Hint. Use the second-order Taylor expansion at $(1, 2)$.

15. Compute $(0.95)^{2.01}$ numerically by using the second-order Taylor approximation to the function $f(x, y) = x^y$ at $(1, 2)$.

Vector-Valued Functions of Two Variables

16

In this section we briefly touch upon the theory of vector-valued functions in several variables. To simplify matters we limit ourselves again to the case of two variables.

First we define vector fields in the plane and extend the notions of *continuity* and *differentiability* to vector-valued functions. Then we discuss Newton's method in two variables. As an application we compute a common zero of two nonlinear functions. Finally, as an extension of Sect. 15.1, we show how smooth surfaces can be described mathematically with the help of parameterisations.

For the required basic notions of vector and matrix algebra we refer to the Appendices A and B.

16.1 Vector Fields and the Jacobian

In the entire section D denotes an open subset of $\mathbb{R}^2$ and

$$\mathbf{F} : D \subset \mathbb{R}^2 \to \mathbb{R}^2 : (x, y) \mapsto \begin{bmatrix} u \\ v \end{bmatrix} = \mathbf{F}(x, y) = \begin{bmatrix} f(x, y) \\ g(x, y) \end{bmatrix}$$

a *vector-valued* function of two variables with values in $\mathbb{R}^2$. Such functions are also called *vector fields* since they assign a vector to every point in the plane. Important applications are provided in physics. For example, the velocity field of a flowing liquid or the gravitational field are mathematically described as vector fields.

In the previous chapter we have already encountered a vector field, namely the gradient of a scalar-valued function of two variables $f : D \to \mathbb{R} : (x, y) \mapsto f(x, y)$.

© Springer Nature Switzerland AG 2018
M. Oberguggenberger and A. Ostermann, *Analysis for Computer Scientists*,
Undergraduate Topics in Computer Science,
https://doi.org/10.1007/978-3-319-91155-7_16

For a partially differentiable function f the gradient

$$\mathbf{F} = \nabla f : D \to \mathbb{R}^2 : (x, y) \mapsto \begin{bmatrix} \dfrac{\partial f}{\partial x}(x, y) \\ \dfrac{\partial f}{\partial y}(x, y) \end{bmatrix}$$

is obviously a vector field.

Continuity and differentiability of vector fields are defined *componentwise*.

Definition 16.1 The function

$$\mathbf{F} : D \subset \mathbb{R}^2 \to \mathbb{R}^2 : (x, y) \mapsto \mathbf{F}(x, y) = \begin{bmatrix} f(x, y) \\ g(x, y) \end{bmatrix}$$

is called continuous (or partially differentiable or Fréchet differentiable, respectively) if and only if its two components $f : D \to \mathbb{R}$ and $g : D \to \mathbb{R}$ have the corresponding property, i.e. they are continuous (or partially differentiable or Fréchet differentiable, respectively).

If both f and g are Fréchet differentiable, one has the linearisations

$$f(x, y) = f(a, b) + \left[\frac{\partial f}{\partial x}(a, b), \frac{\partial f}{\partial y}(a, b) \right] \begin{bmatrix} x - a \\ y - b \end{bmatrix} + R_1(x, y; a, b),$$

$$g(x, y) = g(a, b) + \left[\frac{\partial g}{\partial x}(a, b), \frac{\partial g}{\partial y}(a, b) \right] \begin{bmatrix} x - a \\ y - b \end{bmatrix} + R_2(x, y; a, b)$$

for (x, y) close to (a, b) with remainder terms R_1 and R_2. If one combines these two formulas to one formula using matrix-vector notation, one obtains

$$\begin{bmatrix} f(x, y) \\ g(x, y) \end{bmatrix} = \begin{bmatrix} f(a, b) \\ g(a, b) \end{bmatrix} + \begin{bmatrix} \dfrac{\partial f}{\partial x}(a, b) & \dfrac{\partial f}{\partial y}(a, b) \\ \dfrac{\partial g}{\partial x}(a, b) & \dfrac{\partial g}{\partial y}(a, b) \end{bmatrix} \begin{bmatrix} x - a \\ y - b \end{bmatrix} + \begin{bmatrix} R_1(x, y; a, b) \\ R_2(x, y; a, b) \end{bmatrix},$$

or in shorthand notation

$$\mathbf{F}(x, y) = \mathbf{F}(a, b) + \mathbf{F}'(a, b) \begin{bmatrix} x - a \\ y - b \end{bmatrix} + \mathbf{R}(x, y; a, b)$$

with the remainder term $\mathbf{R}(x, y; a, b)$ and the (2×2)-*Jacobian*

$$\mathbf{F}'(a, b) = \begin{bmatrix} \dfrac{\partial f}{\partial x}(a, b) & \dfrac{\partial f}{\partial y}(a, b) \\ \dfrac{\partial g}{\partial x}(a, b) & \dfrac{\partial g}{\partial y}(a, b) \end{bmatrix}.$$

The linear mapping defined by this matrix is called *(Fréchet) derivative* of the function $\mathbf{F}$ at the point (a, b). The remainder term $\mathbf{R}$ has the property

$$\lim_{(x,y)\to(a,b)} \frac{\sqrt{R_1(x, y; a, b)^2 + R_2(x, y; a, b)^2}}{\sqrt{(x - a)^2 + (y - b)^2}} = 0.$$

Example 16.2 (Polar coordinates) The mapping

$$\mathbf{F} : \mathbb{R}^2 \to \mathbb{R}^2 : (r, \phi) \mapsto \begin{bmatrix} x \\ y \end{bmatrix} = \begin{bmatrix} r \cos \varphi \\ r \sin \varphi \end{bmatrix}$$

is (everywhere) differentiable with derivative (Jacobian)

$$\mathbf{F}'(r, \varphi) = \begin{bmatrix} \cos \varphi & -r \sin \varphi \\ \sin \varphi & r \cos \varphi \end{bmatrix}.$$

16.2 Newton's Method in Two Variables

The linearisation

$$\mathbf{F}(x, y) \approx \mathbf{F}(a, b) + \mathbf{F}'(a, b) \begin{bmatrix} x - a \\ y - b \end{bmatrix}$$

is the key for solving nonlinear equations in two (or more) unknowns. In this section, we derive Newton's method for determining the zeros of a function

$$\mathbf{F}(x, y) = \begin{bmatrix} f(x, y) \\ g(x, y) \end{bmatrix}$$

of two variables and two components.

Example 16.3 (Intersection of a circle with a hyperbola) Consider the circle $x^2 + y^2 = 4$ and the hyperbola $xy = 1$. The points of intersection are the zeros of the vector equation $\mathbf{F}(x, y) = \mathbf{0}$ with

$$\mathbf{F} : \mathbb{R}^2 \to \mathbb{R}^2 : \mathbf{F}(x, y) = \begin{bmatrix} f(x, y) \\ g(x, y) \end{bmatrix} = \begin{bmatrix} x^2 + y^2 - 4 \\ xy - 1 \end{bmatrix}.$$

The level curves $f(x, y) = 0$ and $g(x, y) = 0$ are sketched in Fig. 16.1.

Fig. 16.1 Intersection of a
circle with a hyperbola

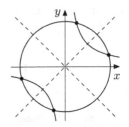

Newton's method for determining the zeros is based on the following idea. For a
starting value (x_0, y_0) which is sufficiently close to the solution, one computes an
improved value by replacing the function by its linear approximation at (x_0, y_0)

$$\mathbf{F}(x, y) \approx \mathbf{F}(x_0, y_0) + \mathbf{F}'(x_0, y_0) \begin{bmatrix} x - x_0 \\ y - y_0 \end{bmatrix}.$$

The zero of the linearisation

$$\mathbf{F}(x_0, y_0) + \mathbf{F}'(x_0, y_0) \begin{bmatrix} x - x_0 \\ y - y_0 \end{bmatrix} = \begin{bmatrix} 0 \\ 0 \end{bmatrix}$$

is taken as improved approximation (x_1, y_1), so

$$\mathbf{F}'(x_0, y_0) \begin{bmatrix} x_1 - x_0 \\ y_1 - y_0 \end{bmatrix} = -\mathbf{F}(x_0, y_0),$$

and

$$\begin{bmatrix} x_1 \\ y_1 \end{bmatrix} = \begin{bmatrix} x_0 \\ y_0 \end{bmatrix} - \left(\mathbf{F}'(x_0, y_0) \right)^{-1} \mathbf{F}(x_0, y_0),$$

respectively. This can only be carried out if the Jacobian is invertible, i.e. its determinant is not equal to zero. In the example above the Jacobian is

$$\mathbf{F}'(x, y) = \begin{bmatrix} 2x & 2y \\ y & x \end{bmatrix}$$

with determinant $\det \mathbf{F}'(x, y) = 2x^2 - 2y^2$. Thus it is singular on the straight lines
$x = \pm y$. These lines are plotted as dashed lines in Fig. 16.1.

The idea now is to iterate the procedure, i.e. to repeat Newton's step with the
improved value as new starting value

$$\begin{bmatrix} x_{k+1} \\ y_{k+1} \end{bmatrix} = \begin{bmatrix} x_k \\ y_k \end{bmatrix} - \begin{bmatrix} \dfrac{\partial f}{\partial x}(x_k, y_k) & \dfrac{\partial f}{\partial y}(x_k, y_k) \\ \dfrac{\partial g}{\partial x}(x_k, y_k) & \dfrac{\partial g}{\partial y}(x_k, y_k) \end{bmatrix}^{-1} \begin{bmatrix} f(x_k, y_k) \\ g(x_k, y_k) \end{bmatrix}$$

for $k = 1, 2, 3, \ldots$ until the desired accuracy is reached. The procedure generally converges rapidly as is shown in the following proposition. For a proof, see [23, Chap. 7, Theorem 7.1].

Proposition 16.4 *Let* $\mathbf{F} : D \to \mathbb{R}^2$ *be twice continuously differentiable with* $\mathbf{F}(a, b) = \mathbf{0}$ *and* $\det \mathbf{F}'(a, b) \neq 0$. *If the starting value* (x_0, y_0) *lies sufficiently close to the solution* (a, b) *then Newton's method converges quadratically.*

One often sums up this fact under the term *local quadratic convergence of Newton's method.*

Example 16.5 The intersection points of the circle and the hyperbola can also be computed analytically. Since

$$xy = 1 \quad \Leftrightarrow \quad x = \frac{1}{y}$$

we may insert $x = 1/y$ into the equation $x^2 + y^2 = 4$ to obtain the biquadratic equation

$$y^4 - 4y^2 + 1 = 0.$$

By substituting $y^2 = u$ the equation is easily solvable. The intersection point with the largest x-component has the coordinates

$$x = \sqrt{2 + \sqrt{3}} = 1.93185165257813657\ldots$$
$$y = \sqrt{2 - \sqrt{3}} = 0.51763809020504152\ldots$$

Application of Newton's method with starting values $x_0 = 2$ and $y_0 = 1$ yields the above solution in 5 steps with 16 digits accuracy. The quadratic convergence can be observed from the fact that the number of correct digits doubles with each step.

x	y	Error
2.000000000000000	1.000000000000000	4.871521418175E-001
2.000000000000000	5.000000000000000E-001	7.039388810410E-002
1.933333333333333	5.166666666666667E-001	1.771734052060E-003
1.931852741096439	5.176370548219287E-001	1.502295005704E-006
1.931851652578934	5.176380902042443E-001	1.127875985998E-012
1.931851652578136	5.176380902050416E-001	2.220446049250E-016

Experiment 16.6 Using the MATLAB programs `mat16_1.m` and `mat16_2.m` compute the intersection points from Example 16.3. Experiment with different starting values, and this way try to determine all four solutions to the problem. What happens if the starting value is chosen to be $(x_0, y_0) = (1, 1)$?

16.3 Parametric Surfaces

In Sect. 15.1 we investigated surfaces as graphs of functions $f : D \subset \mathbb{R}^2 \to \mathbb{R}$. However, similar to the case of curves, this concept is too narrow to represent more complicated surfaces. The remedy is to use parameterisations like it was done for curves.

The starting point for the construction of a parametric surface is a (componentwise) continuous mapping

$$(u, v) \mapsto \mathbf{x}(u, v) = \begin{bmatrix} x(u, v) \\ y(u, v) \\ z(u, v) \end{bmatrix}$$

of a parameter domain $D \subset \mathbb{R}^2$ to $\mathbb{R}^3$. By fixing one parameter $u = u_0$ or $v = v_0$ at a time one obtains coordinate curves in space

$$u \mapsto \mathbf{x}(u, v_0) \quad \ldots \quad u\text{-curve}$$
$$v \mapsto \mathbf{x}(u_0, v) \quad \ldots \quad v\text{-curve}$$

Definition 16.7 A regular parametric surface is defined by a mapping $D \subset \mathbb{R}^2 \to \mathbb{R}^3 : (u, v) \mapsto \mathbf{x}(u, v)$ which satisfies the following conditions

(a) the mapping $(u, v) \mapsto \mathbf{x}(u, v)$ is injective;

(b) the u-curves and the v-curves are continuously differentiable;

(c) the tangent vectors to the u-curves and v-curves are linearly independent at every point (thus always span a plane).

These conditions guarantee that the parametric surface is indeed a two-dimensional smooth subset of $\mathbb{R}^3$.

For a regular surface, the tangent vectors

$$\frac{\partial \mathbf{x}}{\partial u}(u, v) = \begin{bmatrix} \dfrac{\partial x}{\partial u}(u, v) \\ \dfrac{\partial y}{\partial u}(u, v) \\ \dfrac{\partial z}{\partial u}(u, v) \end{bmatrix}, \qquad \frac{\partial \mathbf{x}}{\partial v}(u, v) = \begin{bmatrix} \dfrac{\partial x}{\partial v}(u, v) \\ \dfrac{\partial y}{\partial v}(u, v) \\ \dfrac{\partial z}{\partial v}(u, v) \end{bmatrix}$$

span the *tangent plane* at $\mathbf{x}(u, v)$. The tangent plane has the parametric representation

$$\mathbf{p}(\lambda, \mu) \ = \ \mathbf{x}(u, v) + \lambda \frac{\partial \mathbf{x}}{\partial u}(u, v) + \mu \frac{\partial \mathbf{x}}{\partial v}(u, v), \qquad \lambda, \mu \in \mathbb{R}.$$

The regularity condition (c) is equivalent to the assertion that

$$\frac{\partial \mathbf{x}}{\partial u} \times \frac{\partial \mathbf{x}}{\partial v} \neq \mathbf{0}.$$

The cross product constitutes a normal vector to the (tangent plane of the) surface.

Example 16.8 (Surfaces of rotation)   By rotation of the graph of a continuously differentiable, positive function $z \mapsto h(z)$, $a < z < b$, around the z-axis, one obtains a surface of rotation with parametrisation

$$D = (a, b) \times (0, 2\pi), \qquad \mathbf{x}(u, v) = \begin{bmatrix} h(u) \cos v \\ h(u) \sin v \\ u \end{bmatrix}.$$

The v-curves are horizontal circles, the u-curves are the generator lines. Note that the generator line corresponding to the angle $v = 0$ has been removed to ensure condition (a). To verify condition (c) we compute the cross product of the tangent vectors to the u- and the v-curves

$$\frac{\partial \mathbf{x}}{\partial u} \times \frac{\partial \mathbf{x}}{\partial v} = \begin{bmatrix} h'(u) \cos v \\ h'(u) \sin v \\ 1 \end{bmatrix} \times \begin{bmatrix} -h(u) \sin v \\ h(u) \cos v \\ 0 \end{bmatrix} = \begin{bmatrix} -h(u) \cos v \\ -h(u) \sin v \\ h(u) h'(u) \end{bmatrix} \neq \mathbf{0}.$$

Due to $h(u) > 0$ this vector is not zero; the two tangent vectors are hence not collinear.

Figure 16.2 shows the surface of rotation which is generated by $h(u) = 0.4 + \cos(4\pi u)/3$, $u \in (0, 1)$. In MATLAB one advantageously uses the command `cylinder` in combination with the command `mesh` for the representation of such surfaces.

Example 16.9 (The sphere)   The sphere of radius R is obtained by the parametrisation

$$D = (0, \pi) \times (0, 2\pi), \qquad \mathbf{x}(u, v) = \begin{bmatrix} R \sin u \cos v \\ R \sin u \sin v \\ R \cos u \end{bmatrix}.$$

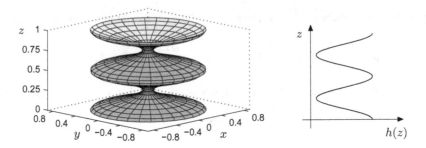

Fig. 16.2 Surface of rotation, generated by rotation of a graph $h(z)$ about the z-axis. The underlying graph $h(z)$ is represented on the right

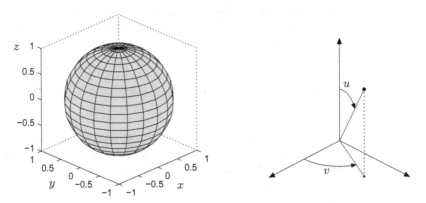

Fig. 16.3 Unit sphere as parametric surface. The interpretation of the parameters u, v as angles is given in the picture on the right

The v-curves are the circles of latitude, the u-curves the meridians. The meaning of the parameters u, v as angles can be seen from Fig. 16.3.

16.4 Exercises

1. Compute the Jacobian of the mapping

$$\begin{bmatrix} u \\ v \end{bmatrix} = \mathbf{F}(x, y) = \begin{bmatrix} x^2 + y^2 \\ x^2 - y^2 \end{bmatrix}.$$

For which values of x and y is the Jacobian invertible?

2. Program Newton's method in several variables and test the program on the problem

$$x^2 + \sin y = 4$$
$$xy = 1$$

with starting values $x = 2$ and $y = 1$. If you are working in MATLAB, you can solve this question by modifying `mat16_2.m`.

3. Compute the tangent vectors $\frac{\partial \mathbf{x}}{\partial u}$, $\frac{\partial \mathbf{x}}{\partial v}$ and the normal vector $\frac{\partial \mathbf{x}}{\partial u} \times \frac{\partial \mathbf{x}}{\partial v}$ to the sphere of radius R (Example 16.9). What can you observe about the direction of the normal vector?

4. Sketch the surface of revolution

$$\mathbf{x}(u, v) = \begin{bmatrix} \cos u \cos v \\ \cos u \sin v \\ u \end{bmatrix}, \quad -1 < u < 1,\ 0 < v < 2\pi.$$

Compute the tangent vectors $\frac{\partial \mathbf{x}}{\partial u}$, $\frac{\partial \mathbf{x}}{\partial v}$ and the normal vector $\frac{\partial \mathbf{x}}{\partial u} \times \frac{\partial \mathbf{x}}{\partial v}$. Determine the equation of the tangent plane at the point $(1/\sqrt{2}, 1/\sqrt{2}, 0)$.

5. Sketch the paraboloid

$$\mathbf{x}(u, v) = \begin{bmatrix} u \cos v \\ u \sin v \\ 1 - u^2 \end{bmatrix}, \quad 0 < u < 1,\ 0 < v < 2\pi$$

and plot some of the u- and v-curves. Compute the tangent vectors $\frac{\partial \mathbf{x}}{\partial u}$, $\frac{\partial \mathbf{x}}{\partial v}$ and the normal vector $\frac{\partial \mathbf{x}}{\partial u} \times \frac{\partial \mathbf{x}}{\partial v}$.

6. Plot some of the u- and v-curves for the helicoid

$$\mathbf{x}(u, v) = \begin{bmatrix} u \cos v \\ u \sin v \\ v \end{bmatrix}, \quad 0 < u < 1,\ 0 < v < 2\pi$$

What kind of curves are they? Try to sketch the surface.

7. A planar vector field (see also Sect. 20.1)

$$(x, y) \mapsto \mathbf{F}(x, y) = \begin{bmatrix} f(x, y) \\ g(x, y) \end{bmatrix}$$

can be visualised by plotting a grid of points (x_i, y_j) in the plane and attaching the vector $\mathbf{F}(x_i, y_j)$ to each grid point. Sketch the vector fields

$$\mathbf{F}(x, y) = \frac{1}{\sqrt{x^2 + y^2}} \begin{bmatrix} x \\ y \end{bmatrix} \quad \text{and} \quad \mathbf{G}(x, y) = \frac{1}{\sqrt{x^2 + y^2}} \begin{bmatrix} -y \\ x \end{bmatrix}$$

in this way.

Integration of Functions of Two Variables

17

In Sect. 11.3 we have shown how to calculate the volume of solids of revolution. If there is no rotational symmetry, however, one needs an extension of integral calculus to functions of two variables. This arises, for example, if one wants to find the volume of a solid that lies between a domain D in the (x, y)-plane and the graph of a non-negative function $z = f(x, y)$. In this section we will extend the notion of Riemann integrals from Chap. 11 to double integrals of functions of two variables. Important tools for the computation of double integrals are their representation as iterated integrals and the transformation formula (change of coordinates). The integration of functions of several variables occurs in numerous applications, a few of which we will discuss.

17.1 Double Integrals

We start with the integration of a real-valued function $z = f(x, y)$ which is defined on a rectangle $R = [a, b] \times [c, d]$. More general domains of integration $D \subset \mathbb{R}^2$ will be discussed below. Since we know from Sect. 11.1 that Riemann integrable functions are necessarily bounded, we assume in the whole section that f is bounded. If f is non-negative, the integral should be interpretable as the volume of the solid with base R and top surface given by the graph of f (see Fig. 17.2). This motivates the following approach in which the solid is approximated by a sum of cuboids.

We place a rectangular grid G over the domain R by partitioning the intervals $[a, b]$ and $[c, d]$ like in Sect. 11.1:

$$Z_x : a = x_0 < x_1 < x_2 < \cdots < x_{n-1} < x_n = b,$$
$$Z_y : c = y_0 < y_1 < y_2 < \cdots < y_{m-1} < y_m = d.$$

© Springer Nature Switzerland AG 2018
M. Oberguggenberger and A. Ostermann, *Analysis for Computer Scientists*,
Undergraduate Topics in Computer Science,
https://doi.org/10.1007/978-3-319-91155-7_17

The rectangular grid is made up of the small rectangles

$$[x_{i-1}, x_i] \times [y_{j-1}, y_j], \quad i = 1, \ldots, n, \ j = 1, \ldots, m.$$

The *mesh size* $\Phi(G)$ is the length of the largest subinterval involved:

$$\Phi(G) = \max \left(|x_i - x_{i-1}|, |y_j - y_{j-1}| \ ; \ i = 1, \ldots, n, \ j = 1, \ldots, m \right).$$

Finally we choose an arbitrary intermediate point $\mathbf{p}_{ij} = (\xi_{ij}, \eta_{ij})$ in each of the rectangles of the grid, see Fig. 17.1.

The double sum

$$S = \sum_{i=1}^{n} \sum_{j=1}^{m} f(\xi_{ij}, \eta_{ij})(x_i - x_{i-1})(y_j - y_{j-1})$$

is again called a *Riemann sum*. Since the volume of a cuboid with base $[x_{i-1}, x_i] \times [y_{j-1}, y_j]$ and height $f(\xi_{ij}, \eta_{ij})$ is

$$f(\xi_{ij}, \eta_{ij})(x_i - x_{i-1})(y_j - y_{j-1}),$$

the above Riemann sum is an approximation to the volume under the graph of f (Fig. 17.2).

Like in Sect. 11.1, the integral is now defined as a limit of Riemann sums. We consider a sequence $G_1, G_2, G_3, \ldots$ of grids whose mesh size $\Phi(G_N)$ tends to zero as $N \to \infty$ and the corresponding Riemann sums S_N.

Fig. 17.1 Partitioning the rectangle R

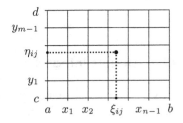

Fig. 17.2 Volume and approximation by cuboids

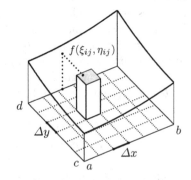

Definition 17.1 A bounded function $z = f(x, y)$ is called *Riemann integrable* on $R = [a, b] \times [c, d]$ if for arbitrary sequences of grids $(G_N)_{N \geq 1}$ with $\Phi(G_N) \to 0$ the corresponding Riemann sums $(S_N)_{N \geq 1}$ tend to the same limit $I(f)$, independently of the choice of intermediate points. This limit

$$I(f) = \iint_R f(x, y) \, d(x, y)$$

is called the *double integral* of f on R.

Experiment 17.2 Study the M-file `mat17_1.m` and experiment with different randomly chosen Riemann sums for the function $z = x^2 + y^2$ on the rectangle $[0, 1] \times [0, 1]$. What happens if you choose finer and finer grids?

As in the case of one variable, one may use the definition of the double integral for obtaining a numerical approximation to the integral. However, it is of little use for the analytic evaluation of integrals. In Sect. 11.1 the fundamental theorem of calculus has proven helpful, here the representation as *iterated integral* does. In this way the computation of double integrals is reduced to the integration of functions in one variable.

Proposition 17.3 (The double integral as iterated integral) *If a bounded function f and its partial functions $x \mapsto f(x, y)$, $y \mapsto f(x, y)$ are Riemann integrable on $R = [a, b] \times [c, d]$, then the mappings $x \mapsto \int_c^d f(x, y) \, dy$ and $y \mapsto \int_a^b f(x, y) \, dx$ are Riemann integrable as well and*

$$\iint_R f(x, y) \, d(x, y) = \int_a^b \left(\int_c^d f(x, y) \, dy \right) dx = \int_c^d \left(\int_a^b f(x, y) \, dx \right) dy.$$

Outline of the proof. If one chooses intermediate points in the Riemann sums of the special form $\mathbf{p}_{ij} = (\xi_i, \eta_j)$ with $\xi_i \in [x_{i-1}, x_i]$, $\eta_j \in [y_{j-1}, y_j]$, then

$$\iint_R f(x, y) \, d(x, y) \approx \sum_{i=1}^n \left(\sum_{j=1}^m f(\xi_i, \eta_j)(y_j - y_{j-1}) \right) (x_i - x_{i-1})$$

$$\approx \sum_{i=1}^n \left(\int_c^d f(\xi_i, y) \, dy \right) (x_i - x_{i-1}) \approx \int_a^b \left(\int_c^d f(x, y) \, dy \right) dx$$

and likewise for the second statement by changing the order. For a rigorous proof of this argument, we refer to the literature, for instance [4, Theorem 8.13 and Corollary].   □

Figure 17.3 serves to illustrate Proposition 17.3. The volume is approximated by summation of thin slices parallel to the axis instead of small cuboids. Proposition 17.3

Fig. 17.3 The double integral as iterated integral

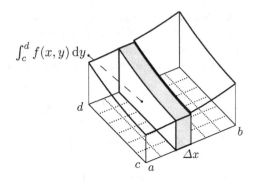

$$\int_c^d f(x, y)\, dy$$

states that the volume of the solid is obtained by integration over the area of the cross sections (perpendicular to the x- or y-axis). In this form Proposition 17.3 is called *Cavalieri's principle.*[1] In general integration theory one also speaks of *Fubini's theorem.*[2] Since in the case of integrability the order of integration does not matter, one often omits the brackets and writes

$$\iint_R f(x, y)\, d(x, y) = \iint_R f(x, y)\, dx\, dy = \int_a^b \int_c^d f(x, y)\, dy\, dx.$$

Example 17.4 Let $R = [0, 1] \times [0, 1]$. The volume of the body

$$B = \{(x, y, z) \in \mathbb{R}^3 : (x, y) \in R,\ 0 \le z \le x^2 + y^2\}$$

is obtained using Proposition 17.3 as follows, see also Fig. 17.4:

$$\iint_R (x^2 + y^2)\, d(x, y) = \int_0^1 \left(\int_0^1 (x^2 + y^2)\, dy \right) dx$$

$$= \int_0^1 \left(x^2 y + \frac{y^3}{3} \right) \Big|_{y=0}^{y=1} dx = \int_0^1 \left(x^2 + \frac{1}{3} \right) dx = \left(\frac{x^3}{3} + \frac{x}{3} \right) \Big|_{x=0}^{x=1} = \frac{2}{3}.$$

Fig. 17.4 The body B

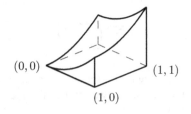

$(0,0)$

$(1,1)$

$(1,0)$

[1] B. Cavalieri, 1598–1647.
[2] G. Fubini, 1879–1943.

Fig. 17.5 Area as volume of the cylinder of height one

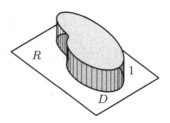

We now turn to the integration over more general (bounded) domains $D \subset \mathbb{R}^2$. The *indicator function* of the domain D is

$$\mathbb{1}_D(x, y) = \begin{cases} 1, & (x, y) \in D, \\ 0, & (x, y) \notin D. \end{cases}$$

We can enclose the bounded domain D in a rectangle R ($D \subset R$). If the Riemann integral of the indicator function of D exists, then it represents the volume of the cylinder of height one and base D and thus the area of D (Fig. 17.5). The result obviously does not depend on the size of the surrounding rectangle since the indicator function assumes the value zero outside the domain D.

Definition 17.5 Let D be a bounded domain and R an enclosing rectangle.

(a) If the indicator function of D is Riemann integrable then the domain D is called *measurable* and one sets

$$\iint_D d(x, y) = \iint_R \mathbb{1}_D(x, y) \, d(x, y).$$

(b) A subset $N \subset \mathbb{R}^2$ is called *set of measure zero*, if $\iint_N d(x, y) = 0$.

(c) For a bounded function $z = f(x, y)$, its integral over a measurable domain D is defined as

$$\iint_D f(x, y) \, d(x, y) = \iint_R f(x, y) \mathbb{1}_D(x, y) \, d(x, y),$$

if $f(x, y) \mathbb{1}_D(x, y)$ is Riemann integrable.

Sets of measure zero are, for example, single points, straight line segments or segments of differentiable curves in the plane. Item (c) of the definition states that the integral of a function f over a domain D is determined by continuing f to a larger rectangle R and assigning the value zero outside D.

Remark 17.6 (a) If D is a measurable domain, N a set of measure zero and f is integrable over the respective domains then

$$\iint_D f(x, y)\, d(x, y) = \iint_{D\setminus N} f(x, y)\, d(x, y).$$

(b) Let $D = D_1 \cup D_2$. If $D_1 \cap D_2$ is a set of measure zero then

$$\iint_D f(x, y)\, d(x, y) = \iint_{D_1} f(x, y)\, d(x, y) + \iint_{D_2} f(x, y)\, d(x, y).$$

The integral over the entire domain D is thus obtained as sum of the integrals over subdomains. The proof of this statement can easily be obtained by working with Riemann sums.

An important class of domains D on which integration is simple are the so-called *normal domains*.

Definition 17.7 (a) A subset $D \subset \mathbb{R}^2$ is called *normal domain of type I* if

$$D = \{(x, y) \in \mathbb{R}^2 \;;\; a \le x \le b, \; v(x) \le y \le w(x)\}$$

with certain continuously differentiable lower and upper bounding functions $x \mapsto v(x), x \mapsto w(x)$.

(b) A subset $D \subset \mathbb{R}^2$ is called *normal domain of type II*

$$D = \{(x, y) \in \mathbb{R}^2 \;;\; c \le y \le d, \; l(y) \le x \le r(y)\}$$

with certain continuously differentiable left and right bounding functions $x \mapsto l(x)$, $x \mapsto r(x)$.

Figure 17.6 shows examples of normal domains.

Proposition 17.8 (Integration over normal domains) *Let D be a normal domain and $f : D \to \mathbb{R}$ continuous. For normal domains of type I, one has*

$$\iint_D f(x, y)\, d(x, y) = \int_a^b \left(\int_{v(x)}^{w(x)} f(x, y)\, dy \right) dx$$

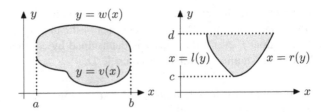

Fig. 17.6 Normal domains of type I and II

and for normal domains of type II

$$\iint_D f(x, y)\, d(x, y) = \int_c^d \left(\int_{l(y)}^{r(y)} f(x, y)\, dx \right) dy.$$

Proof The statements follow from Proposition 17.3. We recall that f is extended by zero outside of D. For details we refer to the remark at the end of [4, Chap. 8.3]. $\square$

Example 17.9 For the calculation of the volume of the body lying between the triangle $D = \{(x, y) \,;\, 0 \leq x \leq 1, 0 \leq y \leq 1 - x\}$ and the graph of $z = x^2 + y^2$, we interpret D as normal domain of type I with the boundaries $v(x) = 0, w(x) = 1 - x$. Consequently

$$\iint_D \left(x^2 + y^2\right) d(x, y) = \int_0^1 \left(\int_0^{1-x} \left(x^2 + y^2\right) dy \right) dx$$

$$= \int_0^1 \left(x^2 y + \frac{y^3}{3}\right) \bigg|_{y=0}^{y=1-x} dx = \int_0^1 \left(x^2(1 - x) + \frac{(1-x)^3}{3}\right) dx = \frac{1}{6},$$

as can be seen by multiplying out and integrating term by term.

17.2 Applications of the Double Integral

For modelling purposes it is useful to introduce a simplified notation for Riemann sums. In the case of equidistant partitions Z_x, Z_y where all subintervals have the same lengths, one writes

$$\Delta x = x_i - x_{i-1}, \quad \Delta y = y_j - y_{j-1}$$

and calls

$$\Delta A = \Delta x\, \Delta y$$

the *area element of the grid* G. If one then takes the right upper corner $\mathbf{p}_{ij} = (x_i, y_j)$ of the subrectangle $[x_{i-1}, x_i] \times [y_{j-1}, y_j]$ as an intermediate point, the corresponding Riemann sum reads

$$S = \sum_{i=1}^n \sum_{j=1}^m f(x_i, y_j)\, \Delta A = \sum_{i=1}^n \sum_{j=1}^m f(x_i, y_j)\, \Delta x\, \Delta y.$$

Application 17.10 (Mass as integral of the density)   A thin plane object D has density $\rho(x, y)$ [mass/unit area] at the point (x, y). If the density ρ is constant everywhere then its total mass is simply the product of density and area. In the case of variable density (e.g. due to a change of the material properties from point to point), we partition D in smaller rectangles with sides Δx, Δy. The mass contained in such a small rectangle around (x, y) is approximately equal to $\rho(x, y) \Delta x \Delta y$. The total mass is thus approximately equal to

$$\sum_{i=1}^{n} \sum_{j=1}^{m} \rho(x_i, y_j) \Delta x \Delta y.$$

However, this is just a Riemann sum for

$$M = \iint_D \rho(x, y) \, dx \, dy.$$

This consideration shows that the integral of the density function is a feasible model for representing the total mass of a two-dimensional object.

Application 17.11 (Centre of gravity)   We consider a two-dimensional flat object D as in Application 17.10. The two statical moments of a small rectangle close to (x, y) with respect to a point (x^*, y^*) are

$$(x - x^*)\rho(x, y) \Delta x \Delta y, \quad (y - y^*)\rho(x, y) \Delta x \Delta y,$$

see Fig. 17.7.

The relevance of the statical moments can be seen if one considers the object under the influence of gravity. Multiplied by the gravitational acceleration g one obtains the moments of force with respect to the axes through (x^*, y^*) in direction of the coordinates (force times lever arm). The *centre of gravity* of the two-dimensional object D is the point (x_S, y_S) with respect to which the total statical moments vanish:

$$\sum_{i=1}^{n} \sum_{j=1}^{m} (x_i - x_S)\,\rho(x_i, y_j)\,\Delta x \Delta y \approx 0, \quad \sum_{i=1}^{n} \sum_{j=1}^{m} (y_j - y_S)\,\rho(x_i, y_j)\,\Delta x \Delta y \approx 0.$$

In the limit, as the mesh size of the grid tends to zero, one obtains

$$\iint_D (x - x_S)\,\rho(x, y)\,dx\,dy = 0, \quad \iint_D (y - y_S)\,\rho(x, y)\,dx\,dy = 0$$

Fig. 17.7 The statical moments

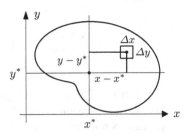

Fig. 17.8 Centre of gravity
of the quarter circle

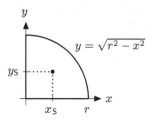

as defining equations for the centre of gravity; i.e.,

$$x_S = \frac{1}{M} \iint_D x\,\rho(x, y)\,dx\,dy, \quad y_S = \frac{1}{M} \iint_D y\,\rho(x, y)\,dx\,dy,$$

where M denotes the total mass as in Application 17.10.

For the special case of a constant density $\rho(x, y) \equiv 1$ one obtains the *geometric centre of gravity* of the domain D.

Example 17.12 (Geometric centre of gravity of a quarter circle) Let D be the quarter circle of radius r about $(0, 0)$ in the first quadrant; i.e., $D = \{(x, y) \; ; \; 0 \leq x \leq r, \; 0 \leq y \leq \sqrt{r^2 - x^2}\}$ (Fig. 17.8). With density $\rho(x, y) \equiv 1$ one obtains the area M as $r^2\pi/4$. The first statical moment is

$$\iint_D x\,dx\,dy = \int_0^r \left(\int_0^{\sqrt{r^2-x^2}} x\,dy \right) dx = \int_0^r \left(xy \Big|_{y=0}^{y=\sqrt{r^2-x^2}} \right) dx$$

$$= \int_0^r x\sqrt{r^2 - x^2}\,dx = -\frac{1}{3}\left(r^2 - x^2\right)^{3/2} \Big|_{x=0}^{x=r} = \frac{1}{3}r^3.$$

The x-coordinate of the centre of gravity is thus given by $x_S = \frac{4}{r^2\pi} \cdot \frac{1}{3}r^3 = \frac{4r}{3\pi}$. For reasons of symmetry, one has $y_S = x_S$.

17.3 The Transformation Formula

Similar to the substitution rule for one-dimensional integrals (Sect. 10.2), the transformation formula for double integrals makes it possible to change coordinates on the domain D of integration. For the purpose of this section it is convenient to assume that D is an open subset of $\mathbb{R}^2$ (see Definition 9.1).

Definition 17.13 A bijective, differentiable mapping $\mathbf{F} : D \to B = \mathbf{F}(D)$ between two open subsets $D, B \subset \mathbb{R}^2$ is called a *diffeomorphism* if the inverse mapping $\mathbf{F}^{-1}$ is also differentiable.

Fig. 17.9 Transformation of
a planar domain

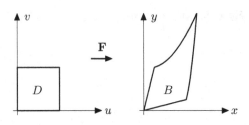

We use the following notation for the variables:

$$\mathbf{F} : D \to B : \begin{bmatrix} u \\ v \end{bmatrix} \mapsto \begin{bmatrix} x \\ y \end{bmatrix} = \begin{bmatrix} x(u, v) \\ y(u, v) \end{bmatrix}.$$

Figure 17.9 shows the image B of the domain $D = (0, 1) \times (0, 1)$ under the transformation

$$\mathbf{F} : \begin{bmatrix} u \\ v \end{bmatrix} \mapsto \begin{bmatrix} x \\ y \end{bmatrix} = \begin{bmatrix} u + v/4 \\ u/4 + v + u^2 v^2 \end{bmatrix}.$$

The aim is to transform the integral of a real-valued function f over the domain B to one over D.

For this purpose we lay a grid G over the domain D in the (u, v)-plane and select a rectangle, for instance with the left lower corner (u, v) and sides spanned by the vectors

$$\begin{bmatrix} \Delta u \\ 0 \end{bmatrix}, \begin{bmatrix} 0 \\ \Delta v \end{bmatrix}.$$

The image of this rectangle under the transformation $\mathbf{F}$ will in general have a curvilinear boundary. In a first approximation we replace it by a parallelogram. In linear approximation (see Sect. 15.4) we have the following:

$$\mathbf{F}(u + \Delta u, v) \approx \mathbf{F}(u, v) + \mathbf{F}'(u, v) \begin{bmatrix} \Delta u \\ 0 \end{bmatrix},$$

$$\mathbf{F}(u, v + \Delta v) \approx \mathbf{F}(u, v) + \mathbf{F}'(u, v) \begin{bmatrix} 0 \\ \Delta v \end{bmatrix}.$$

The approximating parallelogram is thus spanned by the vectors

$$\begin{bmatrix} \dfrac{\partial x}{\partial u}(u, v) \\ \dfrac{\partial y}{\partial u}(u, v) \end{bmatrix} \Delta u, \quad \begin{bmatrix} \dfrac{\partial x}{\partial v}(u, v) \\ \dfrac{\partial y}{\partial v}(u, v) \end{bmatrix} \Delta v$$

and has the area (see Appendix A.5)

$$\left| \det \begin{bmatrix} \dfrac{\partial x}{\partial u}(u, v) & \dfrac{\partial x}{\partial v}(u, v) \\ \dfrac{\partial y}{\partial u}(u, v) & \dfrac{\partial y}{\partial v}(u, v) \end{bmatrix} \Delta u \, \Delta v \right| = \left| \det \mathbf{F}'(u, v) \right| \Delta u \, \Delta v.$$

Fig. 17.10 Transformation
of an area element

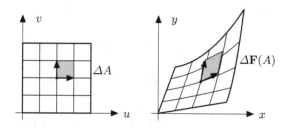

In short, the area element $\Delta A = \Delta u \, \Delta v$ is changed by the transformation $\mathbf{F}$ to the
area element $\Delta \mathbf{F}(A) = \left| \det \mathbf{F}'(u, v) \right| \Delta u \, \Delta v$ (see Fig. 17.10).

Proposition 17.14 (Transformation formula for double integrals) *Let D, B be
open, bounded subsets of $\mathbb{R}^2$, $\mathbf{F} : D \to B$ a diffeomorphism and $f : B \to \mathbb{R}$ a
bounded mapping. Then*

$$\iint_B f(x, y) \, dx \, dy = \iint_D f\big(\mathbf{F}(u, v)\big) \left| \det \mathbf{F}'(u, v) \right| du \, dv,$$

as long as the functions f and $f(\mathbf{F}) \left| \det \mathbf{F}' \right|$ are Riemann integrable.

Outline of the proof. We use Riemann sums on the transformed grid and obtain

$$\iint_B f(x, y) \, dx \, dy \approx \sum_{i=1}^{n} \sum_{j=1}^{m} f(x_i, y_j) \, \Delta \mathbf{F}(A)$$

$$\approx \sum_{i=1}^{n} \sum_{j=1}^{m} f\big(x(u_i, v_j), y(u_i, v_j)\big) \left| \det \mathbf{F}'(u_i, v_j) \right| \Delta u \, \Delta v$$

$$\approx \iint_D f\big(x(u, v), y(u, v)\big) \left| \det \mathbf{F}'(u, v) \right| du \, dv.$$

A rigorous proof is tedious and requires a careful study of the boundary of the domain
D and the behaviour of the transformation $\mathbf{F}$ near the boundary (see for instance [3,
Chap. 19, Theorem 4.7]). $\square$

Example 17.15 The area of the domain B from Fig. 17.9 can be calculated using the
transformation formula with $f(x, y) = 1$ as follows. We have

$$\mathbf{F}'(u, v) = \begin{bmatrix} 1 & 1/4 \\ 1/4 + 2uv^2 & 1 + 2u^2 v \end{bmatrix},$$

$$\left| \det \mathbf{F}'(u, v) \right| = \left| \frac{15}{16} + 2u^2 v - \frac{1}{2} u v^2 \right|$$

and thus

$$\iint_B dx\, dy = \iint_D \left|\det \mathbf{F}'(u, v)\right| du\, dv$$

$$= \int_0^1 \left(\int_0^1 \left(\frac{15}{16} + 2u^2 v - \frac{1}{2} u v^2 \right) dv \right) du$$

$$= \int_0^1 \left(\frac{15}{16} + u^2 - \frac{1}{6} u \right) du = \frac{15}{16} + \frac{1}{3} - \frac{1}{12} = \frac{19}{16}.$$

Example 17.16 (Volume of a hemisphere in polar coordinates) We represent a hemisphere of radius R by the three-dimensional domain

$$\{(x, y, z)\ ;\ 0 \le x^2 + y^2 \le R^2, 0 \le z \le \sqrt{R^2 - x^2 - y^2}\}.$$

Its volume is obtained by integration of the function $f(x, y) = \sqrt{R^2 - x^2 - y^2}$ over the base $B = \{(x, y)\ ;\ 0 \le x^2 + y^2 \le R^2\}$. In polar coordinates

$$\mathbf{F} : \mathbb{R}^2 \to \mathbb{R}^2 : \begin{bmatrix} r \\ \varphi \end{bmatrix} \mapsto \begin{bmatrix} x \\ y \end{bmatrix} = \begin{bmatrix} r \cos \varphi \\ r \sin \varphi \end{bmatrix}$$

the area B can be represented as the image $\mathbf{F}(D)$ of the rectangle $D = [0, R] \times [0, 2\pi]$. However, in order to fulfil the assumptions of Proposition 17.14 we have to switch to open domains on which $\mathbf{F}$ is a diffeomorphism. We can obtain this, for instance, by removing the boundary and the half ray $\{(x, y)\ ;\ 0 \le x \le R,\ y = 0\}$ of the circle B and the boundary of the rectangle D. On the smaller domains D', B' obtained in this way, $\mathbf{F}$ is a diffeomorphism. However, since B differs from B' and D differs from D' by sets of measure zero, the value of the integral is not changed if one replaces B by B' and D by D', see Remark 17.6. We have

$$\mathbf{F}'(r, \varphi) = \begin{bmatrix} \cos \varphi & -r \sin \varphi \\ \sin \varphi & r \cos \varphi \end{bmatrix}, \quad \left|\det \mathbf{F}'(r, \varphi)\right| = r.$$

Substituting $x = r \cos \varphi$, $y = r \sin \varphi$ results in $x^2 + y^2 = r^2$ and we obtain the volume from the transformation formula as

$$\iint_B \sqrt{R^2 - x^2 - y^2}\, dx\, dy = \int_0^R \int_0^{2\pi} \sqrt{R^2 - r^2}\, r\, d\varphi\, dr$$

$$= \int_0^R 2\pi r \sqrt{R^2 - r^2}\, dr$$

$$= -\frac{2\pi}{3} \left(R^2 - r^2 \right)^{3/2} \Big|_{r=0}^{r=R} = \frac{2\pi}{3} R^3,$$

which coincides with the known result from elementary geometry.

17.4 Exercises

1. Compute the volume of the parabolic dome $z = 2 - x^2 - y^2$ above the quadratic domain $D : -1 \leq x \leq 1, -1 \leq y \leq 1$.

2. (From statics) Compute the axial moment of inertia $\iint_D y^2 \, dx \, dy$ of a rectangular cross section $D : 0 \leq x \leq b, -h/2 \leq y \leq h/2$, where $b > 0, h > 0$.

3. Compute the volume of the body bounded by the plane $z = x + y$ above the domain $D : 0 \leq x \leq 1, 0 \leq y \leq \sqrt{1 - x^2}$.

4. Compute the volume of the body bounded by the plane $z = 6 - x - y$ above the domain D, which is bounded by the y-axis and the straight lines $x + y = 6$, $x + 3y = 6 \, (x \geq 0, y \geq 0)$.

5. Compute the geometric centre of gravity of the domain $D : 0 \leq x \leq 1, 0 \leq y \leq 1 - x^2$.

6. Compute the area and the geometric centre of gravity of the semi-ellipse

$$\frac{x^2}{a^2} + \frac{y^2}{b^2} \leq 1, \quad y \geq 0.$$

Hint. Introduce elliptic coordinates $x = ar \cos \varphi, y = br \sin \varphi, 0 \leq r \leq 1, 0 \leq \varphi \leq \pi$, compute the Jacobian and use the transformation formula.

7. (From statics) Compute the axial moment of inertia of a ring with inner radius R_1 and outer radius R_2 with respect to the central axis, i.e. the integral $\iint_D (x^2 + y^2) \, dx \, dy$ over the domain $D : R_1 \leq \sqrt{x^2 + y^2} \leq R_2$.

8. Modify the M-file `mat17_1.m` so that it can evaluate Riemann sums over equidistant partitions with $\Delta x \neq \Delta y$.

9. Let the domain D be bounded by the curves

$$y = x \quad \text{and} \quad y = x^2, \quad 0 \leq x \leq 1.$$

   (a) Sketch D.
   (b) Compute the area of D by means of the double integral $F = \iint_D d(x, y)$.
   (c) Compute the statical moments $\iint_D x \, d(x, y)$ und $\iint_D y \, d(x, y)$.

10. Compute the statical moment $\iint_D y \, d(x, y)$ of the half-disk

$$D = \{(x, y) \in \mathbb{R}^2; \; -1 \leq x \leq 1, \; 0 \leq y \leq \sqrt{1 - x^2}\}$$

   (a) as a double integral, writing D as a normal domain of type I;
   (b) by transformation to polar coordinates.

11. The following integral is written in terms of a normal domain of type II:

$$\int\limits_0^1 \int\limits_y^{y^2+1} x^2 y \, dx \, dy.$$

(a) Compute the integral.
(b) Sketch the domain and represent it as a normal domain of type I.
(c) Interchange the order of integration and recompute the integral.
Hint. In (c) two summands are needed.

Linear Regression

18

Linear regression is one of the most important methods of data analysis. It serves the determination of model parameters, model fitting, assessing the importance of influencing factors, and prediction, in all areas of human, natural and economic sciences. Computer scientists who work closely with people from these areas will definitely come across regression models.

The aim of this chapter is a first introduction into the subject. We deduce the coefficients of the regression models using the method of least squares to minimise the errors. We will only employ methods of descriptive data analysis. We do not touch upon the more advanced probabilistic approaches which are topics of statistics. For that, as well as for nonlinear regression, we refer to the specialised literature.

We start with simple (or univariate) linear regression—a model with a single input and a single output quantity—and explain the basic ideas of analysis of variance for model evaluation. Then we turn to multiple (or multivariate) linear regression with several input quantities. The chapter closes with a descriptive approach to determine the influence of the individual coefficients.

18.1 Simple Linear Regression

A first glance at the basic idea of linear regression was already given in Sect. 8.3. In extension to this, we will now allow more general models, in particular regression lines with nonzero intercept.

Consider pairs of data $(x_1, y_1), \ldots, (x_n, y_n)$, obtained as observations or measurements. Geometrically they form a scatter plot in the plane. The values x_i and y_i may appear repeatedly in this list of data. In particular, for a given x_i there can be data points with different values $y_{i1}, \ldots, y_{ip}$. The general task of *linear regression*

© Springer Nature Switzerland AG 2018
M. Oberguggenberger and A. Ostermann, *Analysis for Computer Scientists*,
Undergraduate Topics in Computer Science,
https://doi.org/10.1007/978-3-319-91155-7_18

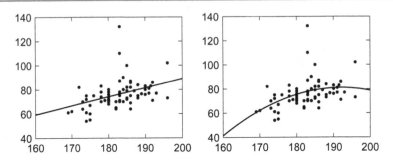

Fig. 18.1 Scatter plot height/weight, line of best fit, best parabola

is to fit the graph of a function

$$y = \beta_0\varphi_0(x) + \beta_1\varphi_1(x) + \cdots + \beta_m\varphi_m(x)$$

to the n data points $(x_1, y_1), \ldots, (x_n, y_n)$. Here the shape functions $\varphi_j(x)$ are given and the (unknown) coefficients β_j are to be determined such that the sum of squares of the errors is minimal (*method of least squares*):

$$\sum_{i=1}^{n}\left(y_i - \beta_0\varphi_0(x_i) - \beta_1\varphi_1(x_i) - \cdots - \beta_m\varphi_m(x_i)\right)^2 \to \min$$

The regression is called *linear* because the function y depends linearly on the unknown coefficients β_j. The choice of the shape functions ensues either from a possible theoretical model or empirically, where different possibilities are subjected to statistical tests. The choice is made, for example, according to the proportion of data variability which is explained by the regression—more about that in Sect. 18.4. The standard question of (simple or univariate) linear regression is to fit a *linear model*

$$y = \beta_0 + \beta_1 x$$

to the data, i.e., to find the *line of best fit* or *regression line* through the scatter plot.

Example 18.1 A sample of $n = 70$ computer science students at the University of Innsbruck in 2002 yielded the data depicted in Fig. 18.1. Here x denotes the height [cm] and y the weight [kg] of the students. The left picture in Fig. 18.1 shows the regression line $y = \beta_0 + \beta_1 x$, the right one a fitted quadratic parabola of the form

$$y = \beta_0 + \beta_1 x + \beta_2 x^2.$$

Note the difference to Fig. 8.8 where the *line of best fit through the origin* was used; i.e., the intercept β_0 was set to zero in the linear model.

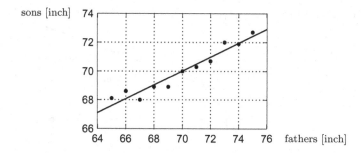

Fig. 18.2 Scatter plot height of fathers/height of the sons, regression line

A variant of the standard problem is obtained by considering the linear model

$$\eta = \beta_0 + \beta_1 \xi$$

for the transformed variables

$$\xi = \varphi(x), \ \eta = \psi(y).$$

Formally this problem is identical to the standard problem of linear regression, however, with transformed data

$$(\xi_i, \eta_i) = \big(\varphi(x_i), \psi(y_i)\big).$$

A typical example is given by the *loglinear regression* with $\xi = \log x$, $\eta = \log y$

$$\log y = \beta_0 + \beta_1 \log x,$$

which in the original variables amounts to the *exponential model*

$$y = e^{\beta_0} x^{\beta_1}.$$

If the variable x itself has several components which enter linearly in the model, then one speaks of *multiple linear regression*. We will deal with it in Sect. 18.3.

The notion of *regression* was introduced by Galton[1] who observed, while investigating the height of sons/fathers, a tendency of *regressing* to the average size. The data taken from [15] clearly show this effect, see Fig. 18.2. The method of least squares goes back to Gauss.

After these introductory remarks about the general concept of linear regression, we turn to *simple linear regression*. We start with setting up the model. The postulated relationship between x and y is linear

$$y = \beta_0 + \beta_1 x$$

---

[1]F. Galton, 1822–1911.

Fig. 18.3 Linear model and
error ε_i

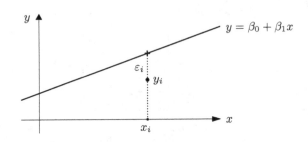

with unknown coefficients β_0 and β_1. In general, the given data will not exactly lie
on a straight line but deviate by ε_i, i.e.,

$$y_i = \beta_0 + \beta_1 x_i + \varepsilon_i,$$

as represented in Fig. 18.3.

From the given data we want to obtain estimated values $\widehat{\beta}_0, \widehat{\beta}_1$ for β_0, β_1. This is
achieved through minimising the sum of squares of the errors

$$L(\beta_0, \beta_1) = \sum_{i=1}^{n} \varepsilon_i^2 = \sum_{i=1}^{n}(y_i - \beta_0 - \beta_1 x_i)^2,$$

so that $\widehat{\beta}_0, \widehat{\beta}_1$ solve the minimisation problem

$$L(\widehat{\beta}_0, \widehat{\beta}_1) = \min\left(L(\beta_0, \beta_1) \; ; \; \beta_0 \in \mathbb{R}, \beta_1 \in \mathbb{R} \right).$$

We obtain $\widehat{\beta}_0$ and $\widehat{\beta}_1$ by setting the partial derivatives of L with respect to β_0 and β_1
to zero:

$$\frac{\partial L}{\partial \beta_0}(\widehat{\beta}_0, \widehat{\beta}_1) = -2 \sum_{i=1}^{n}(y_i - \widehat{\beta}_0 - \widehat{\beta}_1 x_i) = 0,$$

$$\frac{\partial L}{\partial \beta_1}(\widehat{\beta}_0, \widehat{\beta}_1) = -2 \sum_{i=1}^{n} x_i(y_i - \widehat{\beta}_0 - \widehat{\beta}_1 x_i) = 0.$$

This leads to a linear system of equations for $\widehat{\beta}_0, \widehat{\beta}_1$, the so-called *normal equations*

$$n\,\widehat{\beta}_0 + \left(\sum x_i\right)\widehat{\beta}_1 = \sum y_i,$$
$$\left(\sum x_i\right)\widehat{\beta}_0 + \left(\sum x_i^2\right)\widehat{\beta}_1 = \sum x_i y_i.$$

Proposition 18.2 *Assume that at least two x-values in the data set (x_i, y_i), $i = 1, \ldots, n$ are different. Then the normal equations have a unique solution*

$$\widehat{\beta}_0 = \left(\tfrac{1}{n}\sum y_i\right) - \left(\tfrac{1}{n}\sum x_i\right)\widehat{\beta}_1, \qquad \widehat{\beta}_1 = \frac{\sum x_i y_i - \frac{1}{n}\sum x_i \sum y_i}{\sum x_i^2 - \frac{1}{n}\left(\sum x_i\right)^2}$$

which minimises the sum of squares $L(\beta_0, \beta_1)$ of the errors.

Fig. 18.4 Linear model, prediction, residual

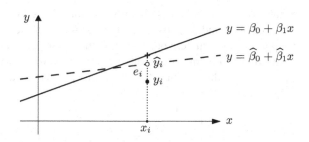

Proof With the notations $\mathbf{x} = (x_1, \ldots, x_n)$ and $\mathbf{1} = (1, \ldots, 1)$ the determinant of the normal equations is $n \sum x_i^2 - (\sum x_i)^2 = \|\mathbf{x}\|^2 \|\mathbf{1}\|^2 - \langle \mathbf{x}, \mathbf{1} \rangle^2$. For vectors of length $n = 2$ and $n = 3$ we know that $\langle \mathbf{x}, \mathbf{1} \rangle = \|\mathbf{x}\| \|\mathbf{1}\| \cdot \cos \angle(\mathbf{x}, \mathbf{1})$, see Appendix A.4, and thus $\|\mathbf{x}\| \|\mathbf{1}\| \geq |\langle \mathbf{x}, \mathbf{1} \rangle|$. This relation, however, is valid in any dimension n (see for instance [2, Chap. VI, Theorem 1.1]), and equality can only occur if $\mathbf{x}$ is parallel to $\mathbf{1}$, so all components x_i are equal. As this possibility was excluded, the determinant of the normal equations is greater than zero and the solution formula is obtained by a simple calculation.

In order to show that this solution minimises $L(\beta_0, \beta_1)$, we compute the Hessian matrix

$$H_L = \begin{bmatrix} \frac{\partial^2 L}{\partial \beta_0^2} & \frac{\partial^2 L}{\partial \beta_0 \partial \beta_1} \\ \frac{\partial^2 L}{\partial \beta_1 \partial \beta_0} & \frac{\partial^2 L}{\partial \beta_1^2} \end{bmatrix} = 2 \begin{bmatrix} n & \sum x_i \\ \sum x_i & \sum x_i^2 \end{bmatrix} = 2 \begin{bmatrix} \|\mathbf{1}\|^2 & \langle \mathbf{x}, \mathbf{1} \rangle \\ \langle \mathbf{x}, \mathbf{1} \rangle & \|\mathbf{x}\|^2 \end{bmatrix}.$$

The entry $\partial^2 L / \partial \beta_0^2 = 2n$ and $\det H_L = 4 \left(\|\mathbf{x}\|^2 \|\mathbf{1}\|^2 - \langle \mathbf{x}, \mathbf{1} \rangle^2 \right)$ are both positive. According to Proposition 15.28, L has an isolated local minimum at the point $(\widehat{\beta}_0, \widehat{\beta}_1)$. Due to the uniqueness of the solution, this is the only minimum of L. $\square$

The assumption that there are at least two different x_i-values in the data set is not a restriction since otherwise the regression problem is not meaningful. The result of the regression is the *predicted regression line*

$$y = \widehat{\beta}_0 + \widehat{\beta}_1 x.$$

The *values predicted by the model* are then

$$\widehat{y}_i = \widehat{\beta}_0 + \widehat{\beta}_1 x_i, \quad i = 1, \ldots, n.$$

Their deviations from the data values y_i are called *residuals*

$$e_i = y_i - \widehat{y}_i = y_i - \widehat{\beta}_0 - \widehat{\beta}_1 x_i, \quad i = 1, \ldots, n.$$

The meaning of these quantities can be seen in Fig. 18.4.

With the above specifications, the *deterministic regression model* is completed. In the *statistical regression model* the errors ε_i are interpreted as random variables with mean zero. Under further probabilistic assumptions the model is made accessible to statistical tests and diagnostic procedures. As mentioned in the introduction, we will not pursue this path here but remain in the framework of descriptive data analysis.

In order to obtain a more lucid representation, we will reformulate the normal equations. For this we introduce the following vectors and matrices:

$$\mathbf{y} = \begin{bmatrix} y_1 \\ y_2 \\ \vdots \\ y_n \end{bmatrix}, \quad \mathbf{X} = \begin{bmatrix} 1 & x_1 \\ 1 & x_2 \\ \vdots & \vdots \\ 1 & x_n \end{bmatrix}, \quad \boldsymbol{\beta} = \begin{bmatrix} \beta_0 \\ \beta_1 \end{bmatrix}, \quad \boldsymbol{\varepsilon} = \begin{bmatrix} \varepsilon_1 \\ \varepsilon_2 \\ \vdots \\ \varepsilon_n \end{bmatrix}.$$

By this, the relations

$$y_i = \beta_0 + \beta_1 x_i + \varepsilon_i, \quad i = 1, \ldots, n,$$

can be written simply as

$$\mathbf{y} = \mathbf{X}\boldsymbol{\beta} + \boldsymbol{\varepsilon}.$$

Further

$$\mathbf{X}^\mathsf{T}\mathbf{X} = \begin{bmatrix} 1 & 1 & \ldots & 1 \\ x_1 & x_2 & \ldots & x_n \end{bmatrix} \begin{bmatrix} 1 & x_1 \\ 1 & x_2 \\ \vdots & \vdots \\ 1 & x_n \end{bmatrix} = \begin{bmatrix} n & \sum x_i \\ \sum x_i & \sum x_i^2 \end{bmatrix},$$

$$\mathbf{X}^\mathsf{T}\mathbf{y} = \begin{bmatrix} 1 & 1 & \ldots & 1 \\ x_1 & x_2 & \ldots & x_n \end{bmatrix} \begin{bmatrix} y_i \\ y_2 \\ \vdots \\ y_n \end{bmatrix} = \begin{bmatrix} \sum y_i \\ \sum x_i y_i \end{bmatrix},$$

so that the normal equations take the form

$$\mathbf{X}^\mathsf{T}\mathbf{X}\widehat{\boldsymbol{\beta}} = \mathbf{X}^\mathsf{T}\mathbf{y}$$

with solution

$$\widehat{\boldsymbol{\beta}} = (\mathbf{X}^\mathsf{T}\mathbf{X})^{-1}\mathbf{X}^\mathsf{T}\mathbf{y}.$$

The predicted values and residuals are

$$\widehat{\mathbf{y}} = \mathbf{X}\widehat{\boldsymbol{\beta}}, \quad \mathbf{e} = \mathbf{y} - \widehat{\mathbf{y}}.$$

Example 18.3 (Continuation of Example 18.1) The data for x = height and y = weight can be found in the M-file mat08_3.m; the matrix $\mathbf{X}$ is generated in MATLAB by

```
X = [ones(size(x)), x];
```

the regression coefficients are obtained by

```
beta = inv(X' * X) * X' * y;
```

The command $\texttt{beta = X\backslash y}$ permits a more stable calculation in MATLAB. In our case the result is

$$\widehat{\beta}_0 = -85.02,$$
$$\widehat{\beta}_1 = 0.8787.$$

This gives the regression line depicted in Fig. 18.1.

18.2 Rudiments of the Analysis of Variance

First indications for the quality of fit of the linear model can be obtained from the *analysis of variance* (ANOVA), which also forms the basis for more advanced statistical test procedures.

The arithmetic mean of the y-values $y_1, \ldots, y_n$ is

$$\bar{y} = \frac{1}{n} \sum_{i=1}^{n} y_i.$$

The deviation of the measured value y_i from the mean value $\bar{y}$ is $y_i - \bar{y}$. The *total sum of squares* or *total variability* of the data is

$$S_{yy} = \sum_{i=1}^{n} (y_i - \bar{y})^2.$$

The total variability is split into two components in the following way:

$$\sum_{i=1}^{n} (y_i - \bar{y})^2 = \sum_{i=1}^{n} (\widehat{y}_i - \bar{y})^2 + \sum_{i=1}^{n} (y_i - \widehat{y}_i)^2.$$

The validity of this relationship will be proven in Proposition 18.4 below. It is interpreted as follows: $\widehat{y}_i - \bar{y}$ is the deviation of the predicted value from the mean value, and

$$SS_R = \sum_{i=1}^{n} (\widehat{y}_i - \bar{y})^2$$

the *regression sum of squares*. This is interpreted as the part of the data variability accounted for by the model. On the other hand $e_i = y_i - \widehat{y}_i$ are the residuals, and

$$SS_E = \sum_{i=1}^{n} (y_i - \widehat{y}_i)^2$$

is the *error sum of squares* which is interpreted as the part of the variability that remains unexplained by the linear model. These notions are best explained by considering the two extremal cases.

(a) The data values y_i themselves already lie on a straight line. Then all $\widehat{y}_i = y_i$ and thus $S_{yy} = SS_R$, $SS_E = 0$, and the regression model describes the data record exactly.

(b) The data values are in no linear relation. Then the line of best fit is the horizontal line through the mean value (see Exercise 13 of Chap. 8), so $\widehat{y}_i = \bar{y}$ for all i and hence $S_{yy} = SS_E$, $SS_R = 0$. This means that the regression model does not offer any indication for a linear relation between the values.

The basis of these considerations is the validity of the following formula.

Proposition 18.4 (Partitioning of total variability) $S_{yy} = SS_R + SS_E$.

Proof In the following we use matrix and vector notation. In particular, we employ the formulas

$$\mathbf{a}^\mathsf{T}\mathbf{b} = \mathbf{b}^\mathsf{T}\mathbf{a} = \sum a_i b_i, \quad \mathbf{1}^\mathsf{T}\mathbf{a} = \mathbf{a}^\mathsf{T}\mathbf{1} = \sum a_i = n\bar{a}, \quad \mathbf{a}^\mathsf{T}\mathbf{a} = \sum a_i^2$$

for vectors $\mathbf{a}, \mathbf{b}$, and the matrix identity $(\mathbf{AB})^\mathsf{T} = \mathbf{B}^\mathsf{T}\mathbf{A}^\mathsf{T}$. We have

$$S_{yy} = (\mathbf{y} - \bar{y}\mathbf{1})^\mathsf{T}(\mathbf{y} - \bar{y}\mathbf{1}) = \mathbf{y}^\mathsf{T}\mathbf{y} - \bar{y}(\mathbf{1}^\mathsf{T}\mathbf{y}) - (\mathbf{y}^\mathsf{T}\mathbf{1})\bar{y} + n\bar{y}^2$$
$$= \mathbf{y}^\mathsf{T}\mathbf{y} - n\bar{y}^2 - n\bar{y}^2 + n\bar{y}^2 = \mathbf{y}^\mathsf{T}\mathbf{y} - n\bar{y}^2,$$
$$SS_E = \mathbf{e}^\mathsf{T}\mathbf{e} = (\mathbf{y} - \widehat{\mathbf{y}})^\mathsf{T}(\mathbf{y} - \widehat{\mathbf{y}}) = (\mathbf{y} - \mathbf{X}\widehat{\boldsymbol{\beta}})^\mathsf{T}(\mathbf{y} - \mathbf{X}\widehat{\boldsymbol{\beta}})$$
$$= \mathbf{y}^\mathsf{T}\mathbf{y} - \widehat{\boldsymbol{\beta}}^\mathsf{T}\mathbf{X}^\mathsf{T}\mathbf{y} - \mathbf{y}^\mathsf{T}\mathbf{X}\widehat{\boldsymbol{\beta}} + \widehat{\boldsymbol{\beta}}^\mathsf{T}\mathbf{X}^\mathsf{T}\mathbf{X}\widehat{\boldsymbol{\beta}} = \mathbf{y}^\mathsf{T}\mathbf{y} - \widehat{\boldsymbol{\beta}}^\mathsf{T}\mathbf{X}^\mathsf{T}\mathbf{y}.$$

For the last equality we have used the normal equations $\mathbf{X}^\mathsf{T}\mathbf{X}\widehat{\boldsymbol{\beta}} = \mathbf{X}^\mathsf{T}\mathbf{y}$ and the transposition formula $\widehat{\boldsymbol{\beta}}^\mathsf{T}\mathbf{X}^\mathsf{T}\mathbf{y} = (\mathbf{y}^\mathsf{T}\mathbf{X}\widehat{\boldsymbol{\beta}})^\mathsf{T} = \mathbf{y}^\mathsf{T}\mathbf{X}\widehat{\boldsymbol{\beta}}$. The relation $\widehat{\mathbf{y}} = \mathbf{X}\widehat{\boldsymbol{\beta}}$ implies in particular $\mathbf{X}^\mathsf{T}\widehat{\mathbf{y}} = \mathbf{X}^\mathsf{T}\mathbf{y}$. Since the first line of $\mathbf{X}^\mathsf{T}$ consists of ones only, it follows that $\mathbf{1}^\mathsf{T}\widehat{\mathbf{y}} = \mathbf{1}^\mathsf{T}\mathbf{y}$ and thus

$$SS_R = (\widehat{\mathbf{y}} - \bar{y}\mathbf{1})^\mathsf{T}(\widehat{\mathbf{y}} - \bar{y}\mathbf{1}) = \widehat{\mathbf{y}}^\mathsf{T}\widehat{\mathbf{y}} - \bar{y}(\mathbf{1}^\mathsf{T}\widehat{\mathbf{y}}) - (\widehat{\mathbf{y}}^\mathsf{T}\mathbf{1})\bar{y} + n\bar{y}^2$$
$$= \widehat{\mathbf{y}}^\mathsf{T}\widehat{\mathbf{y}} - n\bar{y}^2 - n\bar{y}^2 + n\bar{y}^2 = \widehat{\boldsymbol{\beta}}^\mathsf{T}(\mathbf{X}^\mathsf{T}\mathbf{X}\widehat{\boldsymbol{\beta}}) - n\bar{y}^2 = \widehat{\boldsymbol{\beta}}^\mathsf{T}\mathbf{X}^\mathsf{T}\mathbf{y} - n\bar{y}^2.$$

Summation of the obtained expressions for SS_E and SS_R results in the sought after formula.   □

The partitioning of total variability

$$S_{yy} = SS_R + SS_E$$

and its above interpretation suggests using the quantity

$$R^2 = \frac{SS_R}{S_{yy}}$$

for the assessment of the goodness of fit. The quantity R^2 is called *coefficient of determination* and measures the fraction of variability explained by the regression. In the limiting case of an exact fit, where the regression line passes through all data points, we have $SS_E = 0$ and thus $R^2 = 1$. A small value of R^2 indicates that the linear model does not fit the data.

Remark 18.5 An essential point in the proof of Proposition 18.4 was the property of $X^\top$ that its first line was composed of ones only. This is a consequence of the fact that β_0 was a model parameter. In the regression where a straight line through the origin is used (see Sect. 8.3) this is not the case. For a regression which does not have β_0 as a parameter the variance partition is not valid and the coefficient of determination is meaningless.

Example 18.6 We continue the investigation of the relation between height and weight from Example 18.1. Using the MATLAB program `mat18_1.m` and entering the data from `mat08_3.m` results in

$$S_{yy} = 9584.9, \quad SS_E = 8094.4, \quad SS_R = 1490.5$$

and

$$R^2 = 0.1555, \quad R = 0.3943.$$

The low value of R^2 is a clear indication that height and weight are not in a linear relation.

Example 18.7 In Sect. 9.1 the fractal dimension $d = d(A)$ of a bounded subset A of $\mathbb{R}^2$ was defined by the limit

$$d = d(A) = -\lim_{\varepsilon \to 0^+} \log N(A, \varepsilon) / \log \varepsilon,$$

where $N(A, \varepsilon)$ denoted the smallest number of squares of side length ε needed to cover A. For the experimental determination of the dimension of a fractal set A, one rasters the plane with different mesh sizes ε and determines the number $N = N(A, \varepsilon)$ of boxes that have a non-empty intersection with the fractal. As explained in Sect. 9.1, one uses the approximation

$$N(A, \varepsilon) \approx C \cdot \varepsilon^{-d}.$$

Applying logarithms results in

$$\log N(A, \varepsilon) \approx \log C + d \log \frac{1}{\varepsilon},$$

which is a linear model

$$y \approx \beta_0 + \beta_1 x$$

for the quantities $x = \log 1/\varepsilon$, $y = \log N(A, \varepsilon)$. The regression coefficient $\widehat{\beta_1}$ can be used as an estimate for the fractal dimension d.

In Exercise 1 of Sect. 9.6 this procedure was applied to the coastline of Great Britain. Assume that the following values were obtained:

$1/\varepsilon$	4	8	12	16	24	32
$N(A, \varepsilon)$	16	48	90	120	192	283

A linear regression through the logarithms $x = \log 1/\varepsilon$, $y = \log N(A, \varepsilon)$ yields the coefficients

$$\widehat{\beta_0} = 0.9849, \quad d \approx \widehat{\beta_1} = 1.3616$$

with the coefficient of determination

$$R^2 = 0.9930.$$

This is very good fit, which is also confirmed by Fig. 18.5. The given data thus indicate that the fractal dimension of the coastline of Great Britain is $d = 1.36$.

A word of caution is in order. Data analysis can only supply indications, but never a proof that a model is correct. Even if we choose among a number of wrong models the one with the largest R^2, this model will not become correct. A healthy amount of skepticism with respect to purely empirically inferred relations is advisable; models should always be critically questioned. Scientific progress arises from the interplay between the invention of models and their experimental validation through data.

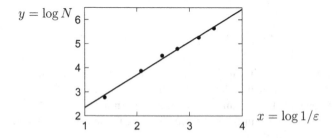

Fig. 18.5 Fractal dimension of the coastline of Great Britain

18.3   Multiple Linear Regression

In multiple (multivariate) linear regression the variable y does not just depend on one regressor variable x, but on several variables, for instance $x_1, x_2, \ldots, x_k$. We emphasise that the notation with respect to Sect. 18.1 is changed; there x_i denoted the ith data value, and now x_i refers to the ith regressor variable. The measurements of the ith regressor variable are now denoted with two indices, namely $x_{i1}, x_{i2}, \ldots, x_{in}$. In total, there are $k \times n$ data values. We again look for a linear model

$$y = \beta_0 + \beta_1 x_1 + \beta_2 x_2 + \cdots + \beta_k x_k$$

with the yet unknown coefficients $\beta_0, \beta_1, \ldots, \beta_k$.

Example 18.8 A vending machine company wants to analyse the delivery time, i.e., the time span y which a driver needs to refill a machine. The most important parameters are the number x_1 of refilled product units and the distance x_2 walked by the driver. The results of an observation of 25 services are given in the M-file mat18_3.m. The data values are taken from [19]. The observations $(x_{11}, x_{21}), (x_{12}, x_{22}), (x_{13}, x_{23}), \ldots, (x_{1,25}, x_{2,25})$ with the corresponding service times $y_1, y_2, y_3, \ldots, y_{25}$ yield a scatter plot in space to which a plane of the form $y = \beta_0 + \beta_1 x_1 + \beta_2 x_2$ should be fitted (Fig. 18.6; use the M-file mat18_4.m for visualisation).

Remark 18.9 A special case of the general multiple linear model $y = \beta_0 + \beta_1 x_1 + \cdots + \beta_k x_k$ is simple linear regression with several nonlinear form functions (as mentioned in Sect. 18.1), i.e.,

$$y = \beta_0 + \beta_1 \varphi_1(x) + \beta_2 \varphi_2(x) + \cdots + \beta_k \varphi_k(x),$$

where $x_1 = \varphi_1(x), x_2 = \varphi_2(x), \cdots, x_k = \varphi_k(x)$ are considered as regressor variables. In particular one can allow polynomial models

$$y = \beta_0 + \beta_1 x + \beta_2 x^2 + \cdots + \beta_k x^k$$

Fig. 18.6 Multiple linear regression through a scatter plot in space

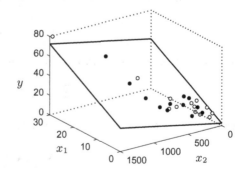

or still more general interactions between several variables, for instance

$$y = \beta_0 + \beta_1 x_1 + \beta_2 x_2 + \beta_3 x_1 x_2.$$

All these cases are treated in the same way as the standard problem of multiple linear regression, after renaming the variables.

The data values for the individual regressor variables are schematically represented as follows:

Variable	y	x_1	x_2	...	x_k
Observation 1	y_1	x_{11}	x_{21}	...	x_{k1}
Observation 2	y_2	x_{12}	x_{22}	...	x_{k2}
$\vdots$	$\vdots$	$\vdots$	$\vdots$		$\vdots$
Observation n	y_n	x_{1n}	x_{2n}	...	x_{kn}

Each value y_i is to be approximated by

$$y_i = \beta_0 + \beta_1 x_{1i} + \beta_2 x_{2i} + \cdots + \beta_k x_{ki} + \varepsilon_i, \quad i = 1, \ldots, n$$

with the errors ε_i. The estimated coefficients $\widehat{\beta}_0, \widehat{\beta}_1, \ldots, \widehat{\beta}_k$ are again obtained as the solution of the minimisation problem

$$L(\beta_0, \beta_1, \ldots, \beta_k) = \sum_{i=1}^{n} \varepsilon_i^2 \to \min$$

Using vector and matrix notation

$$\mathbf{y} = \begin{bmatrix} y_1 \\ y_2 \\ \vdots \\ y_n \end{bmatrix}, \quad \mathbf{X} = \begin{bmatrix} 1 & x_{11} & x_{21} & \cdots & x_{k1} \\ 1 & x_{12} & x_{22} & \cdots & x_{k2} \\ \vdots & \vdots & \vdots & & \vdots \\ 1 & x_{1n} & x_{2n} & \cdots & x_{kn} \end{bmatrix}, \quad \boldsymbol{\beta} = \begin{bmatrix} \beta_0 \\ \beta_1 \\ \vdots \\ \beta_k \end{bmatrix}, \quad \boldsymbol{\varepsilon} = \begin{bmatrix} \varepsilon_1 \\ \varepsilon_2 \\ \vdots \\ \varepsilon_n \end{bmatrix}$$

the linear model can again be written for short as

$$\mathbf{y} = \mathbf{X}\boldsymbol{\beta} + \boldsymbol{\varepsilon}.$$

The coefficients of best fit are obtained as in Sect. 18.1 by the formula

$$\widehat{\boldsymbol{\beta}} = (\mathbf{X}^{\mathsf{T}}\mathbf{X})^{-1}\mathbf{X}^{\mathsf{T}}\mathbf{y}$$

with the predicted values and the residuals

$$\widehat{\mathbf{y}} = \mathbf{X}\widehat{\boldsymbol{\beta}}, \quad \mathbf{e} = \mathbf{y} - \widehat{\mathbf{y}}.$$

The partitioning of total variability

$$S_{yy} = SS_R + SS_E$$

is still valid; the *multiple coefficient of determination*

$$R^2 = SS_R/S_{yy}$$

is an indicator of the goodness of fit of the model.

Example 18.10 We continue the analysis of the delivery times from Example 18.8. Using the MATLAB program `mat18_2.m` and entering the data from `mat18_3.m` results in

$$\widehat{\beta} = \begin{bmatrix} 2.3412 \\ 1.6159 \\ 0.0144 \end{bmatrix}.$$

We obtain the model

$$\widehat{y} = 2.3412 + 1.6159\, x_1 + 0.0144\, x_2$$

with the multiple coefficient of determination of

$$R^2 = 0.9596$$

and the partitioning of total variability

$$S_{yy} = 5784.5, \quad SS_R = 5550.8, \quad SS_E = 233.7$$

In this example merely $(1 - R^2) \cdot 100\% \approx 4\%$ of the variability of the data is not explained by the regression, a very satisfactory goodness of fit.

18.4   Model Fitting and Variable Selection

A recurring problem is to decide which variables should be included in the model. Would the inclusion of $x_3 = x_2^2$ and $x_4 = x_1 x_2$, i.e., the model

$$y = \beta_0 + \beta_1 x_1 + \beta_2 x_2 + \beta_3 x_2^2 + \beta_4 x_1 x_2,$$

lead to better results, and can, e.g., the term $\beta_2 x_2$ be eliminated subsequently? It is not desirable to have too many variables in the model. If there are as many variables as data points, then one can fit the regression exactly through the data and the model would loose its predictive power. A criterion will definitely be to reach a value of R^2 which is as large as possible. Another aim is to eliminate variables that do not contribute essentially to the total variability. An algorithmic procedure for identifying these variables is the sequential partitioning of total variability.

Sequential partitioning of total variability. We include variables stepwise in the model, thus consider the increasing sequence of models with corresponding SS_R:

$$
\begin{aligned}
y &= \beta_0 & SS_R(\beta_0), \\
y &= \beta_0 + \beta_1 x_1 & SS_R(\beta_0, \beta_1), \\
y &= \beta_0 + \beta_1 x_1 + \beta_2 x_2 & SS_R(\beta_0, \beta_1, \beta_2), \\
&\;\;\vdots & \vdots \\
y &= \beta_0 + \beta_1 x_1 + \beta_2 x_2 + \cdots + \beta_k x_k & SS_R(\beta_0, \beta_1, \ldots, \beta_k) = SS_R.
\end{aligned}
$$

Note that $SS_R(\beta_0) = 0$, since in the initial model $\beta_0 = \bar{y}$. The additional explanatory power of the variable x_1 is measured by

$$
SS_R(\beta_1 | \beta_0) = SS_R(\beta_0, \beta_1) - 0,
$$

the power of variable x_2 (if x_1 is already in the model) by

$$
SS_R(\beta_2 | \beta_0, \beta_1) = SS_R(\beta_0, \beta_1, \beta_2) - SS_R(\beta_0, \beta_1),
$$

the power of variable x_k (if $x_1, x_2, \ldots, x_{k-1}$ are in the model) by

$$
SS_R(\beta_k | \beta_0, \beta_1, \ldots, \beta_{k-1}) = SS_R(\beta_0, \beta_1, \ldots, \beta_k) - SS_R(\beta_0, \beta_1, \ldots, \beta_{k-1}).
$$

Obviously,

$$
\begin{aligned}
SS_R(\beta_1 | \beta_0) + SS_R(\beta_2 | \beta_0, \beta_1) + SS_R(\beta_3 | \beta_0, \beta_1, \beta_2) + \cdots \\
+ SS_R(\beta_k | \beta_0, \beta_1, \beta_2, \ldots, \beta_{k-1}) = SS_R.
\end{aligned}
$$

This shows that one can interpret the *sequential, partial coefficient of determination*

$$
\frac{SS_R(\beta_j | \beta_0, \beta_1, \ldots, \beta_{j-1})}{S_{yy}}
$$

as explanatory power of the variables x_j, under the condition that the variables $x_1, x_2, \ldots, x_{j-1}$ are already included in the model. This partial coefficient of determination depends on the order of the added variables. This dependency can be eliminated by averaging over all possible sequences of variables.

Average explanatory power of individual coefficients. One first computes all possible sequential, partial coefficients of determination which can be obtained by adding the variable x_j to all possible combinations of the already included variables. Summing up these coefficients and dividing the result by the total number of possibilities, one obtains a measure for the contribution of the variable x_j to the explanatory power of the model.

Average over orderings was proposed by [16]; further details and advanced considerations can be found, for instance, in [8, 10]. The concept does not use probabilistically motivated indicators. Instead it is based on the data and on combinatorics, thus belongs to descriptive data analysis. Such descriptive methods, in contrast to the commonly used statistical hypothesis testing, do not require additional assumptions which may be difficult to justify.

Example 18.11 We compute the explanatory power of the coefficients in the delivery time problem of Example 18.8. First we fit the two univariate models

$$y = \beta_0 + \beta_1 x_1, \quad y = \beta_0 + \beta_2 x_2$$

and from that obtain

$$SS_R(\beta_0, \beta_1) = 5382.4, \quad SS_R(\beta_0, \beta_2) = 4599.1,$$

with the regression coefficients $\widehat{\beta}_0 = 3.3208$, $\widehat{\beta}_1 = 2.1762$ in the first and $\widehat{\beta}_0 = 4.9612$, $\widehat{\beta}_2 = 0.0426$ in the second case. With the already computed values of the bivariate model

$$SS_R(\beta_0, \beta_1, \beta_2) = SS_R = 5550.8, \quad S_{yy} = 5784.5$$

from Example 18.10 we obtain the two sequences

$$SS_R(\beta_1|\beta_0) = 5382.4 \approx 93.05\% \text{ of } S_{yy}$$
$$SS_R(\beta_2|\beta_0, \beta_1) = 168.4 \approx 2.91\% \text{ of } S_{yy}$$

and

$$SS_R(\beta_2|\beta_0) = 4599.1 \approx 79.51\% \text{ of } S_{yy}$$
$$SS_R(\beta_1|\beta_0, \beta_2) = 951.7 \approx 16.45\% \text{ of } S_{yy}.$$

The average explanatory power of the variable x_1 (or of the coefficient β_1) is

$$\frac{1}{2}\left(93.05 + 16.45\right)\% = 54.75\%,$$

the one of the variable x_2 is

$$\frac{1}{2}\left(2.91 + 79.51\right)\% = 41.21\%;$$

the remaining 4.04% stay unexplained. The result is represented in Fig. 18.7.

Fig. 18.7 Average explanatory powers of the individual variables

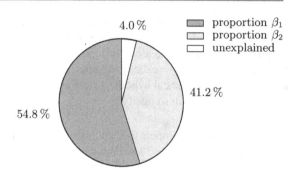

4.0 %

▨ proportion β_1
☐ proportion β_2
☐ unexplained

41.2 %

54.8 %

Numerical calculation of the average explanatory powers. In the case of more than two independent variables one has to take care that all possible sequences (represented by permutations of the variables) are considered. This will be exemplarily shown with three variables x_1, x_2, x_3. In the left column of the table there are the $3! = 6$ permutations of $\{1, 2, 3\}$, the other columns list the sequentially obtained values of SS_R.

$$
\begin{array}{l|lll}
1\ 2\ 3 & SS_R(\beta_1|\beta_0) & SS_R(\beta_2|\beta_0, \beta_1) & SS_R(\beta_3|\beta_0, \beta_1, \beta_2) \\
1\ 3\ 2 & SS_R(\beta_1|\beta_0) & SS_R(\beta_3|\beta_0, \beta_1) & SS_R(\beta_2|\beta_0, \beta_1, \beta_3) \\
2\ 1\ 3 & SS_R(\beta_2|\beta_0) & SS_R(\beta_1|\beta_0, \beta_2) & SS_R(\beta_3|\beta_0, \beta_2, \beta_1) \\
2\ 3\ 1 & SS_R(\beta_2|\beta_0) & SS_R(\beta_3|\beta_0, \beta_2) & SS_R(\beta_1|\beta_0, \beta_2, \beta_3) \\
3\ 1\ 2 & SS_R(\beta_3|\beta_0) & SS_R(\beta_1|\beta_0, \beta_3) & SS_R(\beta_2|\beta_0, \beta_3, \beta_1) \\
3\ 2\ 1 & SS_R(\beta_3|\beta_0) & SS_R(\beta_2|\beta_0, \beta_3) & SS_R(\beta_1|\beta_0, \beta_3, \beta_2)
\end{array}
$$

Obviously the sum of each row is always equal to SS_R, so that the sum of all entries is equal to $6 \cdot SS_R$. Note that amongst the 18 SS_R-values there are actually only 12 different ones.

The average explanatory power of the variable x_1 is defined by M_1/S_{yy}, where

$$
M_1 = \frac{1}{6}\Big(SS_R(\beta_1|\beta_0) + SS_R(\beta_1|\beta_0) + SS_R(\beta_1|\beta_0, \beta_2) + SS_R(\beta_1|\beta_0, \beta_3)
$$
$$
+ SS_R(\beta_1|\beta_0, \beta_2, \beta_3) + SS_R(\beta_1|\beta_0, \beta_3, \beta_2)\Big)
$$

and analogously for the other variables. As remarked above, we have

$$
M_1 + M_2 + M_3 = SS_R,
$$

and thus the total partitioning adds up to one

$$
\frac{M_1}{S_{yy}} + \frac{M_2}{S_{yy}} + \frac{M_3}{S_{yy}} + \frac{SS_E}{S_{yy}} = 1.
$$

For a more detailed analysis of the underlying combinatorics, for the necessary modifications in the case of collinearity of the data (linear dependence of the columns of the matrix X) and for a discussion of the significance of the average explanatory power, we refer to the literature quoted above. The algorithm is implemented in the applet *Linear regression*.

Experiment 18.12 Open the applet *Linear regression* and load data set number 9. It contains experimental data quantifying the influence of different aggregates on a mixture of concrete. The meaning of the output variables x_1 through x_4 and the input variables x_5 through x_{13} is explained in the online description of the applet. Experiment with different selections of the variables of the model. An interesting initial model is obtained, for example, by choosing $x_6, x_8, x_9, x_{10}, x_{11}, x_{12}, x_{13}$ as independent and x_1 as dependent variable; then remove variables with low explanatory power and draw a pie chart.

18.5   Exercises

1. The total consumption of electric energy in Austria 1970–2015 is given in Table 18.1 (from [26, Table 22.13]). The task is to carry out a linear regression of the form $y = \beta_0 + \beta_1 x$ through the data.
   (a) Write down the matrix X explicitly and compute the coefficients $\widehat{\beta} = [\widehat{\beta}_0, \widehat{\beta}_1]^T$ using the MATLAB command beta = X\y.
   (b) Check the goodness of fit by computing R^2. Plot a scatter diagram with the fitted straight line. Compute the forecast $\widehat{y}$ for 2020.

Table 18.1 Electric energy consumption in Austria, year = x_i, consumption = y_i [GWh]

x_i	1970	1980	1990	2000	2005	2010	2013	2014	2015
y_i	23.908	37.473	48.529	58.512	66.083	68.931	69.934	68.918	69.747

2. A sample of $n = 44$ civil engineering students at the University of Innsbruck in the year 1998 gave the values for x = height [cm] and y = weight [kg], listed in the M-file mat18_ex2.m. Compute the regression line $y = \beta_0 + \beta_1 x$, plot the scatter diagram and calculate the coefficient of determination R^2.
3. Solve Exercise 1 using Excel.
4. Solve Exercise 1 using the statistics package SPSS.
   Hint. Enter the data in the worksheet *Data View*; the names of the variables and their properties can be defined in the worksheet *Variable View*. Go to *Analyze* → *Regression* → *Linear*.
5. The stock of buildings in Austria 1869–2011 is given in the M-file mat18_ex5.m (data from [26, Table 12.01]). Compute the regression line $y = \beta_0 + \beta_1 x$ and the regression parabola $y = \alpha_0 + \alpha_1 (x - 1860)^2$ through the data and test which model fits better, using the coefficient of determination R^2.
6. The monthly share index for four breweries from November 1999 to November 2000 is given in the M-file mat18_ex6.m (November 1999 = 100%, from the Austrian magazine *profil* 46/2000). Fit a univariate linear model $y = \beta_0 + \beta_1 x$

to each of the four data sets ($x \ldots$ date, $y \ldots$ share index), plot the results in four equally scaled windows, evaluate the results by computing R^2 and check whether the caption provided by profil is justified by the data. For the calculation you may use the MATLAB program `mat18_1.m`.

Hint. A solution is suggested in the M-file `mat18_exsol6.m`.

7. Continuation of Exercise 5, stock of buildings in Austria. Fit the model

$$y = \beta_0 + \beta_1 x + \beta_2 (x - 1860)^2$$

and compute $SS_R = SS_R(\beta_0, \beta_1, \beta_2)$ and S_{yy}. Further, analyse the increase of explanatory power through adding the respective missing variable in the models of Exercise 5, i.e., compute $SS_R(\beta_2|\beta_0, \beta_1)$ and $SS_R(\beta_1|\beta_0, \beta_2)$ as well as the average explanatory power of the individual coefficients. Compare with the result for data set number 5 in the applet *Linear regression*.

8. The M-file `mat18_ex8.m` contains the mileage per gallon y of 30 cars depending on the engine displacement x_1, the horsepower x_2, the overall length x_3 and the weight x_4 of the vehicle (from: Motor Trend 1975, according to [19]). Fit the linear model

$$y = \beta_0 + \beta_1 x_1 + \beta_2 x_2 + \beta_3 x_3 + \beta_4 x_4$$

and estimate the explanatory power of the individual coefficients through a simple sequential analysis

$$SS_R(\beta_1|\beta_0), \quad SS_R(\beta_2|\beta_0, \beta_1), \quad SS_R(\beta_3|\beta_0, \beta_1, \beta_2), \quad SS_R(\beta_4|\beta_0, \beta_1, \beta_2, \beta_3).$$

Compare your result with the average explanatory power of the coefficients for data set number 2 in the applet *Linear regression*.

Hint. A suggested solution is given in the M-file `mat18_exsol8.m`.

9. Check the results of Exercises 2 and 6 using the applet *Linear regression* (data sets 1 and 4); likewise for the Examples 18.1 and 18.8 with the data sets 8 and 3. In particular, investigate in data set 8 whether height, weight and the risk of breaking a leg are in any linear relation.

10. Continuation of Exercise 14 from Sect. 8.4. A more accurate linear approximation to the relation between shear strength τ and normal stress σ is delivered by Coulomb's model $\tau = c + k\sigma$ where $k = \tan \varphi$ and c [kPa] is interpreted as cohesion. Recompute the regression model of Exercise 14 in Sect. 8.4 with nonzero intercept. Check that the resulting cohesion is indeed small as compared to the applied stresses, and compare the resulting friction angles.

11. (Change point analysis) The consumer prize data from Example 8.21 suggest that there might be a change in the slope of the regression line around the year 2013, see also Fig. 8.9. Given data $(x_1, y_1), \ldots, (x_n, y_n)$ with ordered data points $x_1 < x_2 < \ldots < x_n$, phenomena of this type can be modelled by a piecewise linear regression

$$y = \begin{cases} \alpha_0 + \alpha_1 x, & x \le x_*, \\ \beta_0 + \beta_1 x, & x \ge x_*. \end{cases}$$

If the slopes α_1 and α_2 are different, x_* is called a *change point*. A change point can be detected by fitting models

$$y_i = \begin{cases} \alpha_0 + \alpha_1 x_i, & i = 1, \ldots, m, \\ \beta_0 + \beta_1 x_i, & i = m+1, \ldots, n \end{cases}$$

and varying the index m between 2 and $n-1$ until a two-line model with the smallest total residual sum of squares $SS_R(\alpha_0, \alpha_1) + SS_R(\beta_0, \beta_1)$ is found. The change point x_* is the point of intersection of the two predicted lines. (If the overall one-line model has the smallest SS_R, there is no change point.)

Find out whether there is a change point in the data of Example 8.21. If so, locate it and use the two-line model to predict the consumer price index for 2017.

12. Atmospheric CO_2 concentration has been recorded at Mauna Loa, Hawai, since 1958. The yearly averages (1959–2008) in ppm can be found in the MATLAB program `mat18_ex12.m`; the data are from [14].

(a) Fit an exponential model $y = \alpha_0 \, e^{\alpha_1 x}$ to the data and compare the prediction with the actual data (2017: 406.53 ppm).

Hint. Taking logarithms leads to the linear model $z = \beta_0 + \beta_1 x$ with $z = \log y$, $\beta_0 = \log \alpha_0$, $\beta_1 = \alpha_1$. Estimate the coefficients $\widehat{\beta_0}, \widehat{\beta_1}$ and compute $\widehat{\alpha_0}, \widehat{\alpha_1}$ as well as the prediction for y.

(b) Fit a square exponential model $y = \alpha_0 \, e^{\alpha_1 x + \alpha_2 x^2}$ to the data and check whether this yields a better fit and prediction.

Differential Equations 19

In this chapter we discuss the theory of initial value problems for ordinary differential equations. We limit ourselves to scalar equations here; systems will be discussed in the next chapter.

After presenting the general definition of a differential equation and the geometric significance of its direction field, we start with a detailed discussion of first-order linear equations. As important applications we discuss the modelling of growth and decay processes. Subsequently, we investigate questions of existence and (local) uniqueness of the solution of general differential equations and discuss the method of power series. We also study the qualitative behaviour of solutions close to an equilibrium point. Finally, we discuss the solution of second-order linear problems with constant coefficients.

19.1 Initial Value Problems

Differential equations are equations involving a (sought after) function and its derivative(s). They play a decisive role in modelling time-dependent processes.

Definition 19.1 Let $D \subset \mathbb{R}^2$ be open and $f : D \subset \mathbb{R}^2 \to \mathbb{R}$ continuous. The equation

$$y'(x) = f\big(x, y(x)\big)$$

is called (an ordinary) *first-order differential equation*. A *solution* is a differentiable function $y : I \to D$ which satisfies the equation for all $x \in I$.

© Springer Nature Switzerland AG 2018
M. Oberguggenberger and A. Ostermann, *Analysis for Computer Scientists*,
Undergraduate Topics in Computer Science,
https://doi.org/10.1007/978-3-319-91155-7_19

One often suppresses the *independent variable* x in the notation and writes the above problem for short as

$$y' = f(x, y).$$

The sought after function y in this equation is also called the *dependent variable* (depending on x).

In modelling time-dependent processes, one usually denotes the independent variable by t (for time) and the dependent variable by $x = x(t)$. In this case one writes the first-order differential equation as

$$\dot{x}(t) = f\big(t, x(t)\big)$$

or for short as $\dot{x} = f(t, x)$.

Example 19.2 (Separation of the variables) We want to find all functions $y = y(x)$ satisfying the equation $y'(x) = x \cdot y(x)^2$. In this example one obtains the solutions by *separating the variables*. For $y \neq 0$ one divides the differential equation by y^2 and gets

$$\frac{1}{y^2} \cdot y' = x.$$

The left-hand side of this equation is of the form $g(y) \cdot y'$. Let $G(y)$ be an antiderivative of $g(y)$. According to the chain rule, and recalling that y is a function of x, we obtain

$$\frac{d}{dx} G(y) = \frac{d}{dy} G(y) \cdot \frac{dy}{dx} = g(y) \cdot y'.$$

In our example we have $g(y) = y^{-2}$ and $G(y) = -y^{-1}$, consequently

$$\frac{d}{dx} \left(-\frac{1}{y} \right) = \frac{1}{y^2} \cdot y' = x.$$

Integration of this equation with respect to x results in

$$-\frac{1}{y} = \frac{x^2}{2} + C,$$

where C denotes an arbitrary integration constant. By elementary manipulations we find

$$y = \frac{1}{-x^2/2 - C} = \frac{2}{K - x^2}$$

with the constant $K = -2C$.

The function $y = 0$ is also a solution of the differential equation. Formally, one obtains it from the above solution by setting $K = \infty$. The example shows that differential equations have infinitely many solutions in general. By requiring an additional condition, a unique solution can be selected. For example, setting $y(0) = 1$ gives $y(x) = 2/(2 - x^2)$.

Fig. 19.1 The direction field
of $y' = -2xy/(x^2 + 2y)$

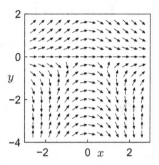

Definition 19.3 The differential equation $y'(x) = f\big(x, y(x)\big)$ together with the additional condition $y(x_0) = y_0$, i.e.,

$$y'(x) = f\big(x, y(x)\big), \quad y(x_0) = y_0,$$

is called *initial value problem*. A solution of an initial value problem is a (continuously) differentiable function $y(x)$, which satisfies the differential equation and the *initial condition* $y(x_0) = y_0$.

Geometric interpretation of a differential equation. For a given first-order differential equation

$$y' = f(x, y), \quad (x, y) \in D \subset \mathbb{R}^2$$

one searches for a differentiable function $y = y(x)$ whose graph lies in D and whose tangents have the slopes $\tan \varphi = y'(x) = f\big(x, y(x)\big)$ for each x. By plotting short arrows with slopes $\tan \varphi = f(x, y)$ at the points $(x, y) \in D$ one obtains the *direction field* of the differential equation. The direction field is *tangential* to the solution curves and offers a good visual impression of their shapes. Figure 19.1 shows the direction field of the differential equation

$$y' = -\frac{2xy}{x^2 + 2y}.$$

The right-hand side has singularities along the curve $y = -x^2/2$ which is reflected by the behaviour of the arrows in the lower part of the figure.

Experiment 19.4 Visualise the direction field of the above differential equation with the applet *Dynamical systems in the plane*.

19.2 First-Order Linear Differential Equations

Let $a(x)$ and $g(x)$ be functions defined on some interval. The equation

$$y' + a(x)\, y = g(x)$$

is called a *first-order linear differential equation*. The function a is the *coefficient*, the right-hand side g is called *inhomogeneity*. The differential equation is called *homogeneous*, if $g = 0$, otherwise *inhomogeneous*. First we state the following important result.

Proposition 19.5 (Superposition principle) *If y and z are solutions of a linear differential equation with possibly different inhomogeneities*

$$y'(x) + a(x)\, y(x) = g(x),$$
$$z'(x) + a(x)\, z(x) = h(x),$$

then their linear combination

$$w(x) = \alpha y(x) + \beta z(x), \qquad \alpha,\ \beta \in \mathbb{R}$$

solves the linear differential equation   .

$$w'(x) + a(x)\, w(x) = \alpha g(x) + \beta h(x).$$

Proof This so-called *superposition principle* follows from the linearity of the derivative and the linearity of the equation.   □

In a first step we compute all solutions of the homogeneous equation. We will use the superposition principle later to find all solutions of the inhomogeneous equation.

Proposition 19.6 *The general solution of the* homogeneous *differential equation*

$$y' + a(x)\, y = 0$$

is

$$y_h(x) = K\mathrm{e}^{-A(x)}$$

with $K \in \mathbb{R}$ and an arbitrary antiderivative $A(x)$ of $a(x)$.

Proof For $y \neq 0$ we separate the variables

$$\frac{1}{y} \cdot y' = -a(x)$$

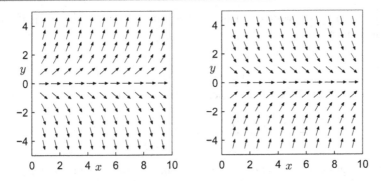

Fig. 19.2 The direction field of $y' = y$ (left) and $y' = -y$ (right)

and use

$$\frac{d}{dy} \log |y| = \frac{1}{y}$$

to obtain

$$\log |y| = -A(x) + C$$

by integrating the equation. From that we infer

$$|y(x)| = e^{-A(x)} e^C.$$

This formula shows that $y(x)$ cannot change sign since the right-hand side is never zero. Thus $K = e^C \cdot \operatorname{sign} y(x)$ is a constant as well, and the formula

$$y(x) = \operatorname{sign} y(x) \cdot |y(x)| = K e^{-A(x)}, \qquad K \in \mathbb{R}$$

yields all solutions of the homogeneous equation.   □

Example 19.7 The linear differential equation

$$\dot{x} = ax$$

with *constant* coefficient a has the general solution

$$x(t) = K e^{at}, \qquad K \in \mathbb{R}.$$

The constant K is determined by $x(0)$, for example.

The direction field of the differential equation $y' = ay$ (depending on the sign of the coefficient) is shown in Fig. 19.2.

Interpretation. Let $x(t)$ be a time-dependent function which describes a growth or decay process (population increase/decrease, change of mass, etc.). We consider a time interval $[t, t + h]$ with $h > 0$. For $x(t) \neq 0$ the relative change of x in this time interval is given by

$$\frac{x(t + h) - x(t)}{x(t)} = \frac{x(t + h)}{x(t)} - 1.$$

The relative *rate of change* (change per unit of time) is thus

$$\frac{x(t + h) - x(t)}{t + h - t} \cdot \frac{1}{x(t)} = \frac{x(t + h) - x(t)}{h \cdot x(t)}.$$

For an *ideal* growth process this rate only depends on time t. In the limit $h \to 0$ this leads to the *instantaneous relative rate of change*

$$a(t) = \lim_{h \to 0} \frac{x(t + h) - x(t)}{h \cdot x(t)} = \frac{\dot{x}(t)}{x(t)}.$$

Ideal growth processes thus may be modelled by the linear differential equation

$$\dot{x}(t) = a(t)x(t).$$

Example 19.8 (Radioactive decay) Let $x(t)$ be the concentration of a radioactive substance at time t. In radioactive decay the rate of change does not depend on time and is negative,

$$a(t) \equiv a < 0.$$

The solution of the equation $\dot{x} = ax$ with initial value $x(0) = x_0$ is

$$x(t) = e^{at}x_0.$$

It is exponentially decreasing and $\lim_{t \to \infty} x(t) = 0$, see Fig. 19.3. The *half life* T, the time in which half of the substance has decayed, is obtained from

$$\frac{x_0}{2} = e^{aT}x_0 \quad \text{as} \quad T = -\frac{\log 2}{a}.$$

The half life for $a = -2$ is indicated in Fig. 19.3 by the dotted lines.

Fig. 19.3 Radioactive decay with constants $a = -0.5, -1, -2$ (top to bottom)

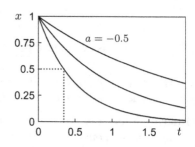

Example 19.9 (Population models) Let $x(t)$ be the size of a population at time t, modelled by $\dot{x} = ax$. If a constant, positive rate of growth $a > 0$ is presumed then the population grows exponentially

$$x(t) = e^{at} x_0, \qquad \lim_{t \to \infty} |x(t)| = \infty.$$

One calls this behaviour *Malthusian law*.[1] In 1839 Verhulst suggested an improved model which also takes limited resources into account

$$\dot{x}(t) = \big(\alpha - \beta x(t)\big) \cdot x(t) \qquad \text{with } \alpha, \beta > 0.$$

The corresponding discrete model was already discussed in Example 5.3, where L denoted the quotient α/β.

The rate of growth in Verhulst's model is population dependent, namely equal to $\alpha - \beta x(t)$, and decreases *linearly* with increasing population. Verhulst's model can be solved by separating the variables (or with maple). One obtains

$$x(t) = \frac{\alpha}{\beta + C\alpha e^{-\alpha t}}$$

and thus, independently of the initial value,

$$\lim_{t \to \infty} x(t) = \frac{\alpha}{\beta},$$

see also Fig. 19.4. The *stationary* solution $x(t) = \alpha/\beta$ is an *asymptotically stable equilibrium point* of Verhulst's model, see Sect. 19.5.

Variation of constants. We now turn to the solution of the *inhomogeneous* equation

$$y' + a(x)y = g(x).$$

We already know the general solution

$$y_h(x) = c \cdot e^{-A(x)}, \qquad c \in \mathbb{R}$$

Fig. 19.4 Population increase according to Malthus and Verhulst

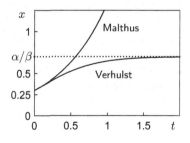

---

[1] T.R. Malthus, 1766–1834.

of the homogeneous equation with the antiderivative

$$A(x) = \int_{x_0}^{x} a(\xi) \, d\xi.$$

We look for a particular solution of the inhomogeneous equation of the form

$$y_p(x) = c(x) \cdot y_h(x) = c(x) \cdot e^{-A(x)},$$

where we allow the constant $c = c(x)$ to be a function of x (variation of constants). Substituting this formula into the inhomogeneous equation and differentiating using the product rule yields

$$\begin{aligned} y_p'(x) + a(x) \, y_p(x) &= c'(x) \, y_h(x) + c(x) \, y_h'(x) + a(x) \, y_p(x) \\ &= c'(x) \, y_h(x) - a(x) \, c(x) \, y_h(x) + a(x) \, y_p(x) \\ &= c'(x) \, y_h(x). \end{aligned}$$

If one equates this expression with the inhomogeneity $g(x)$, one recognises that $c(x)$ fulfils the differential equation

$$c'(x) = e^{A(x)} g(x)$$

which can be solved by integration

$$c(x) = \int_{x_0}^{x} e^{A(\xi)} g(\xi) \, d\xi.$$

We thus obtain the following proposition.

Proposition 19.10 *The differential equation*

$$y' + a(x)y = g(x)$$

has the general solution

$$y(x) = e^{-A(x)} \left(\int_{x_0}^{x} e^{A(\xi)} g(\xi) \, d\xi + K \right)$$

with $A(x) = \int_{x_0}^{x} a(\xi) \, d\xi$ and an arbitrary constant $K \in \mathbb{R}$.

Proof By the above considerations, the function $y(x)$ is a solution of the differential equation $y' + a(x)y = g(x)$. Conversely, let $z(x)$ be any other solution. Then, according to the *superposition principle*, the difference $z(x) - y(x)$ is a solution of the homogeneous equation, so

$$z(x) = y(x) + c \, e^{-A(x)}.$$

Therefore, $z(x)$ also has the form stated in the proposition.   □

Corollary 19.11 *Let y_p be an arbitrary solution of the inhomogeneous linear differential equation*

$$y' + a(x)y = g(x).$$

Then, its general solution can be written as

$$y(x) = y_p(x) + y_h(x) = y_p(x) + K e^{-A(x)}, \qquad K \in \mathbb{R}.$$

Proof This statement follows from the proof of Proposition 19.10 or directly from the superposition principle. □

Example 19.12 We solve the problem $y' + 2y = e^{4x} + 1$. The solution of the homogeneous equation is $y_h(x) = c\, e^{-2x}$. A particular solution can be found by variation of constants. From

$$c(x) = \int_0^x e^{2\xi}\left(e^{4\xi} + 1\right) d\xi = \frac{1}{6} e^{6x} + \frac{1}{2} e^{2x} - \frac{2}{3}$$

it follows that

$$y_p(x) = \frac{1}{6} e^{4x} - \frac{2}{3} e^{-2x} + \frac{1}{2}.$$

The general solution is thus

$$y(x) = y_p(x) + y_h(x) = K e^{-2x} + \frac{1}{6} e^{4x} + \frac{1}{2}.$$

Here, we have combined the two terms containing e^{-2x}. The new constant K can be determined from an additional initial condition $y(0) = \alpha$, namely

$$K = \alpha - \frac{2}{3}.$$

19.3 Existence and Uniqueness of the Solution

Finding analytic solutions of differential equations can be a difficult problem and is often impossible. Apart from some types of differential equations (e.g., linear problems or equations with separable variables), there is no general procedure to determine the solution explicitly. Thus numerical methods are used frequently (see Chap. 21). In the following we discuss the existence and uniqueness of solutions of general initial value problems.

Proposition 19.13 (Peano's theorem[2]) *If the function f is continuous in a neigh-bourhood of* (x_0, y_0), *then the initial value problem*

$$y' = f(x, y), \qquad y(x_0) = y_0$$

has a solution $y(x)$ *for x close to* x_0.

Instead of a proof (see [11, Part I, Theorem 7.6]), we discuss the limitations of this proposition. First it only guarantees the existence of a local solution in the neighbourhood of the initial value. The next example shows that one cannot expect more, in general.

Example 19.14 We solve the differential equation $\dot{x} = x^2$, $x(0) = 1$. Separation of the variables yields

$$\int \frac{dx}{x^2} = \int dt = t + C,$$

and thus

$$x(t) = \frac{1}{1 - t}.$$

This function has a singularity at $t = 1$, where the solution ceases to exist. This behaviour is called *blow up*.

Furthermore, Peano's theorem does not give any information on how many solu-tions an initial value problem has. In general, solutions need not be unique, as it is shown in the following example.

Example 19.15 The initial value problem $y' = 2\sqrt{|y|}$, $y(0) = 0$ has infinitely many solutions

$$y(x) = \begin{cases} (x - b)^2, & b < x, \\ 0, & -a \le x \le b, \\ -(x - a)^2, & x < -a, \end{cases} \qquad a, b \ge 0 \text{ arbitrary.}$$

For example, for $x < -a$, one verifies at once

$$y'(x) = -2(x - a) = 2(a - x) = 2|x - a| = 2\sqrt{(x - a)^2} = 2\sqrt{|y|}.$$

Thus the continuity of f is not sufficient to guarantee the uniqueness of the solution of initial value problems. One needs somewhat more regularity, namely Lipschitz[3] continuity with respect to the second variable (see also Definition C.14).

---

[2]G. Peano, 1858–1932.
[3]R. Lipschitz, 1832–1903.

Definition 19.16 Let $D \subset \mathbb{R}^2$ and $f : D \rightarrow \mathbb{R}$. The function f is said to satisfy a *Lipschitz condition* with *Lipschitz constant* L on D, if the inequality $|f(x, y) - f(x, z)| \leq L |y - z|$ holds for all points $(x, y), (x, z) \in D$.

According to the mean value theorem (Proposition 8.4)

$$f(x, y) - f(x, z) = \frac{\partial f}{\partial y}(x, \xi)(y - z)$$

for every differentiable function. If the derivative is bounded, then the function satisfies a Lipschitz condition. In this case one can choose

$$L = \sup \left| \frac{\partial f}{\partial y}(x, \xi) \right|.$$

Counterexample 19.17 The function $g(x, y) = \sqrt{|y|}$ does not satisfy a Lipschitz condition in any D that contains a point with $y = 0$ because

$$\frac{|g(x, y) - g(x, 0)|}{|y - 0|} = \frac{\sqrt{|y|}}{|y|} = \frac{1}{\sqrt{|y|}} \rightarrow \infty \quad \text{for} \quad y \rightarrow 0.$$

Proposition 19.18 *If the function f satisfies a Lipschitz condition in the neighbourhood of (x_0, y_0), then the initial value problem*

$$y' = f(x, y), \quad y(x_0) = y_0$$

has a unique *solution $y(x)$ for x close to x_0.*

Proof We only show uniqueness, the existence of a solution $y(x)$ on the interval $[x_0, x_0 + H]$ follows (for small H) from Peano's theorem. Uniqueness is proven indirectly. Assume that z is another solution, *different* from y, on the interval $[x_0, x_0 + H]$ with $z(x_0) = y_0$. The number

$$x_1 = \inf\{x \in \mathbb{R} \,; \, x_0 \leq x \leq x_0 + H \text{ and } y(x) \neq z(x)\}$$

is thus well-defined. We infer from the continuity of y and z that $y(x_1) = z(x_1)$. Now we choose $h > 0$ so small that $x_1 + h \leq x_0 + H$ and integrate the differential equation

$$y'(x) = f(x, y(x))$$

from x_1 to $x_1 + h$. This gives

$$y(x_1 + h) - y(x_1) = \int_{x_1}^{x_1+h} y'(x)\, dx = \int_{x_1}^{x_1+h} f(x, y(x))\, dx$$

and

$$z(x_1 + h) - y(x_1) = \int_{x_1}^{x_1+h} f\big(x, z(x)\big)\, \mathrm{d}x.$$

Subtracting the first formula above from the second yields

$$z(x_1 + h) - y(x_1 + h) = \int_{x_1}^{x_1+h} \Big(f\big(x, z(x)\big) - f\big(x, y(x)\big)\Big)\, \mathrm{d}x.$$

The Lipschitz condition on f gives

$$|z(x_1 + h) - y(x_1 + h)| \leq \int_{x_1}^{x_1+h} \big|f\big(x, z(x)\big) - f\big(x, y(x)\big)\big|\, \mathrm{d}x$$

$$\leq L \int_{x_1}^{x_1+h} |z(x) - y(x)|\, \mathrm{d}x.$$

Let now

$$M = \max\big\{|z(x) - y(x)|\,;\ x_1 \leq x \leq x_1 + h\big\}.$$

Due to the continuity of y and z, this maximum exists, see the discussion after Proposition 6.15. After possibly decreasing h this maximum is attained at $x_1 + h$ and

$$M = |z(x_1 + h) - y(x_1 + h)| \leq L \int_{x_1}^{x_1+h} M\, \mathrm{d}x \leq L h\, M.$$

For a sufficiently small h, namely $Lh < 1$, the inequality

$$M \leq L h\, M$$

implies $M = 0$. Since one can choose h arbitrarily small, $y(x) = z(x)$ holds true for $x_1 \leq x \leq x_1 + h$ in contradiction to the definition of x_1. Hence the assumed different solution z does not exist.   □

19.4 Method of Power Series

We have encountered several examples of functions that can be represented as series, e.g. in Chap. 12. Motivated by this we try to solve the initial value problem

$$y' = f(x, y), \qquad y(x_0) = y_0$$

by means of a series

$$y(x) = \sum_{n=0}^{\infty} a_n (x - x_0)^n.$$

We will use the fact that *convergent power series* can be differentiated and rearranged term by term, see for instance [3, Chap. 9, Corollary 7.4].

Example 19.19 We solve once more the linear initial value problem

$$y' = y, \qquad y(0) = 1.$$

For that we differentiate the ansatz

$$y(x) = \sum_{n=0}^{\infty} a_n x^n = a_0 + a_1 x + a_2 x^2 + a_3 x^3 + \cdots$$

term by term with respect to x

$$y'(x) = \sum_{n=1}^{\infty} n a_n x^{n-1} = a_1 + 2a_2 x + 3a_3 x^2 + 4a_4 x^3 + \cdots$$

and substitute the result into the differential equation to get

$$a_1 + 2a_2 x + 3a_3 x^2 + 4a_4 x^3 + \cdots = a_0 + a_1 x + a_2 x^2 + a_3 x^3 + \cdots$$

Since this equation has to hold for all x, the unknowns a_n can be determined by equating the coefficients of same powers of x. This gives

$$a_1 = a_0, \qquad 2a_2 = a_1,$$
$$3a_3 = a_2, \qquad 4a_4 = a_3,$$

and so on. Due to $a_0 = y(0) = 1$ this infinite system of equations can be solved recursively. One obtains

$$a_0 = 1, \quad a_1 = 1, \quad a_2 = \frac{1}{2!}, \quad a_3 = \frac{1}{3!}, \quad \ldots, \quad a_n = \frac{1}{n!}$$

and thus the (expected) solution

$$y(x) = \sum_{n=0}^{\infty} \frac{x^n}{n!} = e^x.$$

Example 19.20 (A particular Riccati differential equation[4]) For the solution of the initial value problem

$$y' = y^2 + x^2, \qquad y(0) = 1,$$

---

[4]J.F. Riccati, 1676–1754.

we make the ansatz

$$y(x) = \sum_{n=0}^{\infty} a_n x^n = a_0 + a_1 x + a_2 x^2 + a_3 x^3 + \cdots$$

The initial condition $y(0) = 1$ immediately gives $a_0 = 1$. First, we compute the product (see also Proposition C.10)

$$
\begin{aligned}
y(x)^2 &= (1 + a_1 x + a_2 x^2 + a_3 x^3 + \cdots)^2 \\
&= 1 + 2a_1 x + (a_1^2 + 2a_2)x^2 + (2a_3 + 2a_2 a_1)x^3 + \cdots
\end{aligned}
$$

and substitute it into the differential equation

$$
\begin{aligned}
a_1 + 2a_2 x &+ 3a_3 x^2 + 4a_4 x^3 + \cdots \\
&= 1 + 2a_1 x + (1 + a_1^2 + 2a_2)x^2 + (2a_3 + 2a_2 a_1)x^3 + \cdots
\end{aligned}
$$

Equating coefficients results in

$$
\begin{aligned}
a_1 &= 1, \\
2a_2 &= 2a_1, & a_2 &= 1 \\
3a_3 &= 1 + a_1^2 + 2a_2, & a_3 &= 4/3 \\
4a_4 &= 2a_3 + 2a_2 a_1, & a_4 &= 7/6, \quad \ldots
\end{aligned}
$$

Thus we obtain a good approximation to the solution for small x

$$y(x) = 1 + x + x^2 + \frac{4}{3}x^3 + \frac{7}{6}x^4 + \mathcal{O}(x^5).$$

The maple command

```
dsolve({diff(y(x),x)=x^2+y(x)^2, y(0)=1}, y(x), series);
```

carries out the above computations.

19.5 Qualitative Theory

Often one can describe the qualitative behaviour of the solutions of differential equations without solving the equations themselves. As the simplest case we discuss the stability of nonlinear differential equations in the neighbourhood of an equilibrium point. A differential equation is called *autonomous*, if its right-hand side does not explicitly depend on the independent variable.

Definition 19.21 The point $y^\star \in \mathbb{R}$ is called an *equilibrium* of the autonomous differential equation $y' = f(y)$, if $f(y^\star) = 0$.

Equilibrium points are particular solutions of the differential equation, so-called stationary solutions.

In order to investigate solutions in the neighbourhood of an equilibrium point, we *linearise* the differential equation at the equilibrium. Let

$$w(x) = y(x) - y^\star$$

denote the distance of the solution $y(x)$ from the equilibrium. Taylor series expansion of f shows that

$$w' = y' = f(y) = f(y) - f(y^\star) = f'(y^\star)w + \mathcal{O}(w^2),$$

hence

$$w'(x) = \big(a + \mathcal{O}(w)\big)w$$

with $a = f'(y^\star)$. It is decisive how solutions of this problem behave for small w. Obviously the value of the coefficient $a + \mathcal{O}(w)$ is crucial. If $a < 0$, then $a + \mathcal{O}(w) < 0$ for sufficiently small w and the function $|w(x)|$ decreases. If on the other hand $a > 0$, then the function $|w(x)|$ increases for small w. With these considerations one has proven the following proposition.

Proposition 19.22 *Let $y^\star$ be an* equilibrium point *of the differential equation $y' = f(y)$ and assume that $f'(y^\star) < 0$. Then all solutions of the differential equation with initial value $w(0)$ close to $y^\star$ satisfy the estimate*

$$|w(x)| \le C \cdot e^{bx} \cdot |w(0)|$$

with constants $C > 0$ and $b < 0$.

Under the conditions of the proposition one calls the equilibrium point *asymptotically stable*. An asymptotically stable equilibrium attracts all solutions in a sufficiently small neighbourhood (exponentially fast), since due to $b < 0$

$$|w(x)| \to 0 \quad \text{as } x \to \infty.$$

Example 19.23 Verhulst's model

$$y' = (\alpha - \beta y)y, \quad \alpha, \beta > 0$$

has two equilibrium points, namely $y_1^\star = 0$ and $y_2^\star = \alpha/\beta$. Due to

$$f'(y_1^\star) = \alpha - 2\beta y_1^\star = \alpha, \quad f'(y_2^\star) = \alpha - 2\beta y_2^\star = -\alpha,$$

$y_1^\star = 0$ is *unstable* and $y_2^\star = \alpha/\beta$ is *asymptotically stable*.

19.6   Second-Order Problems

The equation

$$y''(x) + ay'(x) + by(x) = g(x)$$

is called a second-order linear differential equation with *constant* coefficients a, b and inhomogeneity g.

Example 19.24 (Mass–spring–damper model)   According to Newton's second law of motion, a mass–spring system is modelled by the second-order differential equation

$$y''(x) + ky(x) = 0,$$

where $y(x)$ denotes the position of the mass and k is the stiffness of the spring. The solution of this equation describes a free vibration without damping and excitation. A more realistic model is obtained by adding a viscous damping force $-cy'(x)$ and an external excitation $g(x)$. This results in the differential equation

$$my''(x) + cy'(x) + ky(x) = g(x),$$

which is of the above form.

By introducing the new variable $z(x) = y'(x)$, the homogeneous problem

$$y'' + ay' + by = 0$$

can be rewritten as a system of first-order equations

$$y' = z$$
$$z' = -by - az,$$

see Chap. 20, where this approach is worked out in detail.

Here, we will follow a different idea. Let α and β denote the roots of the quadratic equation

$$\lambda^2 + a\lambda + b = 0,$$

which is called the *characteristic equation* of the homogeneous problem. Then, the second-order problem

$$y''(x) + ay'(x) + by(x) = g(x)$$

can be factorised in the following way:

$$\left(\frac{\mathrm{d}^2}{\mathrm{d}x^2} + a\frac{\mathrm{d}}{\mathrm{d}x} + b \right) y(x) = \left(\frac{\mathrm{d}}{\mathrm{d}x} - \beta \right) \left(\frac{\mathrm{d}}{\mathrm{d}x} - \alpha \right) y(x) = g(x).$$

Setting

$$w(x) = y'(x) - \alpha y(x),$$

we obtain the following first-order linear differential equation for w

$$w'(x) - \beta w(x) = g(x).$$

This problem has the general solution (see Proposition 19.10)

$$w(x) = K_2 e^{\beta(x-x_0)} + \int_{x_0}^{x} e^{\beta(x-\xi)} g(\xi) \, d\xi$$

with some constant K_2. Inserting this expression into the definition of w shows that y is the solution of the first-order problem

$$y'(x) - \alpha y(x) = K_2 e^{\beta(x-x_0)} + \int_{x_0}^{x} e^{\beta(x-\xi)} g(\xi) \, d\xi.$$

Let us assume for a moment that $\alpha \neq \beta$. Applying once more Proposition 19.10 for the solution of this problem gives

$$y(x) = K_1 e^{\alpha(x-x_0)} + \int_{x_0}^{x} e^{\alpha(x-\eta)} w(\eta) \, d\eta$$

$$= K_1 e^{\alpha(x-x_0)} + K_2 \int_{x_0}^{x} e^{\alpha(x-\eta)} e^{\beta(\eta-x_0)} \, d\eta$$

$$+ \int_{x_0}^{x} e^{\alpha(x-\eta)} \int_{x_0}^{\eta} e^{\beta(\eta-\xi)} g(\xi) \, d\xi \, d\eta.$$

Since

$$\int_{x_0}^{x} e^{\alpha(x-\eta)} e^{\beta(\eta-x_0)} \, d\eta = e^{\alpha x - \beta x_0} \int_{x_0}^{x} e^{\eta(\beta-\alpha)} \, d\eta$$

$$= \frac{1}{\beta - \alpha} \left(e^{\beta(x-x_0)} - e^{\alpha(x-x_0)} \right),$$

we finally obtain

$$y(x) = c_1 e^{\alpha(x-x_0)} + c_2 e^{\beta(x-x_0)} + \int_{x_0}^{x} e^{\alpha(x-\eta)} \int_{x_0}^{\eta} e^{\beta(\eta-\xi)} g(\xi) \, d\xi \, d\eta$$

with

$$c_1 = K_1 - \frac{K_2}{\beta - \alpha}, \qquad c_2 = \frac{K_2}{\beta - \alpha}.$$

By setting $g = 0$, one obtains the general solution of the homogeneous problem

$$y_h(x) = c_1 e^{\alpha(x-x_0)} + c_2 e^{\beta(x-x_0)}.$$

The double integral

$$\int_{x_0}^{x} e^{\alpha(x-\eta)} \int_{x_0}^{\eta} e^{\beta(\eta-\xi)} g(\xi) \, d\xi \, d\eta$$

is a particular solution of the inhomogeneous problem. Note that, due to the linearity of the problem, the superposition principle (see Proposition 19.5) is again valid.

Summarising the above calculations gives the following two propositions.

Proposition 19.25 *Consider the homogeneous differential equation*

$$y''(x) + ay'(x) + by(x) = 0$$

and let α and β denote the roots of its characteristic equation

$$\lambda^2 + a\lambda + b = 0.$$

The general (real) solution of this problem is given by

$$y_h(x) = \begin{cases} c_1 e^{\alpha x} + c_2 e^{\beta x} & \text{for } \alpha \neq \beta \in \mathbb{R}, \\ (c_1 + c_2 x) e^{\alpha x} & \text{for } \alpha = \beta \in \mathbb{R}, \\ e^{\rho x} \big(c_1 \cos(\theta x) + c_2 \sin(\theta x) \big) & \text{for } \alpha = \rho + i\theta, \quad \rho, \theta \in \mathbb{R}, \end{cases}$$

for arbitrary real constants c_1 and c_2.

Proof Since the characteristic equation has real coefficients, the roots are either both real or conjugate complex, i.e. $\alpha = \overline{\beta}$. The case $\alpha \neq \beta$ was already considered above. In the complex case where $\alpha = \rho + i\theta$, we use Euler's formula

$$e^{\alpha x} = e^{\rho x} \big(\cos(\theta x) + i \sin(\theta x) \big).$$

This shows that $c_1 e^{\rho x} \cos(\theta x)$ and $c_2 e^{\rho x} \sin(\theta x)$ are the searched for real solutions. Finally, in the case $\alpha = \beta$, the above calculations show

$$y_h(x) = K_1 e^{\alpha(x-x_0)} + K_2 \int_{x_0}^{x} e^{\alpha(x-\eta)} e^{\alpha(\eta-x_0)} \, d\eta$$

$$= (c_1 + c_2 x) e^{\alpha x}$$

with $c_1 = (K_1 - K_2 x_0) e^{-\alpha x_0}$ and $c_2 = K_2 e^{-\alpha x_0}$. $\square$

Proposition 19.26 *Let y_p be an arbitrary solution of the inhomogeneous differential equation*

$$y''(x) + ay'(x) + by(x) = g(x).$$

Then its general solution can be written as

$$y(x) = y_h(x) + y_p(x)$$

where y_h is the general solution of the homogeneous problem.

Proof Superposition principle.   □

Example 19.27 In order to find the general solution of the inhomogeneous differential equation

$$y''(x) - 4y(x) = e^x$$

we first consider the homogeneous part. Its characteristic equation $\lambda^2 - 4 = 0$ has the roots $\lambda_1 = 2$ and $\lambda_2 = -2$. Therefore,

$$y_h(x) = c_1 e^{2x} + c_2 e^{-2x}.$$

A particular solution of the inhomogeneous problem is found by the general formula

$$y_p(x) = \int_0^x e^{2(x-\eta)} \int_0^\eta e^{-2(\eta-\xi)} e^\xi \, d\xi \, d\eta$$

$$= e^{2x} \int_0^x e^{-4\eta} \frac{1}{3} \left(e^{3\eta} - 1 \right) d\eta$$

$$= \frac{1}{3} e^{2x} \left(\left(1 - e^{-x} \right) + \frac{1}{4} \left(e^{-4x} - 1 \right) \right).$$

Comparing this with y_h shows that the choice $y_p(x) = -\frac{1}{3} e^x$ is possible as well, since the other terms solve the homogeneous equation.

   In general, however, it is simpler to use as ansatz for y_p a linear combination of the inhomogeneity and its derivatives. In our case, the ansatz would be $y_p(x) = ae^x$. Inserting this ansatz into the inhomogeneous problem gives $a - 4a = 1$, which results again in $y_p(x) = -\frac{1}{3} e^x$.

Example 19.28 The characteristic equation of the homogeneous problem

$$y''(x) - 10y'(x) + 25y(x) = 0$$

has the double root $\lambda_1 = \lambda_2 = 5$. Therefore, its general solution is

$$y(x) = c_1 e^{5x} + c_2 x e^{5x}.$$

Example 19.29 The characteristic equation of the homogeneous problem

$$y''(x) + 2y'(x) + 2y(x) = 0$$

has the complex conjugate roots $\lambda_1 = -1 + i$ and $\lambda_2 = -1 - i$. The complex form of its general solution is

$$y(x) = c_1 e^{-(1+i)x} + c_2 e^{-(1-i)x}$$

with complex coefficients c_1 and c_2.

The real form is

$$y(x) = e^{-x}\left(d_1 \cos x + d_2 \sin x\right)$$

with real coefficients d_1 and d_2.

19.7 Exercises

1. Find the general solution of the following differential equations and sketch some solution curves

   (a) $\dot{x} = \dfrac{x}{t}$,   (b) $\dot{x} = \dfrac{t}{x}$,   (c) $\dot{x} = \dfrac{-t}{x}$.

   The direction field is most easily plotted with maple, e.g. with DEplot.

2. Using the applet *Dynamical systems in the plane*, solve Exercise 1 by rewriting the respective differential equation as an equivalent autonomous system by adding the equation $\dot{t} = 1$.

   Hint. The variables are denoted by x and y in the applet. For example, Exercise 1(a) would have to be written as $x' = x/y$ and $y' = 1$.

3. According to Newton's law of cooling, the rate of change of the temperature x of an object is proportional to the difference of its temperature and the ambient temperature a. This is modelled by the differential equation

$$\dot{x} = k(a - x),$$

   where k is a proportionality constant. Find the general solution of this differential equation.

   How long does it take to cool down an object from $x(0) = 100°$ to $40°$ at an ambient temperature of $20°$, if it cooled down from $100°$ to $80°$ in 5 minutes?

4. Solve Verhulst's differential equation from Example 19.9 and compute the limit $t \to \infty$ of the solution.

5. A tank contains $100\,l$ of liquid A. Liquid B is added at a rate of $5\,l/s$, while at the same time the mixture is pumped out with a rate of $10\,l/s$. We are interested in the amount $x(t)$ of the liquid B in the tank at time t. From the balance equation $\dot{x}(t) = \text{rate(in)} - \text{rate(out)} = \text{rate(in)} - 10 \cdot x(t)/\text{total amount}(t)$ one obtains the differential equation

$$\dot{x} = 5 - \frac{10x}{100 - 5t}, \quad x(0) = 0.$$

Explain the derivation of this equation in detail and use maple (with dsolve) to solve the initial value problem. When is the tank empty?

6. Solve the differential equations

$$\text{(a)} \quad y' = ay, \qquad \text{(b)} \quad y' = ay + 2$$

with the method of power series.

7. Find the solution of the initial value problem

$$\dot{x}(t) = 1 + x(t)^2$$

with initial value $x(0) = 0$. In which interval does the solution exist?

8. Find the solution of the initial value problem

$$\dot{x}(t) + 2x(t) = e^{4t} + 1$$

with initial value $x(0) = 0$.

9. Find the general solutions of the differential equations

$$\text{(a)} \quad \ddot{x} + 4\dot{x} - 5x = 0, \qquad \text{(b)} \quad \ddot{x} + 4\dot{x} + 5x = 0, \qquad \text{(c)} \quad \ddot{x} + 4\dot{x} = 0.$$

10. Find a particular solution of the problem

$$\ddot{x}(t) + \dot{x}(t) - 6x(t) = t^2 + 2t - 1.$$

Hint. Use the ansatz $y_p(t) = at^2 + bt + c$.

11. Find the general solution of the differential equation

$$y''(x) + 4y(x) = \cos x$$

and specify the solution for the initial data $y(0) = 1$, $y'(0) = 0$.
Hint. Consider the ansatz $y_p(x) = k_1 \cos x + k_2 \sin x$.

12. Find the general solution of the differential equation

$$y''(x) + 4y'(x) + 5y(x) = \cos 2x$$

and specify the solution for the initial data $y(0) = 1$, $y'(0) = 0$.
Hint. Consider the ansatz $y_p(x) = k_1 \cos 2x + k_2 \sin 2x$.

13. Find the general solution of the homogeneous equation

$$y''(x) + 2y'(x) + y(x) = 0.$$

Systems of Differential Equations

20

Systems of differential equations, often called differentiable dynamical systems, play a vital role in modelling time-dependent processes in mechanics, meteorology, biology, medicine, economics and other sciences. We limit ourselves to two-dimensional systems, whose solutions (trajectories) can be graphically represented as curves in the plane. The first section introduces linear systems, which can be solved analytically as will be shown. In many applications, however, nonlinear systems are required. In general, their solution cannot be given explicitly. Here it is of primary interest to understand the qualitative behaviour of solutions. In the second section of this chapter, we touch upon the rich qualitative theory of dynamical systems. The third section is devoted to analysing the mathematical pendulum in various ways. Numerical methods will be discussed in Chap. 21.

20.1 Systems of Linear Differential Equations

We start with the description of various situations which lead to systems of differential equations. In Chap. 19 Malthus' population model was presented, where the rate of change of a population $x(t)$ was assumed proportional to the existing population:

$$\dot{x}(t) = ax(t).$$

The presence of a second population $y(t)$ could result in a decrease or increase of the rate of change of $x(t)$. Conversely, the population $x(t)$ could also affect the rate of change of $y(t)$. This results in a coupled system of equations

$$\dot{x}(t) = ax(t) + by(t),$$
$$\dot{y}(t) = cx(t) + dy(t),$$

© Springer Nature Switzerland AG 2018
M. Oberguggenberger and A. Ostermann, *Analysis for Computer Scientists*,
Undergraduate Topics in Computer Science,
https://doi.org/10.1007/978-3-319-91155-7_20

with positive or negative coefficients b and c, which describe the interaction of the populations. This is the general form of a *linear system of differential equations* in two unknowns, written for short as

$$\dot{x} = ax + by,$$
$$\dot{y} = cx + dy.$$

Refined models are obtained, if one takes into account the dependence of the rate of growth on food supply, for instance. For one species this would result in an equation of the form

$$\dot{x} = (v - n)x,$$

where v denotes the available food supply and n a threshold value. So, the population is increasing if the available quantity of food is larger than n, and is otherwise decreasing. In the case of a predator–prey relationship of species x to species y, in which y is the food for x, the relative rates of change are not constant. A common assumption is that these rates contain a term that depends linearly on the other species. Under this assumption, one obtains the nonlinear system

$$\dot{x} = (ay - n)x,$$
$$\dot{y} = (d - cx)y.$$

This is the famous predator–prey model of Lotka[1] and Volterra[2] (for a detailed derivation we refer to [13, Chap. 12.2]).

The general form of a *system of nonlinear differential equations* is

$$\dot{x} = f(x, y),$$
$$\dot{y} = g(x, y).$$

Geometrically this can be interpreted in the following way. The right-hand side defines a vector field

$$(x, y) \mapsto \begin{bmatrix} f(x, y) \\ g(x, y) \end{bmatrix}$$

on $\mathbb{R}^2$; the left-hand side is the velocity vector of a plane curve

$$t \mapsto \begin{bmatrix} x(t) \\ y(t) \end{bmatrix}.$$

The solutions are thus plane curves whose velocity vectors are given by the vector field.

---

[1] A.J. Lotka, 1880–1949
[2] V. Volterra, 1860–1940

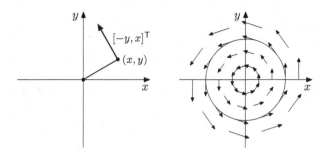

Fig. 20.1 Vector field and solution curves

Example 20.1 (Rotation of the plane)   The vector field

$$(x, y) \mapsto \begin{bmatrix} -y \\ x \end{bmatrix}$$

is perpendicular to the corresponding position vectors $[x, y]^\mathsf{T}$, see Fig. 20.1. The solutions of the system of differential equations

$$\dot{x} = -y,$$
$$\dot{y} = x$$

are the circles (Fig. 20.1)

$$x(t) = R \cos t,$$
$$y(t) = R \sin t,$$

where the radius R is given by the initial values, for instance, $x(0) = R$ and $y(0) = 0$.

Remark 20.2 The geometrical, two-dimensional representation is made possible by the fact that the right-hand side of the system does not dependent on time t explicitly. Such systems are called *autonomous*. A representation which includes the time axis (like in Chap. 19) would require a three-dimensional plot with a three-dimensional direction field

$$(x, y, t) \mapsto \begin{bmatrix} f(x, y) \\ g(x, y) \\ 1 \end{bmatrix}.$$

The solutions are represented as spatial curves

$$t \mapsto \begin{bmatrix} x(t) \\ y(t) \\ t \end{bmatrix},$$

Fig. 20.2 Direction field and
space-time diagram for
$\dot{x} = -y,\ \dot{y} = x$

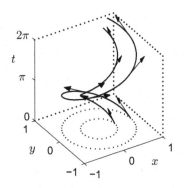

see the space-time diagram in Fig. 20.2.

Example 20.3 Another type of example which demonstrates the meaning of the vector field and the solution curves is obtained from the flow of ideal fluids. For example,

$$\dot{x} = 1 - \frac{x^2 - y^2}{(x^2 + y^2)^2},$$

$$\dot{y} = \frac{-2xy}{(x^2 + y^2)^2}$$

describes a plane, stationary potential flow around the cylinder $x^2 + y^2 \le 1$ (Fig. 20.3). The right-hand side describes the flow velocity at the point (x, y). The solution curves follow the streamlines

$$y\left(1 - \frac{1}{x^2 + y^2}\right) = C.$$

Fig. 20.3 Plane potential
flow around a cylinder

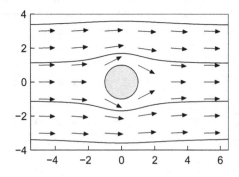

Here C denotes a constant. This can be checked by differentiating the above relation with respect to t and substituting $\dot{x}$ and $\dot{y}$ by the right-hand side of the differential equation.

Experiment 20.4 Using the applet *Dynamical systems in the plane*, study the vector field and the solution curves of the system of differential equations from Examples 20.1 and 20.3. In a similar way, study the systems of differential equations

$$\dot{x} = y, \qquad \dot{x} = y, \qquad \dot{x} = -y, \qquad \dot{x} = x, \qquad \dot{x} = y,$$
$$\dot{y} = -x, \qquad \dot{y} = x, \qquad \dot{y} = -x, \qquad \dot{y} = x, \qquad \dot{y} = y$$

and try to understand the behaviour of the solution curves.

Before turning to the solution theory of planar linear systems of differential equations, it is useful to introduce a couple of notions that serve to describe the qualitative behaviour of solution curves. The system of differential equations

$$\dot{x}(t) = f\big(x(t), y(t)\big),$$
$$\dot{y}(t) = g\big(x(t), y(t)\big)$$

together with prescribed values at $t = 0$

$$x(0) = x_0, \quad y(0) = y_0,$$

is again called an *initial value problem*. In this chapter we assume the functions f and g to be at least continuous. By a *solution curve* or a *trajectory* we mean a continuously differentiable curve $t \mapsto [x(t)\ y(t)]^{\mathsf{T}}$ whose components fulfil the system of differential equations.

For the case of a single differential equation the notion of an equilibrium point was introduced in Definition 19.21. For systems of differential equations one has an analogous notion.

Definition 20.5 (Equilibrium point) A point (x^*, y^*) is called *equilibrium point* or *equilibrium* of the system of differential equations, if $f(x^*, y^*) = 0$ and $g(x^*, y^*) = 0$.

The name comes from the fact that a solution with initial value $x_0 = x^*$, $y_0 = y^*$ remains at (x^*, y^*) for all times; in other words, if (x^*, y^*) is an equilibrium point, then $x(t) = x^*$, $y(t) = y^*$ is a solution to the system of differential equations since both the left- and right-hand sides will be zero.

From Chap. 19 we know that solutions of differential equations do not have to exist for large times. However, if solutions with initial values in a neighbourhood of an equilibrium point exist for all times then the following notions are meaningful.

Definition 20.6 Let (x^*, y^*) be an equilibrium point. If there is a neighbourhood U of (x^*, y^*) so that all trajectories with initial values (x_0, y_0) in U converge to the equilibrium point (x^*, y^*) as $t \to \infty$, then this equilibrium is called *asymptotically stable*. If for every neighbourhood V of (x^*, y^*) there is a neighbourhood W of (x^*, y^*) so that all trajectories with initial values (x_0, y_0) in W stay entirely in V, then the equilibrium (x^*, y^*) is called *stable*. An equilibrium point which is not stable is called *unstable*.

In short, stability means that trajectories that start close to the equilibrium point remain close to it; asymptotic stability means that the trajectories are *attracted* by the equilibrium point. In the case of an unstable equilibrium point there are trajectories that move away from it; in linear systems these trajectories are unbounded, and in the nonlinear case they can also converge to another equilibrium or a periodic solution (for instance, see the discussion of the mathematical pendulum in Sect. 20.3 or [13]).

In the following we determine the solution to the initial value problem

$$\dot{x} = ax + by, \qquad x(0) = x_0,$$
$$\dot{y} = cx + dy, \qquad y(0) = y_0.$$

This is a two-dimensional system of first-order linear differential equations. For this purpose we first discuss the three basic types of such systems and then show how arbitrary systems can be transformed to a system of basic type.

We denote the coefficient matrix by

$$\mathbf{A} = \begin{bmatrix} a & b \\ c & d \end{bmatrix}.$$

The decisive question is whether $\mathbf{A}$ is similar to a matrix of type I, II or III, as described in Appendix B.2. A matrix of type I has real eigenvalues and is similar to a diagonal matrix. A matrix of type II has a double real eigenvalue; its canonical form, however, contains a nilpotent part. The case of two complex conjugate eigenvalues is finally covered by type III.

Type I—real eigenvalues, diagonalisable matrix. In this case the standard form of the system is

$$\dot{x} = \alpha x, \qquad x(0) = x_0,$$
$$\dot{y} = \beta y, \qquad y(0) = y_0.$$

We know from Example 19.7 that the solutions are given by

$$x(t) = x_0 e^{\alpha t}, \quad y(t) = y_0 e^{\beta t}$$

and in particular exist for all times $t \in \mathbb{R}$. Obviously $(x^*, y^*) = (0, 0)$ is an equilibrium point. If $\alpha < 0$ and $\beta < 0$, then all solution curves approach the equilibrium $(0, 0)$ as $t \to \infty$; this equilibrium is asymptotically stable. If $\alpha \geq 0, \beta \geq 0$ (not both

Fig. 20.4 Real eigenvalues, unstable equilibrium

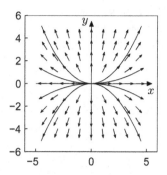

equal to zero), then the solution curves leave every neighbourhood of $(0, 0)$ and the equilibrium is unstable. Similarly, instability is present in the case where $\alpha > 0$, $\beta < 0$ (or vice versa). One calls such an equilibrium a *saddle point*.

If $\alpha \neq 0$ and $x_0 \neq 0$, then one can solve for t and represent the solution curves as graphs of functions:

$$
e^t = \left(\frac{x}{x_0}\right)^{1/\alpha}, \qquad y = y_0 \left(\frac{x}{x_0}\right)^{\beta/\alpha}.
$$

Example 20.7 The three systems

$$
\begin{array}{lll}
\dot{x} = x, & \dot{x} = -x, & \dot{x} = x, \\
\dot{y} = 2y, & \dot{y} = -2y, & \dot{y} = -2y
\end{array}
$$

have the solutions

$$
\begin{array}{lll}
x(t) = x_0 e^t, & x(t) = x_0 e^{-t}, & x(t) = x_0 e^t, \\
y(t) = y_0 e^{2t}, & y(t) = y_0 e^{-2t}, & y(t) = y_0 e^{-2t},
\end{array}
$$

respectively. The vector fields and some solutions are shown in Figs. 20.4, 20.5 and 20.6. One recognises that all coordinate half axes are solutions curves.

Type II—double real eigenvalue, not diagonalisable. The case of a double real eigenvalue $\alpha = \beta$ is a special case of type I, if the coefficient matrix is diagonalisable. There is, however, the particular situation of a double eigenvalue and a nilpotent part. Then the standard form of the system is

$$
\begin{array}{ll}
\dot{x} = \alpha x + y, & x(0) = x_0, \\
\dot{y} = \alpha y, & y(0) = y_0.
\end{array}
$$

We compute the solution component

$$
y(t) = y_0 e^{\alpha t},
$$

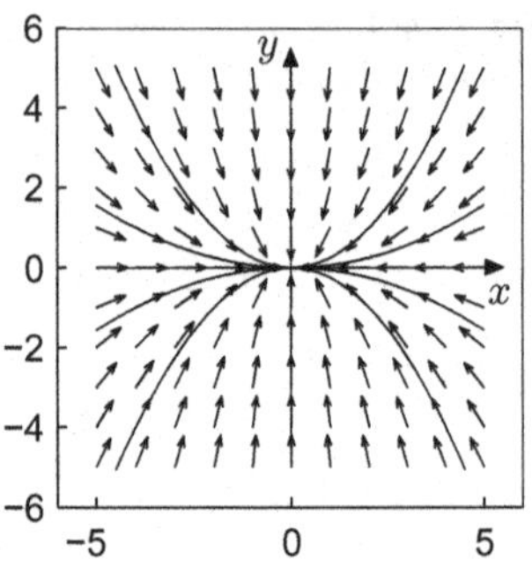

Fig. 20.5 Real eigenvalues, asymptotically stable equilibrium

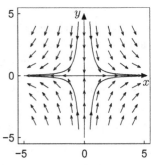

Fig. 20.6 Real eigenvalues, saddle point

Fig. 20.7 Double real eigenvalue, matrix not diagonalisable

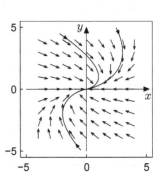

substitute it into the first equation

$$\dot{x}(t) = \alpha x(t) + y_0 e^{\alpha t}, \quad x(0) = x_0$$

and apply the variation of constants formula from Chap. 19:

$$x(t) = e^{\alpha t}\left(x_0 + \int_0^t e^{-\alpha s} y_0 e^{\alpha s}\, \mathrm{d}s\right) = e^{\alpha t}\left(x_0 + t y_0\right).$$

The vector fields and some solution curves for the case $\alpha = -1$ are depicted in Fig. 20.7.

Type III—complex conjugate eigenvalues. In this case the standard form of the system is

$$\dot{x} = \alpha x - \beta y, \quad x(0) = x_0,$$
$$\dot{y} = \beta x + \alpha y, \quad y(0) = y_0.$$

By introducing the complex variable z and the complex coefficients γ, z_0 as

$$z = x + iy, \quad \gamma = \alpha + i\beta, \quad z_0 = x_0 + iy_0,$$

we see that the above system represents the real and the imaginary parts of the equation

$$(\dot{x} + i\dot{y}) = (\alpha + i\beta)(x + iy), \quad x(0) + iy(0) = x_0 + iy_0.$$

From the complex formulation

$$\dot{z} = \gamma z, \quad z(0) = z_0,$$

the solutions can be derived immediately:

$$z(t) = z_0 e^{\gamma t}.$$

Splitting the left- and right-hand sides into real and imaginary parts, one obtains

$$x(t) + iy(t) = (x_0 + iy_0) e^{(\alpha + i\beta)t}$$
$$= (x_0 + iy_0) e^{\alpha t}(\cos \beta t + i \sin \beta t).$$

From that we get (see Sect. 4.2)

$$x(t) = x_0 e^{\alpha t} \cos \beta t - y_0 e^{\alpha t} \sin \beta t,$$
$$y(t) = x_0 e^{\alpha t} \sin \beta t + y_0 e^{\alpha t} \cos \beta t.$$

The point $(x^*, y^*) = (0, 0)$ is again an equilibrium point. In the case $\alpha < 0$ it is asymptotically stable; for $\alpha > 0$ it is unstable; for $\alpha = 0$ it is stable but not asymptotically stable. Indeed the solution curves are circles and hence bounded, but are not attracted by the origin as $t \to \infty$.

Example 20.8 The vector fields and solutions curves for the two systems

$$\dot{x} = \tfrac{1}{10}x - y, \quad \dot{x} = -\tfrac{1}{10}x - y,$$
$$\dot{y} = x + \tfrac{1}{10}y, \quad \dot{y} = x - \tfrac{1}{10}y$$

Fig. 20.8 Complex
eigenvalues, unstable

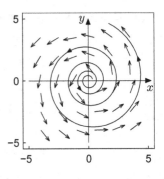

Fig. 20.9 Complex
eigenvalues, asymptotically
stable

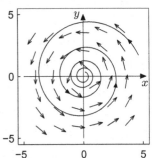

are given in Figs. 20.8 and 20.9. For the stable case $\dot{x} = -y$, $\dot{y} = x$ we refer to
Fig. 20.1.

General solution of a linear system of differential equations. The similarity transformation from Appendix B allows us to solve arbitrary linear systems of differential
equations by reduction to the three standard cases.

Proposition 20.9 *For an arbitrary* (2×2)*-matrix* **A***, the initial value problem*

$$\begin{bmatrix} \dot{x}(t) \\ \dot{y}(t) \end{bmatrix} = \mathbf{A} \begin{bmatrix} x(t) \\ y(t) \end{bmatrix}, \quad \begin{bmatrix} x(0) \\ y(0) \end{bmatrix} = \begin{bmatrix} x_0 \\ y_0 \end{bmatrix}$$

has a unique solution that exists for all times $t \in \mathbb{R}$. *This solution can be computed
explicitly by transformation to one of the types I, II or III.*

Proof According to Appendix B.2 there is an invertible matrix **T** such that

$$\mathbf{T}^{-1}\mathbf{A}\mathbf{T} = \mathbf{B},$$

where **B** belongs to one of the standard types I, II, III. We set

$$\begin{bmatrix} u \\ v \end{bmatrix} = \mathbf{T}^{-1} \begin{bmatrix} x \\ y \end{bmatrix}$$

Fig. 20.10 Example 20.10

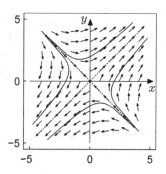

and obtain the transformed system

$$\begin{bmatrix} \dot{u} \\ \dot{v} \end{bmatrix} = \mathbf{T}^{-1} \begin{bmatrix} \dot{x} \\ \dot{y} \end{bmatrix} = \mathbf{T}^{-1} \mathbf{A} \begin{bmatrix} x \\ y \end{bmatrix} = \mathbf{T}^{-1} \mathbf{A} \mathbf{T} \begin{bmatrix} u \\ v \end{bmatrix} = \mathbf{B} \begin{bmatrix} u \\ v \end{bmatrix}, \quad \begin{bmatrix} u(0) \\ v(0) \end{bmatrix} = \mathbf{T}^{-1} \begin{bmatrix} x_0 \\ y_0 \end{bmatrix}.$$

We solve this system of differential equations depending on its type, as explained above. Each of these systems in standard form has a unique solution which exists for all times. The reverse transformation

$$\begin{bmatrix} x \\ y \end{bmatrix} = \mathbf{T} \begin{bmatrix} u \\ v \end{bmatrix}$$

yields the solution of the original system.   □

Thus, modulo a linear transformation, the types I, II, III actually comprise all cases that can occur.

Example 20.10 We study the solution curves of the system

$$\dot{x} = x + 2y,$$
$$\dot{y} = 2x + y.$$

The corresponding coefficient matrix

$$\mathbf{A} = \begin{bmatrix} 1 & 2 \\ 2 & 1 \end{bmatrix}$$

has the eigenvalues $\lambda_1 = 3$ and $\lambda_2 = -1$ with respective eigenvectors $\mathbf{e}_1 = [1\ 1]^\mathsf{T}$ and $\mathbf{e}_2 = [-1\ 1]^\mathsf{T}$. It is of type I, and the origin is a saddle point. The vector field and some solutions can be seen in Fig. 20.10.

Remark 20.11 The proof of Proposition 20.9 shows the structure of the general solution of a linear system of differential equations. Assume, for example, that the roots λ_1, λ_2 of the characteristic polynomial of the coefficient matrix are real and distinct, so the system is of type I. The general solution in transformed coordinates is given by

$$u(t) = C_1 e^{\lambda_1 t}, \quad v(t) = C_2 e^{\lambda_2 t}.$$

If we denote the columns of the transformation matrix by $\mathbf{t}_1, \mathbf{t}_2$, then the solution in the original coordinates is

$$\begin{bmatrix} x(t) \\ y(t) \end{bmatrix} = \mathbf{t}_1 u(t) + \mathbf{t}_2 v(t) = \begin{bmatrix} t_{11} C_1 e^{\lambda_1 t} + t_{12} C_2 e^{\lambda_2 t} \\ t_{21} C_1 e^{\lambda_1 t} + t_{22} C_2 e^{\lambda_2 t} \end{bmatrix}.$$

Every component is a particular linear combination of the transformed solutions $u(t), v(t)$. In the case of complex conjugate roots $\mu \pm i\nu$ (type III) the components of the general solution are particular linear combinations of the functions $e^{\mu t} \cos \nu t$ and $e^{\mu t} \sin \nu t$. In the case of a double root α (type II), the components are given as linear combinations of the functions $e^{\alpha t}$ and $t e^{\alpha t}$.

20.2 Systems of Nonlinear Differential Equations

In contrast to linear systems of differential equations, the solutions to nonlinear systems can generally not be expressed by explicit formulas. Apart from numerical methods (Chap. 21) the qualitative theory is of interest. It describes the behaviour of solutions without knowing them explicitly. In this section we will demonstrate this with the help of an example from population dynamics.

The Lotka–Volterra model. In Sect. 20.1 the predator–prey model of Lotka and Volterra was introduced. In order to simplify the presentation, we set all coefficients equal to one. Thus the system becomes

$$\dot{x} = x(y - 1),$$
$$\dot{y} = y(1 - x).$$

The equilibrium points are $(x^*, y^*) = (1, 1)$ and $(x^{**}, y^{**}) = (0, 0)$. Obviously, the coordinate half axes are solution curves given by

$$\begin{aligned} x(t) &= x_0 e^{-t}, \quad & x(t) &= 0, \\ y(t) &= 0, \quad & y(t) &= y_0 e^t. \end{aligned}$$

The equilibrium $(0, 0)$ is thus a saddle point (unstable); we will later analyse the type of equilibrium $(1, 1)$. In the following we will only consider the first quadrant $x \geq 0, y \geq 0$, which is relevant in biological models. Along the straight line $x = 1$ the vector field is horizontal, and along the straight line $y = 1$ it is vertical. It looks

Fig. 20.11 Vector field of
the Lotka–Volterra model

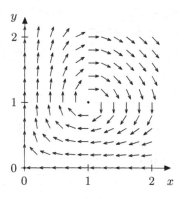

as if the solution curves rotate about the equilibrium point $(1, 1)$, see Fig. 20.11.

In order to be able to verify this conjecture we search for a function $H(x, y)$ which is constant along the solution curves:

$$H(x(t), y(t)) = C.$$

Such a function is called a *first integral, invariant* or *conserved quantity* of the system of differential equations. Consequently, we have

$$\frac{\mathrm{d}}{\mathrm{d}t} H(x(t), y(t)) = 0$$

or by the chain rule for functions in two variables (Proposition 15.16)

$$\frac{\partial H}{\partial x} \dot{x} + \frac{\partial H}{\partial y} \dot{y} = 0.$$

With the ansatz

$$H(x, y) = F(x) + G(y),$$

we should have

$$F'(x)\dot{x} + G'(y)\dot{y} = 0.$$

Inserting the differential equations we obtain

$$F'(x) x(y - 1) + G'(y) y(1 - x) = 0,$$

and a separation of the variables yields

$$\frac{x\, F'(x)}{x - 1} = \frac{y\, G'(y)}{y - 1}.$$

Since the variables x and y are independent of each other, this is only possible if both sides are constant:

$$\frac{x\,F'(x)}{x-1} = C, \quad \frac{y\,G'(y)}{y-1} = C.$$

It follows that

$$F'(x) = C\left(1 - \frac{1}{x}\right), \quad G'(y) = C\left(1 - \frac{1}{y}\right)$$

and thus

$$H(x, y) = C(x - \log x + y - \log y) + D.$$

This function has a global minimum at $(x^*, y^*) = (1, 1)$, as can also be seen in Fig. 20.12.

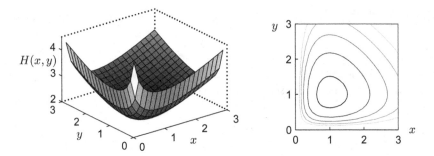

Fig. 20.12 First integral and level sets

The solution curves of the Lotka–Volterra system lie on the level sets

$$x - \log x + y - \log y = \text{const.}$$

These level sets are obviously closed curves. The question arises whether the solution curves are also closed, and the solutions thus are periodic. In the following proposition we will answer this question affirmatively. Periodic, closed solution curves are called *periodic orbits*.

Proposition 20.12 *For initial values* $x_0 > 0$, $y_0 > 0$ *the solution curves of the Lotka–Volterra system are periodic orbits and* $(x^*, y^*) = (1, 1)$ *is a stable equilibrium point.*

Outline of proof The proof of the fact that the solution

$$t \mapsto \begin{bmatrix} x(t) \\ y(t) \end{bmatrix}, \quad \begin{bmatrix} x(0) \\ y(0) \end{bmatrix} = \begin{bmatrix} x_0 \\ y_0 \end{bmatrix}$$

exists (and is unique) for all initial values $x_0 \geq 0$, $y_0 \geq 0$ and all times $t \in \mathbb{R}$ requires methods that go beyond the scope of this book. The interested reader is referred to [13, Chap. 8]. In order to prove periodicity, we take initial values $(x_0, y_0) \neq (1, 1)$ and show that the corresponding solution curves return to the initial value after finite time $\tau > 0$. For that we split the first quadrant $x > 0$, $y > 0$ into four regions

$$Q_1 : x > 1, y > 1; \qquad Q_2 : x < 1, y > 1;$$
$$Q_3 : x < 1, y < 1; \qquad Q_4 : x > 1, y < 1$$

and show that every solution curve moves (clockwise) through all four regions in finite time. For instance, consider the case $(x_0, y_0) \in Q_3$, so $0 < x_0 < 1, 0 < y_0 < 1$. We want to show that the solution curve reaches the region Q_2 in finite time; i.e. $y(t)$ assumes the value one. From the differential equations it follows that

$$\dot{x} = x(y - 1) < 0, \qquad \dot{y} = y(1 - x) > 0$$

in region Q_3 and thus

$$x(t) < x_0, \quad y(t) > y_0, \quad \dot{y}(t) > y_0(1 - x_0),$$

as long as $(x(t), y(t))$ stays in region Q_3. If $y(t)$ were less than one for all times $t > 0$, then the following inequalities would hold:

$$1 > y(t) = y_0 + \int_0^t \dot{y}(s)\,ds > y_0 + \int_0^t y_0(1 - x_0)\,ds = y_0 + ty_0(1 - x_0).$$

However, the latter expression diverges to infinity as $t \to \infty$, a contradiction. Consequently, $y(t)$ has to reach the value 1 and thus the region Q_2 in finite time. Likewise one reasons for the other regions. Thus there exists a time $\tau > 0$ such that $(x(\tau), y(\tau)) = (x_0, y_0)$.

From that the periodicity of the orbit follows. Since the system of differential equations is autonomous, $t \mapsto (x(t + \tau), y(t + \tau))$ is a solution as well. As just shown, both solutions have the same initial value at $t = 0$. The uniqueness of the solution of initial value problems implies that the two solutions are identical, so

$$x(t) = x(t + \tau), \quad y(t) = y(t + \tau)$$

is fulfilled for all times $t \in \mathbb{R}$. However, this proves that the solution $t \mapsto (x(t), y(t))$ is periodic with period τ.

All solution curves in the first quadrant with the exception of the equilibrium are thus periodic orbits. Solution curves that start close to $(x^*, y^*) = (1, 1)$ stay close, see Fig. 20.12. The point $(1, 1)$ is thus a stable equilibrium. $\qquad\square$

Fig. 20.13 Solution curves
of the Lotka–Volterra model

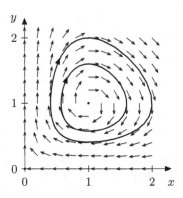

Figure 20.13 shows some solution curves. The populations of predator and prey thus increase and decrease periodically and in opposite direction. For further population models we refer to [6].

20.3 The Pendulum Equation

As a second example of a nonlinear system we consider the *mathematical pendulum*. It models an object of mass m that is attached to the origin with a (massless) cord of length l and moves under the gravitational force $-mg$, see Fig. 20.14. The variable $x(t)$ denotes the angle of deflection from the vertical direction, measured in counterclockwise direction. The tangential acceleration of the object is equal to $l\ddot{x}(t)$, and the tangential component of the gravitational force is $-mg\sin x(t)$. According to Newton's law, force = mass × acceleration, we have

$$-mg\sin x = ml\ddot{x}$$

or

$$ml\ddot{x} + mg\sin x = 0.$$

This is a second-order nonlinear differential equation. We will later reduce it to a first-order system, but for a start, we wish to derive a conserved quantity.

Conservation of energy. Multiplying the pendulum equation by $l\dot{x}$ gives

$$ml^2\dot{x}\ddot{x} + mgl\dot{x}\sin x = 0.$$

We identify $\dot{x}\ddot{x}$ as the derivative of $\frac{1}{2}\dot{x}^2$ and $\dot{x}\sin x$ as the derivative of $1 - \cos x$ and arrive at a conserved quantity, which we denote by $H(x, \dot{x})$:

$$\frac{d}{dt}H(x, \dot{x}) = \frac{d}{dt}\left(\frac{1}{2}ml^2\dot{x}^2 + mgl\big(1 - \cos x\big)\right) = 0;$$

Fig. 20.14 Derivation of the pendulum equation

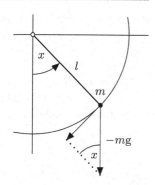

that is, $H(x(t), \dot{x}(t))$ is constant when $x(t)$ is a solution of the pendulum equation. Recall from mechanics that the *kinetic energy* of the moving mass is given by

$$T(\dot{x}) = \frac{1}{2}ml^2\dot{x}^2.$$

The *potential energy* is defined as the work required to move the mass from its height $-l$ at rest to position $-l\cos x$, that is

$$U(x) = \int_{-l}^{-l\cos x} mg \; d\xi = mgl(1 - \cos x).$$

Thus the conserved quantity is identified as the *total energy*

$$H(x, \dot{x}) = T(\dot{x}) + U(x),$$

in accordance with the well-known mechanical principle of conservation of total energy.

Note that the linearisation

$$\sin x = x + \mathcal{O}(x^3) \approx x$$

for small angles x leads to the approximation

$$ml\ddot{x} + mgx = 0.$$

For convenience, we will cancel m in the equation and set $g/l = 1$. Then the pendulum equation reads

$$\ddot{x} = -\sin x,$$

with the conserved quantity

$$H(x, \dot{x}) = \frac{1}{2}\dot{x}^2 + 1 - \cos x,$$

while the linearised pendulum equation reads

$$\ddot{x} = -x.$$

Reduction to a first-order system. Every explicit second-order differential equation $\ddot{x} = f(x, \dot{x})$ can be reduced to a first-order system by introducing the new variable $y = \dot{x}$, resulting in the system

$$\dot{x} = y,$$
$$\dot{y} = f(x, y).$$

Applying this procedure to the pendulum equation and adjoining initial data leads to the system

$$\dot{x} = y, \qquad x(0) = x_0,$$
$$\dot{y} = -\sin x, \qquad y(0) = y_0$$

for the mathematical pendulum. Here x denotes the angle of deflection and y the angular velocity of the object.

The linearised pendulum equation can be written as the system

$$\dot{x} = y, \qquad x(0) = x_0,$$
$$\dot{y} = -x, \qquad y(0) = y_0.$$

Apart from the change in sign this system of differential equations coincides with that of Example 20.1. It is a system of type III; hence its solution is given by

$$x(t) = x_0 \cos t + y_0 \sin t,$$
$$y(t) = -x_0 \sin t + y_0 \cos t.$$

The first line exhibits the solution to the second-order linearised equation $\ddot{x} = -x$ with initial data $x(0) = x_0$, $\dot{x}(0) = y_0$. The same result can be obtained directly by the methods of Sect. 19.6.

Solution trajectories of the nonlinear pendulum. In the coordinates (x, y), the total energy reads

$$H(x, y) = \frac{1}{2}y^2 + 1 - \cos x.$$

As was shown above, it is a conserved quantity; hence solution curves for prescribed initial values (x_0, y_0) lie on the level sets $H(x, y) = C$; i.e.

$$\frac{1}{2}y^2 + 1 - \cos x = \frac{1}{2}y_0^2 + 1 - \cos x_0,$$

$$y = \pm\sqrt{y_0^2 - 2\cos x_0 + 2\cos x}.$$

Fig. 20.15 Solution curves, mathematical pendulum

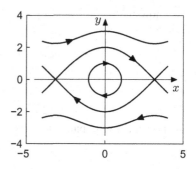

Figure 20.15 shows some solution curves. There are unstable equilibria at $y = 0$, $x = \ldots, -3\pi, -\pi, \pi, 3\pi, \ldots$ which are connected by limit curves. One of the two limit curves passes through $x_0 = 0$, $y_0 = 2$. The solution with these initial values lies on the limit curve and approaches the equilibrium $(\pi, 0)$ as $t \to \infty$, and $(-\pi, 0)$ as $t \to -\infty$. Initial values that lie between these limit curves (for instance the values $x_0 = 0$, $|y_0| < 2$) give rise to periodic solutions of small amplitude (less than π). The solutions outside represent large oscillations where the pendulum loops. We remark that effects of friction are not taken into account in this model.

Power series solutions. The method of power series for solving differential equations has been introduced in Chap. 19. We have seen that the linearised pendulum equation $\ddot{x} = -x$ can be solved explicitly by the methods of Sects. 19.6 and 20.1. Also, the nonlinear pendulum equation can be solved explicitly with the aid of certain higher transcendental functions, the Jacobian elliptic functions. Nevertheless, it is of interest to analyse the solutions of these equations by means of powers series, especially in view of the fact that they can be readily obtained in maple.

Example 20.13 (Power series for the linearised pendulum)   As an example, we solve the initial value problem

$$\ddot{x} = -x, \qquad x(0) = a, \quad \dot{x}(0) = 0$$

by means of the power series ansatz

$$x(t) = \sum_{n=0}^{\infty} c_n t^n = c_0 + c_1 t + c_2 t^2 + c_3 t^3 + c_4 t^4 + \cdots$$

We have

$$\dot{x}(t) = \sum_{n=1}^{\infty} n c_n t^{n-1} = c_1 + 2c_2 t + 3c_3 t^2 + 4c_4 t^3 + \cdots$$

$$\ddot{x}(t) = \sum_{n=2}^{\infty} n(n-1) c_n t^{n-2} = 2c_2 + 6c_3 t + 12c_4 t^2 + \cdots$$

We know that $c_0 = a$ and $c_1 = 0$. Equating $\ddot{x}(t)$ with $-x(t)$ gives, up to second degree,

$$2c_2 + 6c_3 t + 12c_4 t^2 + \cdots = -a - c_2 t^2 - \cdots$$

thus

$$c_2 = -\frac{a}{2}, \quad c_3 = 0, \quad c_4 = -\frac{c_2}{12} = \frac{a}{24}, \quad \ldots$$

The power series expansion starts with

$$x(t) = a\left(1 - \frac{1}{2}t^2 + \frac{1}{24}t^4 \mp \ldots\right)$$

and seemingly coincides with the Taylor series of the known solution $x(t) = a \cos t$.

Example 20.14 (Power series for the nonlinear pendulum) We turn to the initial value problem for the nonlinear pendulum equation

$$\ddot{x} = -\sin x, \qquad x(0) = a, \quad \dot{x}(0) = 0,$$

making the same power series ansatz as in Example 20.13. Developing the sine function into its Taylor series, inserting the lowest order terms of the power series of $x(t)$ and noting that $c_0 = a$, $c_1 = 0$ yields

$$
\begin{aligned}
-\sin x(t) &= -\left(x(t) - \frac{1}{3!}x(t)^3 + \frac{1}{5!}x(t)^5 + \ldots\right) \\
&= -\left(a + c_2 t^2 + \cdots\right) + \frac{1}{3!}\left(a + c_2 t^2 + \ldots\right)^3 - \frac{1}{5!}\left(a + c_2 t^2 + \cdots\right)^5 \\
&= -\left(a + c_2 t^2 + \cdots\right) + \frac{1}{6}\left(a^3 + 3a^2 c_2 t^2 + \cdots\right) \\
&\quad - \frac{1}{120}\left(a^5 + 5a^4 c_2 t^2 + \cdots\right),
\end{aligned}
$$

where we have used the binomial formulas. Equating the last line with

$$\ddot{x}(t) = 2c_2 + 6c_3 t + 12c_4 t^2 + \cdots$$

shows that

$$
\begin{aligned}
2c_2 &= -a + \frac{1}{6}a^3 - \frac{1}{120}a^5 \pm \ldots \\
6c_3 &= 0 \\
12c_4 &= c_2\left(-1 + \frac{3}{6}a^2 - \frac{5}{120}a^4 \pm \ldots\right)
\end{aligned}
$$

which suggests that

$$c_2 = -\frac{1}{2}\sin a, \quad c_4 = \frac{1}{24}\sin a \cos a.$$

Collecting terms and factoring a out finally results in the expansion

$$x(t) = a \left(1 - \frac{1}{2} \frac{\sin a}{a} t^2 + \frac{1}{24} \frac{\sin a \cos a}{a} t^4 \pm \ldots \right).$$

The expansion can be checked by means of the maple command

```
ode:=diff(x(t),[t$2])=-sin(x(t))
ics:=x(0)=a,D(x)(0)=0
dsolve({ode,ics}, x(t), series);
```

If the initial deflection $x_0 = a$ is sufficiently small, then

$$\frac{\sin a}{a} \approx 1, \qquad \cos a \approx 1,$$

see Proposition 6.10, and so the solution $x(t)$ is close to the solution $a \cos t$ of the linearised pendulum equation, as expected.

20.4   Exercises

1. The space-time diagram of a two-dimensional system of differential equations (Remark 20.2) can be obtained by introducing time as third variable $z(t) = t$ and passing to the three-dimensional system

$$\begin{bmatrix} \dot{x} \\ \dot{y} \\ \dot{z} \end{bmatrix} = \begin{bmatrix} f(x, y) \\ g(x, y) \\ 1 \end{bmatrix}.$$

   Use this observation to visualise the systems from Examples 20.1 and 20.3. Study the time-dependent solution curves with the applet *Dynamical systems in space*.

2. Compute the general solutions of the following three systems of differential equations by transformation to standard form:

$$\dot{x} = \tfrac{3}{5}x - \tfrac{4}{5}y, \qquad \dot{x} = -3y, \qquad \dot{x} = \tfrac{7}{4}x - \tfrac{5}{4}y,$$
$$\dot{y} = -\tfrac{4}{5}x - \tfrac{3}{5}y, \qquad \dot{y} = x, \qquad \dot{y} = \tfrac{5}{4}x + \tfrac{1}{4}y.$$

   Visualise the solution curves with the applet *Dynamical systems in the plane*.

3. Small, undamped oscillations of an object of mass m attached to a spring are described by the differential equation $m\ddot{x} + kx = 0$. Here, $x = x(t)$ denotes the displacement from the position of rest and k is the spring stiffness. Introduce the variable $y = \dot{x}$ and rewrite the second-order differential equation as a linear system of differential equations. Find the general solution.

4. A company deposits its profits in an account with continuous interest rate $a\%$. The balance is denoted by $x(t)$. Simultaneously the amount $y(t)$ is withdrawn continuously from the account, where the rate of withdrawal is equal to $b\%$ of the account balance. With $r = a/100$, $s = b/100$ this leads to the linear system of differential equations

$$\dot{x}(t) = r(x(t) - y(t)),$$
$$\dot{y}(t) = s\,x(t).$$

Find the solution $(x(t), y(t))$ for the initial values $x(0) = 1$, $y(0) = 0$ and analyse how big s can be in comparison with r so that the account balance $x(t)$ is increasing for all times without oscillations.

5. A national economy has two sectors (for instance industry and agriculture) with the production volumes $x_1(t)$, $x_2(t)$ at time t. If one assumes that the investments are proportional to the respective growth rate, then the classical model of Leontief[3] [24, Chap. 9.5] states

$$x_1(t) = a_{11}x_1(t) + a_{12}x_2(t) + b_1\dot{x}_1(t) + c_1(t),$$
$$x_2(t) = a_{21}x_1(t) + a_{22}x_2(t) + b_2\dot{x}_2(t) + c_2(t).$$

Here a_{ij} denotes the required amount of goods from sector i to produce one unit of goods in sector j. Further $b_i\dot{x}_i(t)$ are the investments, and $c_i(t)$ is the consumption in sector i. Under the simplifying assumptions $a_{11} = a_{22} = 0$, $a_{12} = a_{21} = a$ ($0 < a < 1$), $b_1 = b_2 = 1$, $c_1(t) = c_2(t) = 0$ (no consumption) one obtains the system of differential equations

$$\dot{x}_1(t) = x_1(t) - ax_2(t),$$
$$\dot{x}_2(t) = -ax_1(t) + x_2(t).$$

Find the general solution and discuss the result.

6. Use the applet *Dynamical systems in the plane* to analyse the solution curves of the differential equations of the mathematical pendulum and translate the mathematical results to statements about the mechanical behaviour.

7. Derive the conserved quantity $H(x, y) = \frac{1}{2}y^2 + 1 - \cos x$ of the pendulum equation by means of the ansatz $H(x, y) = F(x) + G(y)$ as for the Lotka–Volterra system.

8. Using maple, find the power series solution to the nonlinear pendulum equation $\ddot{x} = -\sin x$ with initial data

$$x(0) = a, \quad \dot{x}(0) = 0 \quad \text{and} \quad x(0) = 0, \quad \dot{x}(0) = b.$$

Check by how much its coefficients differ from the ones of the power series solution of the corresponding linearised pendulum equation $\ddot{x} = -x$ for various values of a, b between 0 and 1.

---

[3] W. Leontief, 1906–1999.

9. The differential equation $m\ddot{x}(t) + kx(t) + 2cx^3(t) = 0$ describes a nonlinear mass–spring system where $x(t)$ is the displacement of the mass m, k is the stiffness of the spring and the term cx^3 models nonlinear effects ($c > 0 \ldots$ hardening, $c < 0 \ldots$ softening).

(a) Show that

$$H(x, \dot{x}) = \frac{1}{2}\left(m\dot{x}^2 + kx^2 + cx^4\right)$$

is a conserved quantity.

(b) Assume that $m = 1$, $k = 1$ and $x(0) = 0$, $\dot{x}(0) = 1$. Reduce the second-order equation to a first-order system. Making use of the conserved quantity, plot the solution curves for the values of $c = 0, c = -0.2, c = 0.2$ and $c = 5$. *Hint.* A typical maple command is

```
with(plots,implicitplot); c:=5;
implicitplot(y^2+x^2+c*x^4=1,x=-1.5..1.5,y=-1.5..1.5);
```

10. Using maple, find the power series solution to the nonlinear differential equation $\ddot{x}(t) + x(t) + 2cx^3(t) = 0$ with initial data $x(0) = a$, $\dot{x}(0) = b$. Compare it to the solution with $c = 0$.

Numerical Solution of Differential Equations

21

As we have seen in the last two chapters, only particular classes of differential equations can be solved analytically. Especially for nonlinear problems one has to rely on numerical methods.

In this chapter we discuss several variants of Euler's method as a prototype. Motivated by the Taylor expansion of the analytical solution we deduce Euler approximations and study their stability properties. In this way we introduce the reader to several important aspects of the numerical solution of differential equations. We point out, however, that for most real-life applications one has to use more sophisticated numerical methods.

21.1 The Explicit Euler Method

The differential equation

$$y'(x) = f\big(x, y(x)\big)$$

defines the slope of the tangent to the solution curve $y(x)$. Expanding the solution at the point $x + h$ into a Taylor series

$$y(x + h) = y(x) + hy'(x) + \mathcal{O}(h^2)$$

and inserting the above value for $y'(x)$, one obtains

$$y(x + h) = y(x) + hf\big(x, y(x)\big) + \mathcal{O}(h^2)$$

© Springer Nature Switzerland AG 2018
M. Oberguggenberger and A. Ostermann, *Analysis for Computer Scientists*,
Undergraduate Topics in Computer Science,
https://doi.org/10.1007/978-3-319-91155-7_21

and consequently for small h the approximation

$$y(x + h) \approx y(x) + hf\big(x, y(x)\big).$$

This observation motivates the (explicit) *Euler method*.

Euler's method. For the numerical solution of the initial value problem

$$y'(x) = f\big(x, y(x)\big), \qquad y(a) = y_0$$

on the interval $[a, b]$ we first divide the interval into N parts of length $h = (b - a)/N$ and define the grid points $x_j = x_0 + jh$, $0 \le j \le N$, see Fig. 21.1.

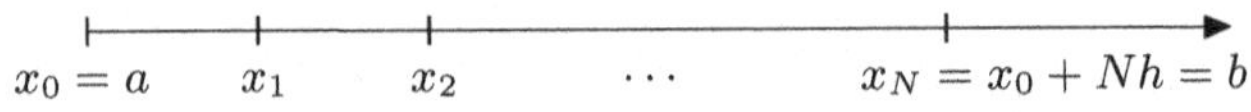

Fig. 21.1 Equidistant grid points $x_j = x_0 + jh$

The distance h between two grid points is called *step size*. We look for a numerical approximation y_n to the exact solution $y(x_n)$ at x_n, i.e. $y_n \approx y(x_n)$. According to the considerations above we should have

$$y(x_{n+1}) \approx y(x_n) + hf\big(x_n, y(x_n)\big).$$

If one replaces the exact solution by the numerical approximation and $\approx$ by $=$, then one obtains the explicit Euler method

$$y_{n+1} = y_n + hf(x_n, y_n),$$

which *defines* the approximation y_{n+1} as a function of y_n.

Starting from the initial value y_0 one computes from this recursion the approximations $y_1, y_2, \ldots, y_N \approx y(b)$. The points (x_i, y_i) are the vertices of a polygon which approximates the graph of the exact solution $y(x)$. Figure 21.2 shows the exact solution of the differential equation $y' = y$, $y(0) = 1$ as well as polygons defined by Euler's method for three different step sizes.

Euler's method is convergent of order 1, see [11, Chap. II.3]. On bounded intervals $[a, b]$ one thus has the uniform error estimate

$$|y(x_n) - y_n| \le Ch$$

for all $n \ge 1$ and sufficiently small h with $0 \le nh \le b - a$. The constant C depends on the length of the interval and the solution $y(x)$, however, it does not depend on n and h.

Fig. 21.2 Euler
approximation to
$y' = y$, $y(0) = 1$

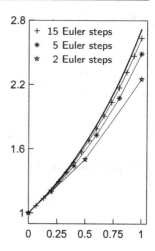

Example 21.1 The solution of the initial value problem $y' = y$, $y(0) = 1$ is $y(x) = e^x$. For $nh = 1$ the numerical solution y_n approximates the exact solution at $x = 1$. Due to

$$y_n = y_{n-1} + h y_{n-1} = (1 + h) y_{n-1} = \cdots = (1 + h)^n y_0$$

we have

$$y_n = (1 + h)^n = \left(1 + \frac{1}{n}\right)^n \approx e.$$

The convergence of Euler's method thus implies

$$e = \lim_{n \to \infty} \left(1 + \frac{1}{n}\right)^n.$$

This formula for e was already deduced in Example 7.11.

In commercial software packages, methods of higher order are used for the numerical integration, for example Runge–Kutta or multi-step methods. All these methods are refinements of Euler's method. In modern implementations of these algorithms the error is automatically estimated and the step size adaptively adjusted to the problem. For more details, we refer to [11, 12].

Experiment 21.2 In MATLAB you can find information on the numerical solution of differential equations by calling help funfun. For example, one can solve the initial value problem

$$y' = y^2, \qquad y(0) = 0.9$$

on the interval [0, 1] with the command

```
[x,y] = ode23('qfun', [0,1], 0.9);
```

The file `qfun.m` has to contain the definition of the function

```
function yp = f(x,y)
yp = y.^2;
```

For a plot of the solution, one sets the option

```
myopt = odeset('OutputFcn','odeplot')
```

and calls the solver by

```
[x,y] = ode23('qfun', [0,1], 0.9, myopt);
```

Start the program with different initial values and observe the *blow up* for $y(0) \geq 1$.

21.2   Stability and Stiff Problems

The linear differential equation

$$y' = ay, \quad y(0) = 1$$

has the solution

$$y(x) = e^{ax}.$$

For $a \leq 0$ this solution has the following qualitative property, independent of the size of a:

$$|y(x)| \leq 1 \quad \text{for all } x \geq 0.$$

We are investigating whether numerical methods preserve this property. For that we solve the differential equation with the explicit Euler method and obtain

$$y_n = y_{n-1} + hay_{n-1} = (1+ha)y_{n-1} = \cdots = (1+ha)^n y_0 = (1+ha)^n.$$

For $-2 \leq ha \leq 0$ the numerical solution obeys the same bound

$$|y_n| = \left|(1+ha)^n\right| = \left|1+ha\right|^n \leq 1$$

as the exact solution. However, for $ha < -2$ a dramatic instability occurs although the exact solution is harmless. In fact, all explicit methods have the same difficulties in this situation: The solution is only stable under very restrictive conditions on the step size. For the explicit Euler method the condition for stability is

$$-2 \le ha \le 0.$$

For $a \ll 0$ this implies a drastic restriction on the step size, which eventually makes the method in this situation inefficient.

In this case a remedy is offered by implicit methods, for example, the *implicit Euler method*

$$y_{n+1} = y_n + hf(x_{n+1}, y_{n+1}).$$

It differs from the explicit method by the fact that the slope of the tangent is now taken at the endpoint. For the determination of the numerical solution, a nonlinear equation has to be solved in general. Therefore, such methods are called implicit. The implicit Euler method has the same accuracy as the explicit one, but by far better stability properties, as the following analysis shows. If one applies the implicit Euler method to the initial value problem

$$y' = ay, \quad y(0) = 1, \quad \text{with } a \le 0,$$

one obtains

$$y_n = y_{n-1} + hf(x_n, y_n) = y_{n-1} + hay_n,$$

and therefore

$$y_n = \frac{1}{1 - ha} y_{n-1} = \cdots = \frac{1}{(1 - ha)^n} y_0 = \frac{1}{(1 - ha)^n}.$$

The procedure is thus stable, i.e. $|y_n| \le 1$, if

$$\left| (1 - ha)^n \right| \ge 1.$$

However, for $a \le 0$ this is fulfilled for all $h \ge 0$. Thus the procedure is stable for *arbitrarily large* step sizes.

Remark 21.3 A differential equation is called *stiff*, if for its solution the *implicit Euler* method is more efficient (often dramatically more efficient) than the *explicit* method.

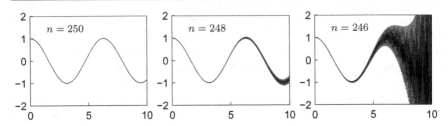

Fig. 21.3 Instability of the explicit Euler method. In each case the picture shows the exact solution and the approximating polygons of Euler's method with n steps

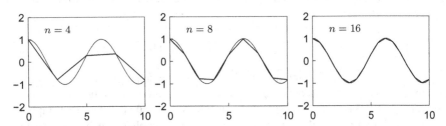

Fig. 21.4 Stability of the implicit Euler method. In each case the picture shows the exact solution and the approximating polygons of Euler's method with n steps

Example 21.4 (From [12, Chap. IV.1]) We integrate the initial value problem

$$y' = -50(y - \cos x), \qquad y(0) = 0.997.$$

Its exact solution is

$$y(x) = \frac{2500}{2501} \cos x + \frac{50}{2501} \sin x - \frac{6503}{250100} \, e^{-50x}$$
$$\approx \cos(x - 0.02) - 0.0026 \, e^{-50x}.$$

The solution looks quite harmless and resembles $\cos x$, but the equation is stiff with $a = -50$. Warned by the analysis above we expect difficulties for explicit methods.

We integrate this differential equation numerically on the interval $[0, 10]$ with the explicit Euler method and step sizes $h = 10/n$ with $n = 250$, 248 and 246. For $n < 250$, i.e. $h > 1/25$, exponential instabilities occur, see Fig. 21.3. This is consistent with the considerations above because the product ah satisfies $ah \le -2$ for $h > 1/25$.

However, if one integrates the differential equation with the implicit Euler method, then even for very large step sizes no instabilities arise, see Fig. 21.4. The implicit Euler method is more costly than the explicit one, as the computation of y_{n+1} from

$$y_{n+1} = y_n + hf(x_{n+1}, y_{n+1})$$

generally requires the solution of a nonlinear equation.

21.3 Systems of Differential Equations

For the derivation of a simple numerical method for solving systems of differential equations

$$\dot{x}(t) = f\big(t, x(t), y(t)\big), \qquad\qquad x(t_0) = x_0,$$
$$\dot{y}(t) = g\big(t, x(t), y(t)\big), \qquad\qquad y(t_0) = y_0,$$

one again starts from the Taylor expansion of the analytic solution

$$x(t + h) = x(t) + h\dot{x}(t) + \mathcal{O}(h^2),$$
$$y(t + h) = y(t) + h\dot{y}(t) + \mathcal{O}(h^2)$$

and replaces the derivatives by the right-hand sides of the differential equations. For small step sizes h this motivates the explicit Euler method

$$x_{n+1} = x_n + hf\big(t_n, x_n, y_n\big),$$
$$y_{n+1} = y_n + hg\big(t_n, x_n, y_n\big).$$

One interprets x_n and y_n as numerical approximations to the exact solution $x(t_n)$ and $y(t_n)$ at time $t_n = t_0 + nh$.

Example 21.5 In Sect. 20.2 we have investigated the Lotka–Volterra model

$$\dot{x} = x(y - 1),$$
$$\dot{y} = y(1 - x).$$

In order to compute the periodic orbit through the point $(x_0, y_0) = (2, 2)$ numerically, we apply the explicit Euler method and obtain the recursion

$$x_{n+1} = x_n + hx_n(y_n - 1),$$
$$y_{n+1} = y_n + hy_n(1 - x_n).$$

Starting from the initial values $x_0 = 2$ and $y_0 = 2$ this recursion determines the numerical solution for $n \geq 0$. The results for three different step sizes are depicted in Fig. 21.5. Note the linear convergence of the numerical solution for $h \to 0$.

This numerical experiment shows that one has to choose a very small step size in order to obtain the periodicity of the true orbit in the numerical solution. Alternatively, one can use numerical methods of higher order or—in the present example—also the following modification of Euler's method

$$x_{n+1} = x_n + hx_n(y_n - 1),$$
$$y_{n+1} = y_n + hy_n(1 - x_{n+1}).$$

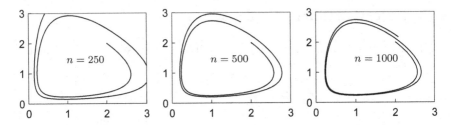

Fig. 21.5 Numerical computation of a periodic orbit of the Lotka–Volterra model. The system was integrated on the interval $0 \leq t \leq 14$ with Euler's method and constant step sizes $h = 14/n$ for $n = 250, 500$ and 1000

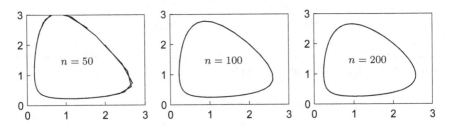

Fig. 21.6 Numerical computation of a periodic orbit of the Lotka–Volterra model. The system was integrated on the interval $0 \leq t \leq 14$ with the modified Euler method with constant step sizes $h = 14/n$ for $n = 50, 100$ and 200

In this method one uses instead of x_n the updated value x_{n+1} for the computation of y_{n+1}. The numerical results, obtained with this modified Euler method, are given in Fig. 21.6. One clearly recognises the superiority of this approach compared to the original one. Clearly, the *geometric* structure of the solution was better captured.

21.4 Exercises

1. Solve the special Riccati equation $y' = x^2 + y^2$, $y(0) = -4$ for $0 \leq x \leq 2$ with MATLAB.
2. Solve with MATLAB the linear system of differential equations

$$\dot{x} = y, \qquad \dot{y} = -x$$

with initial values $x(0) = 1$ and $y(0) = 0$ on the interval $[0, b]$ for $b = 2\pi, 10\pi$ and 200π. Explain the observations.
   Hint. In MATLAB one can use the command ode23('mat21_1',[0 2*pi], [0 1]), where the file mat21_1.m defines the right-hand side of the differential equation.

3. Solve the Lotka–Volterra system

$$\dot{x} = x(y - 1), \qquad \dot{y} = y(1 - x)$$

for $0 \le t \le 14$ with initial values $x(0) = 2$ and $y(0) = 2$ in MATLAB. Compare your results with Figs. 21.5 and 21.6.

4. Let $y'(x) = f(x, y(x))$. Show by Taylor expansion that

$$y(x + h) = y(x) + hf\left(x + \frac{h}{2}, y(x) + \frac{h}{2}f(x, y(x))\right) + \mathcal{O}(h^3)$$

and deduce from this the numerical scheme

$$y_{n+1} = y_n + hf\left(x_n + \frac{h}{2}, y_n + \frac{h}{2}f(x_n, y_n)\right).$$

Compare the accuracy of this scheme with that of the explicit Euler method applied to the Riccati equation of Exercise 1.

5. Apply the numerical scheme

$$y_{n+1} = y_n + hf\left(x_n + \frac{h}{2}, y_n + \frac{h}{2}f(x_n, y_n)\right)$$

to the solution of the differential equation

$$y' = y, \qquad y(0) = 1$$

and show that

$$y_n = \left(1 + h + \frac{h^2}{2}\right)^n.$$

Deduce from this identity a formula for approximating e. How do the results compare to the corresponding formula obtained with the explicit Euler scheme? *Hint:* Choose $h = 1/n$ for $n = 10, 100, 1000, 10000$.

6. Let $a \le 0$. Apply the numerical scheme

$$y_{n+1} = y_n + hf\left(x_n + \frac{h}{2}, y_n + \frac{h}{2}f(x_n, y_n)\right)$$

to the linear differential equation $y' = ay$, $y(0) = 1$ and find a condition on the step size h such that $|y_n| \le 1$ for all $n \in \mathbb{N}$.

Vector Algebra

A

In various sections of this book we referred to the notion of a vector. We assumed the reader to have a basic knowledge on standard school level. In this appendix we recapitulate some basic notions of vector algebra. For a more detailed presentation we refer to [2].

A.1 Cartesian Coordinate Systems

A *Cartesian coordinate system* in the plane (in space) consists of two (three) real lines *(coordinate axes)* which intersect in right angles at the point O (origin). We always assume that the coordinate system is positively (right-handed) oriented. In a planar right-handed system, the positive y-axis lies to the left in viewing direction of the positive x-axis (Fig. A.1). In a positively oriented three-dimensional coordinate system, the direction of the positive z-axis is obtained by turning the x-axis in the direction of the y-axis according to the *right-hand rule*, see Fig. A.2.

The *coordinates* of a point are obtained by parallel projection of the point onto the coordinate axes. In the case of the plane, the point A has the coordinates a_1 and a_2, and we write

$$A = (a_1, a_2) \in \mathbb{R}^2.$$

In an analogous way a point A in space with coordinates a_1, a_2 and a_3 is denoted as

$$A = (a_1, a_2, a_3) \in \mathbb{R}^3.$$

Thus one has a unique representation of points by pairs or triples of real numbers.

© Springer Nature Switzerland AG 2018
M. Oberguggenberger and A. Ostermann, *Analysis for Computer Scientists*,
Undergraduate Topics in Computer Science,
https://doi.org/10.1007/978-3-319-91155-7

Fig. A.1 Cartesian
coordinate system in the
plane

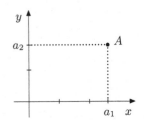

Fig. A.2 Cartesian
coordinate system in space

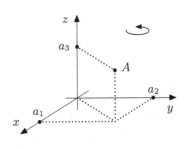

A.2   Vectors

For two points P and Q in the plane (in space) there exists *exactly one* parallel
translation which moves P to Q. This translation is called a *vector*. Vectors are thus
quantities with *direction and length*. The direction is that from P to Q and the length
is the distance between the two points. Vectors are used to model, e.g., forces and
velocities. We always write vectors in boldface.

For a vector $\mathbf{a}$, the vector $-\mathbf{a}$ denotes the parallel translation which undoes the
action of $\mathbf{a}$; the *zero vector* $\mathbf{0}$ does not cause any translation. The composition of
two parallel translations is again a parallel translation. The corresponding operation
for vectors is called *addition* and is performed according to the *parallelogram rule*.
For a real number $\lambda \geq 0$, the vector $\lambda\mathbf{a}$ is the vector which has the same direction
as $\mathbf{a}$, but λ times the length of $\mathbf{a}$. This operation is called *scalar multiplication*. For
addition and scalar multiplication the usual rules of computation apply.

Let $\mathbf{a}$ be the parallel translation from P to Q. The length of the vector $\mathbf{a}$, i.e. the
distance between P and Q, is called *norm* (or *magnitude*) of the vector. We denote
it by $\|\mathbf{a}\|$. A vector $\mathbf{e}$ with $\|\mathbf{e}\| = 1$ is called a *unit vector*.

A.3   Vectors in a Cartesian Coordinate System

In a Cartesian coordinate system with origin O, we denote the three unit vectors in
direction of the three coordinate axes by $\mathbf{e}_1$, $\mathbf{e}_2$, $\mathbf{e}_3$, see Fig. A.3. These three vectors
are called the *standard basis* of $\mathbb{R}^3$. Here $\mathbf{e}_1$ stands for the parallel translation which
moves O to $(1, 0, 0)$, etc.

Fig. A.3 Representation of **a** in components

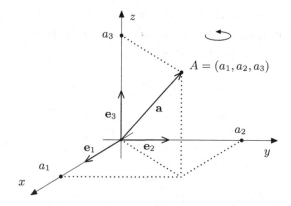

The vector **a** which moves O to A can be decomposed in a unique way as $\mathbf{a} = a_1\mathbf{e}_1 + a_2\mathbf{e}_2 + a_3\mathbf{e}_3$. We denote it by

$$\mathbf{a} = \begin{bmatrix} a_1 \\ a_2 \\ a_3 \end{bmatrix},$$

where the column on the right-hand side is the so-called *coordinate vector* of **a** with respect to the standard basis $\mathbf{e}_1, \mathbf{e}_2, \mathbf{e}_3$. The vector **a** is also called *position vector* of the point A. Since we are always working with the standard basis, we *identify* a vector with its coordinate vector, i.e.

$$\mathbf{e}_1 = \begin{bmatrix} 1 \\ 0 \\ 0 \end{bmatrix}, \quad \mathbf{e}_2 = \begin{bmatrix} 0 \\ 1 \\ 0 \end{bmatrix}, \quad \mathbf{e}_3 = \begin{bmatrix} 0 \\ 0 \\ 1 \end{bmatrix}$$

and

$$\mathbf{a} = a_1\mathbf{e}_1 + a_2\mathbf{e}_2 + a_3\mathbf{e}_3 = \begin{bmatrix} a_1 \\ 0 \\ 0 \end{bmatrix} + \begin{bmatrix} 0 \\ a_2 \\ 0 \end{bmatrix} + \begin{bmatrix} 0 \\ 0 \\ a_3 \end{bmatrix} = \begin{bmatrix} a_1 \\ a_2 \\ a_3 \end{bmatrix}.$$

To distinguish between points and vectors we write the coordinates of points in a row, but use column notation for vectors.

For column vectors the usual rules of computation apply:

$$\begin{bmatrix} a_1 \\ a_2 \\ a_3 \end{bmatrix} + \begin{bmatrix} b_1 \\ b_2 \\ b_3 \end{bmatrix} = \begin{bmatrix} a_1 + b_1 \\ a_2 + b_2 \\ a_3 + b_3 \end{bmatrix}, \quad \lambda \begin{bmatrix} a_1 \\ a_2 \\ a_3 \end{bmatrix} = \begin{bmatrix} \lambda a_1 \\ \lambda a_2 \\ \lambda a_3 \end{bmatrix}.$$

Thus the addition and the scalar multiplication are defined *componentwise*.

The norm of a vector $\mathbf{a} \in \mathbb{R}^2$ with components a_1 and a_2 is computed with Pythagoras' theorem as $\|\mathbf{a}\| = \sqrt{a_1^2 + a_2^2}$. Hence the components of the vector $\mathbf{a}$ have the representation

$$a_1 = \|\mathbf{a}\| \cdot \cos \alpha \quad \text{and} \quad a_2 = \|\mathbf{a}\| \cdot \sin \alpha,$$

and we obtain

$$\mathbf{a} = \|\mathbf{a}\| \cdot \begin{bmatrix} \cos \alpha \\ \sin \alpha \end{bmatrix} = \text{length} \cdot \text{direction},$$

see Fig. A.4. For the norm of a vector $\mathbf{a} \in \mathbb{R}^3$ the analogous formula $\|\mathbf{a}\| = \sqrt{a_1^2 + a_2^2 + a_3^2}$ holds.

Remark A.1 The plane $\mathbb{R}^2$ (and likewise the space $\mathbb{R}^3$) appears in two roles: On the one hand as *point space* (its objects are points which cannot be added) and on the other hand as *vector space* (its objects are vectors that can be added). By parallel translation, $\mathbb{R}^2$ (as vector space) can be attached to every point of $\mathbb{R}^2$ (as point space), see Fig. A.5. In general, however, point space and vector space are different sets, as shown in the following example.

Example A.2 (Particle on a circle) Let P be the position of a particle which moves on a circle and $\mathbf{v}$ its velocity vector. Then the point space is the circle and the vector space the tangent to the circle at the point P, see Fig. A.6.

Fig. A.4 A vector $\mathbf{a}$ with its components a_1 and a_2

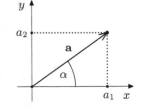

Fig. A.5 Force $\mathbf{F}$ applied at P

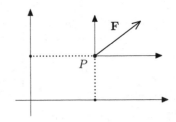

Fig. A.6 Velocity vector is
tangential to the circle

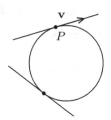

A.4 The Inner Product (Dot Product)

The *angle* $\angle(\mathbf{a}, \mathbf{b})$ between two vectors $\mathbf{a}, \mathbf{b}$ is uniquely determined by the condition $0 \leq \angle(\mathbf{a}, \mathbf{b}) \leq \pi$. One calls a vector $\mathbf{a}$ *orthogonal (perpendicular)* to $\mathbf{b}$ (in symbols: $\mathbf{a} \perp \mathbf{b}$), if $\angle(\mathbf{a}, \mathbf{b}) = \frac{\pi}{2}$. By definition, the zero vector $\mathbf{0}$ is orthogonal to all vectors.

Definition A.3 Let $\mathbf{a}, \mathbf{b}$ be planar (or spatial) vectors. The number

$$\langle \mathbf{a}, \mathbf{b} \rangle = \begin{cases} \|\mathbf{a}\| \cdot \|\mathbf{b}\| \cdot \cos \angle(\mathbf{a}, \mathbf{b}) & \mathbf{a} \neq 0, \ \mathbf{b} \neq 0, \\ 0 & \text{otherwise,} \end{cases}$$

is called the *inner product (dot product)* of $\mathbf{a}$ and $\mathbf{b}$.

For planar vectors $\mathbf{a}, \mathbf{b} \in \mathbb{R}^2$ the inner product is calculated from their components as

$$\langle \mathbf{a}, \mathbf{b} \rangle = \left\langle \begin{bmatrix} a_1 \\ a_2 \end{bmatrix}, \begin{bmatrix} b_1 \\ b_2 \end{bmatrix} \right\rangle = a_1 b_1 + a_2 b_2.$$

For vectors $\mathbf{a}, \mathbf{b} \in \mathbb{R}^3$ the analogous formula holds:

$$\langle \mathbf{a}, \mathbf{b} \rangle = \left\langle \begin{bmatrix} a_1 \\ a_2 \\ a_3 \end{bmatrix}, \begin{bmatrix} b_1 \\ b_2 \\ b_3 \end{bmatrix} \right\rangle = a_1 b_1 + a_2 b_2 + a_3 b_3.$$

Example A.4 The standard basis vectors $\mathbf{e}_i$ have length 1 and are mutually orthogonal, i.e.

$$\langle \mathbf{e}_i, \mathbf{e}_j \rangle = \begin{cases} 1, & i = j, \\ 0, & i \neq j. \end{cases}$$

For vectors $\mathbf{a}, \mathbf{b}, \mathbf{c}$ and a scalar $\lambda \in \mathbb{R}$ the inner product obeys the rules

(a) $\langle \mathbf{a}, \mathbf{b} \rangle = \langle \mathbf{b}, \mathbf{a} \rangle$,

(b) $\langle \mathbf{a}, \mathbf{a} \rangle = \|\mathbf{a}\|^2$,

(c) $\langle \mathbf{a}, \mathbf{b} \rangle = 0 \ \Leftrightarrow \ \mathbf{a} \perp \mathbf{b}$,

(d) $\langle \lambda \mathbf{a}, \mathbf{b} \rangle = \langle \mathbf{a}, \lambda \mathbf{b} \rangle = \lambda \langle \mathbf{a}, \mathbf{b} \rangle$,

(e) $\langle \mathbf{a} + \mathbf{b}, \mathbf{c} \rangle = \langle \mathbf{a}, \mathbf{c} \rangle + \langle \mathbf{b}, \mathbf{c} \rangle$.

Example A.5 For the vectors

$$
\mathbf{a} = \begin{bmatrix} 2 \\ -4 \\ 0 \end{bmatrix}, \quad \mathbf{b} = \begin{bmatrix} 6 \\ 3 \\ 4 \end{bmatrix}, \quad \mathbf{c} = \begin{bmatrix} 1 \\ 0 \\ -1 \end{bmatrix}
$$

we have

$$
\|\mathbf{a}\|^2 = 4 + 16 = 20, \quad \|\mathbf{b}\|^2 = 36 + 9 + 16 = 61, \quad \|\mathbf{c}\|^2 = 1 + 1 = 2,
$$

and

$$
\langle \mathbf{a}, \mathbf{b} \rangle = 12 - 12 = 0, \quad \langle \mathbf{a}, \mathbf{c} \rangle = 2.
$$

From this we conclude that $\mathbf{a}$ is perpendicular to $\mathbf{b}$ and

$$
\cos \angle(\mathbf{a}, \mathbf{c}) = \frac{\langle \mathbf{a}, \mathbf{c} \rangle}{\|\mathbf{a}\| \cdot \|\mathbf{c}\|} = \frac{2}{\sqrt{20}\sqrt{2}} = \frac{1}{\sqrt{10}}.
$$

The value of the angle between $\mathbf{a}$ and $\mathbf{c}$ is thus

$$
\angle(\mathbf{a}, \mathbf{c}) = \arccos \frac{1}{\sqrt{10}} = 1.249 \text{ rad}.
$$

A.5　The Outer Product (Cross Product)

For vectors $\mathbf{a}, \mathbf{b}$ in $\mathbb{R}^2$ one defines

$$
\mathbf{a} \times \mathbf{b} = \begin{bmatrix} a_1 \\ a_2 \end{bmatrix} \times \begin{bmatrix} b_1 \\ b_2 \end{bmatrix} = \det \begin{bmatrix} a_1 & b_1 \\ a_2 & b_2 \end{bmatrix} = a_1 b_2 - a_2 b_1 \in \mathbb{R},
$$

the *cross product* of $\mathbf{a}$ and $\mathbf{b}$. An elementary calculation shows that

$$
|\mathbf{a} \times \mathbf{b}| = \|\mathbf{a}\| \cdot \|\mathbf{b}\| \cdot \sin \angle(\mathbf{a}, \mathbf{b}).
$$

Thus $|\mathbf{a} \times \mathbf{b}|$ is the *area* of the parallelogram spanned by $\mathbf{a}$ and $\mathbf{b}$.

For vectors $\mathbf{a}, \mathbf{b} \in \mathbb{R}^3$ one defines the *cross product* as

$$
\mathbf{a} \times \mathbf{b} = \begin{bmatrix} a_1 \\ a_2 \\ a_3 \end{bmatrix} \times \begin{bmatrix} b_1 \\ b_2 \\ b_3 \end{bmatrix} = \begin{bmatrix} a_2 b_3 - a_3 b_2 \\ a_3 b_1 - a_1 b_3 \\ a_1 b_2 - a_2 b_1 \end{bmatrix} \in \mathbb{R}^3.
$$

This product has the following geometric interpretation: If $\mathbf{a} = \mathbf{0}$ or $\mathbf{b} = \mathbf{0}$ or $\mathbf{a} = \lambda \mathbf{b}$ then $\mathbf{a} \times \mathbf{b} = \mathbf{0}$. Otherwise $\mathbf{a} \times \mathbf{b}$ is the vector

(a) which is *perpendicular* to $\mathbf{a}$ and $\mathbf{b}$: $\langle \mathbf{a} \times \mathbf{b}, \mathbf{a} \rangle = \langle \mathbf{a} \times \mathbf{b}, \mathbf{b} \rangle = 0$;

(b) which is directed such that $\mathbf{a}, \mathbf{b}, \mathbf{a} \times \mathbf{b}$ forms a *right-handed system*;

(c) whose length is equal to the *area* F of the *parallelogram* spanned by $\mathbf{a}$ and $\mathbf{b}$:
$F = \|\mathbf{a} \times \mathbf{b}\| = \|\mathbf{a}\| \cdot \|\mathbf{b}\| \cdot \sin \angle(\mathbf{a}, \mathbf{b})$.

Example A.6 Let E be the plane spanned by the two vectors

$$\mathbf{a} = \begin{bmatrix} 1 \\ -1 \\ 2 \end{bmatrix} \quad \text{and} \quad \mathbf{b} = \begin{bmatrix} 1 \\ 0 \\ 1 \end{bmatrix}.$$

Then

$$\mathbf{a} \times \mathbf{b} = \begin{bmatrix} 1 \\ -1 \\ 2 \end{bmatrix} \times \begin{bmatrix} 1 \\ 0 \\ 1 \end{bmatrix} = \begin{bmatrix} -1 \\ 1 \\ 1 \end{bmatrix}$$

is a vector perpendicular to this plane.

For $\mathbf{a}, \mathbf{b}, \mathbf{c} \in \mathbb{R}^3$ and $\lambda \in \mathbb{R}$ the following rules apply

(a) $\mathbf{a} \times \mathbf{a} = \mathbf{0}, \quad \mathbf{a} \times \mathbf{b} = -(\mathbf{b} \times \mathbf{a})$,

(b) $\lambda(\mathbf{a} \times \mathbf{b}) = (\lambda \mathbf{a}) \times \mathbf{b} = \mathbf{a} \times (\lambda \mathbf{b})$,

(c) $(\mathbf{a} + \mathbf{b}) \times \mathbf{c} = \mathbf{a} \times \mathbf{c} + \mathbf{b} \times \mathbf{c}$.

However, the cross product is *not associative* and

$$\mathbf{a} \times (\mathbf{b} \times \mathbf{c}) \neq (\mathbf{a} \times \mathbf{b}) \times \mathbf{c}$$

for general $\mathbf{a}, \mathbf{b}, \mathbf{c}$. For instance, the standard basis vectors of the $\mathbb{R}^3$ satisfy the following identities

$$\mathbf{e}_1 \times (\mathbf{e}_1 \times \mathbf{e}_2) = \mathbf{e}_1 \times \mathbf{e}_3 = -\mathbf{e}_2,$$
$$(\mathbf{e}_1 \times \mathbf{e}_1) \times \mathbf{e}_2 = \mathbf{0} \times \mathbf{e}_2 = \mathbf{0}.$$

A.6 Straight Lines in the Plane

The general equation of a straight line in the (x, y)-plane is

$$ax + by = c,$$

where at least one of the coefficients a and b must be different from zero. The straight line consists of all points (x, y) which satisfy the above equation,

$$g = \left\{ (x, y) \in \mathbb{R}^2; ax + by = c \right\}.$$

If $b = 0$ (and thus $a \neq 0$) we get

$$x = \frac{c}{a},$$

and thus a line parallel to the y-axis. If $b \neq 0$, one can solve for y and obtains the standard form of a straight line

$$y = -\frac{a}{b}x + \frac{c}{b} = kx + d$$

with *slope k* and *intercept d*.

The *parametric representation* of the straight line is obtained from the general solution of the linear equation

$$ax + by = c.$$

Since this equation is underdetermined, one replaces the independent variable by a parameter and solves for the other variable.

Example A.7 In the equation

$$y = kx + d$$

x is considered as independent variable. One sets $x = \lambda$ and obtains $y = k\lambda + d$ and thus the parametric representation

$$\begin{bmatrix} x \\ y \end{bmatrix} = \begin{bmatrix} 0 \\ d \end{bmatrix} + \lambda \begin{bmatrix} 1 \\ k \end{bmatrix}, \quad \lambda \in \mathbb{R}.$$

Example A.8 In the equation

$$x = 4$$

y is the independent variable (it does not even appear). This straight line in parametric representation is

$$\begin{bmatrix} x \\ y \end{bmatrix} = \begin{bmatrix} 4 \\ 0 \end{bmatrix} + \lambda \begin{bmatrix} 0 \\ 1 \end{bmatrix}.$$

In general, the parametric representation of a straight line is of the form

$$\begin{bmatrix} x \\ y \end{bmatrix} = \begin{bmatrix} p \\ q \end{bmatrix} + \lambda \begin{bmatrix} u \\ v \end{bmatrix}, \quad \lambda \in \mathbb{R}$$

(position vector of a point plus a multiple of a direction vector). A vector perpendicular to this straight line is called a *normal vector*. It is a multiple of

$$\begin{bmatrix} v \\ -u \end{bmatrix}, \quad \text{since} \quad \left\langle \begin{bmatrix} u \\ v \end{bmatrix}, \begin{bmatrix} v \\ -u \end{bmatrix} \right\rangle = 0.$$

The conversion to the nonparametric form is obtained by multiplying the equation in parametric form by a normal vector. Thereby the parameter is eliminated. In the example above one obtains

$$vx - uy = pv - qu.$$

In particular, the coefficients of x and y in the nonparametric form are just the components of a normal vector of the straight line.

A.7 Planes in Space

The general form of a plane in $\mathbb{R}^3$ is

$$ax + by + cz = d,$$

where at least one of the coefficients a, b, c is different from zero. The plane consists of all points which satisfy the above equation, i.e.

$$E = \left\{ (x, y, z) \in \mathbb{R}^3; \, ax + by + cz = d \right\}.$$

Since at least one of the coefficients is nonzero, one can solve the equation for the corresponding unknown.

For example, if $c \neq 0$ one can solve for z to obtain

$$z = -\frac{a}{c}x - \frac{b}{c}y + \frac{d}{c} = kx + ly + e.$$

Here k represents the slope in x-direction, l is the slope in y-direction and e is the intercept on the z-axis (because $z = e$ for $x = y = 0$). By introducing parameters for the independent variables x and y

$$x = \lambda, \quad y = \mu, \quad z = k\lambda + l\mu + e$$

one thus obtains the *parametric representation* of the plane:

$$\begin{bmatrix} x \\ y \\ z \end{bmatrix} = \begin{bmatrix} 0 \\ 0 \\ e \end{bmatrix} + \lambda \begin{bmatrix} 1 \\ 0 \\ k \end{bmatrix} + \mu \begin{bmatrix} 0 \\ 1 \\ l \end{bmatrix}, \quad \lambda, \mu \in \mathbb{R}.$$

In general, the parametric representation of a plane in $\mathbb{R}^3$ is

$$\begin{bmatrix} x \\ y \\ z \end{bmatrix} = \begin{bmatrix} p \\ q \\ r \end{bmatrix} + \lambda \begin{bmatrix} v_1 \\ v_2 \\ v_3 \end{bmatrix} + \mu \begin{bmatrix} w_1 \\ w_2 \\ w_3 \end{bmatrix}$$

with $\mathbf{v} \times \mathbf{w} \neq \mathbf{0}$. If one multiplies this equation with $\mathbf{v} \times \mathbf{w}$ and uses

$$\langle \mathbf{v}, \mathbf{v} \times \mathbf{w} \rangle = \langle \mathbf{w}, \mathbf{v} \times \mathbf{w} \rangle = 0,$$

one again obtains the *nonparametric* form

$$\left\langle \begin{bmatrix} x \\ y \\ z \end{bmatrix}, \mathbf{v} \times \mathbf{w} \right\rangle = \left\langle \begin{bmatrix} p \\ q \\ r \end{bmatrix}, \mathbf{v} \times \mathbf{w} \right\rangle.$$

Example A.9 We compute the nonparametric form of the plane

$$\begin{bmatrix} x \\ y \\ z \end{bmatrix} = \begin{bmatrix} 3 \\ 1 \\ 1 \end{bmatrix} + \lambda \begin{bmatrix} 1 \\ -1 \\ 2 \end{bmatrix} + \mu \begin{bmatrix} 1 \\ 0 \\ 1 \end{bmatrix}.$$

A normal vector to this plane is given by

$$\mathbf{v} \times \mathbf{w} = \begin{bmatrix} 1 \\ -1 \\ 2 \end{bmatrix} \times \begin{bmatrix} 1 \\ 0 \\ 1 \end{bmatrix} = \begin{bmatrix} -1 \\ 1 \\ 1 \end{bmatrix},$$

and thus the equation of the plane is

$$-x + y + z = -1.$$

A.8 Straight Lines in Space

A straight line in $\mathbb{R}^3$ can be seen as the *intersection of two planes*:

$$g : \begin{cases} ax + by + cz = d, \\ ex + fy + gz = h. \end{cases}$$

The straight line is the set of all points (x, y, z) which fulfil this system of equations (two equations in three unknowns). Generically, the solution of the above system can be parametrised by one parameter (this is the case of a straight line). However, it may also happen that the planes are parallel. In this situation they either coincide, or they do not intersect at all.

A straight line can also be represented *parametrically* by the position vector of a point and an arbitrary multiple of a direction vector

$$\begin{bmatrix} x \\ y \\ z \end{bmatrix} = \begin{bmatrix} p \\ q \\ r \end{bmatrix} + \lambda \begin{bmatrix} u \\ v \\ w \end{bmatrix}, \quad \lambda \in \mathbb{R}.$$

The direction vector is obtained as difference of the position vectors of two points on the straight line.

Example A.10 We want to determine the straight line through the points $P = (1, 2, 0)$ and $Q = (3, 1, 2)$. A direction vector $\mathbf{a}$ of this line is given by

$$\mathbf{a} = \begin{bmatrix} 3 \\ 1 \\ 2 \end{bmatrix} - \begin{bmatrix} 1 \\ 2 \\ 0 \end{bmatrix} = \begin{bmatrix} 2 \\ -1 \\ 2 \end{bmatrix}.$$

Thus a parametric representation of the straight line is

$$g : \begin{bmatrix} x \\ y \\ z \end{bmatrix} = \begin{bmatrix} 1 \\ 2 \\ 0 \end{bmatrix} + \lambda \begin{bmatrix} 2 \\ -1 \\ 2 \end{bmatrix}, \quad \lambda \in \mathbb{R}.$$

The conversion from parametric to nonparametric form and vice versa is achieved by *elimination* or *introduction* of a parameter λ. In the example above one computes $z = 2\lambda$ from the last equation and inserts it into the first two equations. This yields the nonparametric form

$$x - z = 1,$$
$$2y + z = 4.$$

Matrices

B

In this book matrix algebra is required in multi-dimensional calculus, for systems of differential equations and for linear regression. This appendix serves to outline the basic notions. A more detailed presentation can be found in [2].

B.1   Matrix Algebra

An $(m \times n)$-*matrix* $\mathbf{A}$ is a rectangular scheme of the form

$$\mathbf{A} = \begin{bmatrix} a_{11} & a_{12} & \dots & a_{1n} \\ a_{21} & a_{22} & \dots & a_{2n} \\ \vdots & \vdots & & \vdots \\ a_{m1} & a_{m2} & \dots & a_{mn} \end{bmatrix}.$$

The *entries (coefficients, elements)* a_{ij}, $i = 1, \dots, m$, $j = 1, \dots, n$ of the matrix $\mathbf{A}$ are real or complex numbers. In this section we restrict ourselves to real numbers. An $(m \times n)$-matrix has m rows and n columns; if $m = n$, and the matrix is called *square*. Vectors of length m can be understood as matrices with one column, i.e as $(m \times 1)$-matrices. In particular, one refers to the columns

$$\mathbf{a}_j = \begin{bmatrix} a_{1j} \\ a_{2j} \\ \vdots \\ a_{mj} \end{bmatrix}, \quad j = 1, \dots, n$$

of a matrix $\mathbf{A}$ as *column vectors* and accordingly also writes

$$\mathbf{A} = [\mathbf{a}_1 \vdots \mathbf{a}_2 \vdots \dots \vdots \mathbf{a}_n]$$

© Springer Nature Switzerland AG 2018
M. Oberguggenberger and A. Ostermann, *Analysis for Computer Scientists*,
Undergraduate Topics in Computer Science,
https://doi.org/10.1007/978-3-319-91155-7

for the matrix. The rows of the matrix are sometimes called *row vectors*.

The *product* of an $(m \times n)$-matrix $\mathbf{A}$ with a vector $\mathbf{x}$ of length n is defined as

$$\mathbf{y} = \mathbf{A}\mathbf{x}, \qquad \begin{bmatrix} y_1 \\ y_2 \\ \vdots \\ y_m \end{bmatrix} = \begin{bmatrix} a_{11}x_1 + a_{12}x_2 + \ldots + a_{1n}x_n \\ a_{21}x_1 + a_{22}x_2 + \ldots + a_{2n}x_n \\ \vdots \\ a_{m1}x_1 + a_{m2}x_2 + \ldots + a_{mn}x_n \end{bmatrix}$$

and results in a vector $\mathbf{y}$ of length m. The kth entry of $\mathbf{y}$ is obtained by the inner product of the kth row vector of the matrix $\mathbf{A}$ (written as a column) with the vector $\mathbf{x}$.

Example B.1 For instance, the product of a (2×3)-matrix with a vector of length 3 is computed as follows:

$$\mathbf{A} = \begin{bmatrix} a & b & c \\ d & e & f \end{bmatrix}, \quad \mathbf{x} = \begin{bmatrix} 3 \\ -1 \\ 2 \end{bmatrix}, \quad \mathbf{A}\mathbf{x} = \begin{bmatrix} 3a - b + 2c \\ 3d - e + 2f \end{bmatrix}.$$

The assignment $\mathbf{x} \mapsto \mathbf{y} = \mathbf{A}\mathbf{x}$ defines a *linear mapping* from $\mathbb{R}^n$ to $\mathbb{R}^m$. The linearity is characterised by the validity of the relations

$$\mathbf{A}(\mathbf{u} + \mathbf{v}) = \mathbf{A}\mathbf{u} + \mathbf{A}\mathbf{v}, \qquad \mathbf{A}(\lambda\mathbf{u}) = \lambda\mathbf{A}\mathbf{u}$$

for all $\mathbf{u}, \mathbf{v} \in \mathbb{R}^n$ and $\lambda \in \mathbb{R}$, which follow immediately from the definition of matrix multiplication. If $\mathbf{e}_j$ is the jth standard basis vector of $\mathbb{R}^n$, then obviously

$$\mathbf{a}_j = \mathbf{A}\mathbf{e}_j.$$

This means that the columns of the matrix $\mathbf{A}$ are just the images of the standard basis vectors under the linear mapping defined by $\mathbf{A}$.

Matrix arithmetic. Matrices of the same format can be added and subtracted by adding or subtracting their components. Multiplication with a number $\lambda \in \mathbb{R}$ is also defined componentwise. The *transpose* $\mathbf{A}^\mathsf{T}$ of a matrix $\mathbf{A}$ is obtained by swapping rows and columns; i.e. the ith row of the matrix $\mathbf{A}^\mathsf{T}$ consists of the elements of the ith column of $\mathbf{A}$:

$$\mathbf{A} = \begin{bmatrix} a_{11} & a_{12} & \ldots & a_{1n} \\ a_{21} & a_{22} & \ldots & a_{2n} \\ \vdots & \vdots & & \vdots \\ a_{m1} & a_{m2} & \ldots & a_{mn} \end{bmatrix}, \quad \mathbf{A}^\mathsf{T} = \begin{bmatrix} a_{11} & a_{21} & \ldots & a_{m1} \\ a_{12} & a_{22} & \ldots & a_{m2} \\ \vdots & \vdots & & \vdots \\ a_{1n} & a_{2n} & \ldots & a_{mn} \end{bmatrix}.$$

By transposition an $(m \times n)$-matrix becomes an $(n \times m)$-matrix. In particular, transposition changes a column vector into a row vector and vice versa.

Example B.2 For the matrix $\mathbf{A}$ and the vector $\mathbf{x}$ from Example B.1 we have:

$$\mathbf{A}^{\mathsf{T}} = \begin{bmatrix} a & d \\ b & e \\ c & f \end{bmatrix}, \quad \mathbf{x}^{\mathsf{T}} = \begin{bmatrix} 3 & -1 & 2 \end{bmatrix}, \quad \mathbf{x} = [3 \; -1 \; 2]^{\mathsf{T}}.$$

If $\mathbf{a}, \mathbf{b}$ are vectors of length n, then one can regard $\mathbf{a}^{\mathsf{T}}$ as a $(1 \times n)$-matrix. Its product with the vector $\mathbf{b}$ is defined as above and coincides with the inner product:

$$\mathbf{a}^{\mathsf{T}}\mathbf{b} = \sum_{i=1}^{n} a_i b_i = \langle \mathbf{a}, \mathbf{b} \rangle.$$

More generally, the *product* of an $(m \times n)$-matrix $\mathbf{A}$ with an $(n \times l)$-matrix $\mathbf{B}$ can be defined by forming the inner products of the row vectors of $\mathbf{A}$ with the column vectors of $\mathbf{B}$. This means that the element c_{ij} in the ith row and jth column of $\mathbf{C} = \mathbf{AB}$ is obtained by inner multiplication of the ith row of $\mathbf{A}$ with the jth column of $\mathbf{B}$:

$$c_{ij} = \sum_{k=1}^{n} a_{ik} b_{kj}.$$

The result is an $(m \times l)$-matrix. The product is only defined if the dimensions match, i.e. if the number of columns n of $\mathbf{A}$ is equal to the number of rows of $\mathbf{B}$. The matrix product corresponds to the composition of linear mappings. If $\mathbf{B}$ is the matrix of a linear mapping $\mathbb{R}^l \to \mathbb{R}^n$ and $\mathbf{A}$ the matrix of a linear mapping $\mathbb{R}^n \to \mathbb{R}^m$, then $\mathbf{AB}$ is just the matrix of the composition of the two mappings $\mathbb{R}^l \to \mathbb{R}^n \to \mathbb{R}^m$. The transposition of the product is given by the formula

$$(\mathbf{AB})^{\mathsf{T}} = \mathbf{B}^{\mathsf{T}}\mathbf{A}^{\mathsf{T}},$$

which can easily be deduced from the definitions.

Square matrices. The entries $a_{11}, a_{22}, \ldots, a_{nn}$ of an $(n \times n)$-matrix $\mathbf{A}$ are called the *diagonal elements*. A square matrix $\mathbf{D}$ is called *diagonal matrix*, if its entries are all zero with the possible exception of the diagonal elements. Special cases are the *zero matrix* and the *unit matrix* of dimension $n \times n$:

$$\mathbf{O} = \begin{bmatrix} 0 & 0 & \ldots & 0 \\ 0 & 0 & \ldots & 0 \\ \vdots & \vdots & \ddots & \vdots \\ 0 & 0 & \ldots & 0 \end{bmatrix}, \quad \mathbf{I} = \begin{bmatrix} 1 & 0 & \ldots & 0 \\ 0 & 1 & \ldots & 0 \\ \vdots & \vdots & \ddots & \vdots \\ 0 & 0 & \ldots & 1 \end{bmatrix}.$$

The unit matrix is the identity with respect to matrix multiplication. For all $(n \times n)$-matrices $\mathbf{A}$ it holds that $\mathbf{IA} = \mathbf{AI} = \mathbf{A}$. If for a given matrix $\mathbf{A}$ there exists a matrix $\mathbf{B}$ with the property

$$\mathbf{BA} = \mathbf{AB} = \mathbf{I},$$

then one calls $\mathbf{A}$ *invertible* or *regular* and $\mathbf{B}$ the *inverse* of $\mathbf{A}$, denoted by

$$\mathbf{B} = \mathbf{A}^{-1}.$$

Let $\mathbf{x} \in \mathbb{R}^n$, $\mathbf{A}$ an invertible $(n \times n)$-matrix and $\mathbf{y} = \mathbf{A}\mathbf{x}$. Then $\mathbf{x}$ can be computed as $\mathbf{x} = \mathbf{A}^{-1}\mathbf{y}$; in particular, $\mathbf{A}^{-1}\mathbf{A}\mathbf{x} = \mathbf{x}$ and $\mathbf{A}\mathbf{A}^{-1}\mathbf{y} = \mathbf{y}$. This shows that the linear mapping $\mathbb{R}^n \to \mathbb{R}^n$ induced by the matrix $\mathbf{A}$ is bijective and $\mathbf{A}^{-1}$ represents the inverse mapping. The bijectivity of $\mathbf{A}$ can be expressed in yet another way. It means that for every $\mathbf{y} \in \mathbb{R}^n$ there is one and only one $\mathbf{x} \in \mathbb{R}^n$ such that

$$\mathbf{A}\mathbf{x} = \mathbf{y}, \quad \text{or} \quad \begin{array}{l} a_{11}x_1 + a_{12}x_2 + \ldots + a_{1n}x_n = y_1, \\ a_{21}x_1 + a_{22}x_2 + \ldots + a_{2n}x_n = y_2, \\ \vdots \qquad \vdots \qquad\qquad \vdots \qquad \vdots \\ a_{m1}x_1 + a_{m2}x_2 + \ldots + a_{mn}x_n = y_n. \end{array}$$

The latter can be considered as a linear system of equations with right-hand side $\mathbf{y}$ and solution $\mathbf{x} = [x_1 \; x_2 \; \ldots \; x_n]^\mathsf{T}$. In other words, the invertibility of a matrix $\mathbf{A}$ is equivalent with the bijectivity of the corresponding linear mapping and equivalent with the unique solvability of the corresponding linear system of equations (for arbitrary right-hand sides).

For the remainder of this appendix we restrict our attention to (2×2)-matrices. Let $\mathbf{A}$ be a (2×2)-matrix with the corresponding system of equations:

$$\mathbf{A} = [\mathbf{a}_1 \vdots \mathbf{a}_2] = \begin{bmatrix} a_{11} & a_{12} \\ a_{21} & a_{22} \end{bmatrix}, \quad \begin{array}{l} a_{11}x_1 + a_{12}x_2 = y_1, \\ a_{21}x_1 + a_{22}x_2 = y_2. \end{array}$$

An important role is played by the *determinant* of the matrix $\mathbf{A}$. In the (2×2)-case it is defined as the cross product of the column vectors:

$$\det \mathbf{A} = \mathbf{a}_1 \times \mathbf{a}_2 = a_{11}a_{22} - a_{21}a_{12}.$$

Since $\mathbf{a}_1 \times \mathbf{a}_2 = \|\mathbf{a}_1\| \|\mathbf{a}_2\| \sin \angle(\mathbf{a}_1, \mathbf{a}_2)$, the column vectors $\mathbf{a}_1, \mathbf{a}_2$ are linearly dependent (so—in $\mathbb{R}^2$—multiples of each other), if and only if $\det \mathbf{A} = 0$. The following theorem characterises invertibility in the (2×2)-case completely.

Proposition B.3 *For (2×2)-matrices $\mathbf{A}$ the following statements are equivalent:*

(a) $\mathbf{A}$ is invertible.

(b) The linear mapping $\mathbb{R}^2 \to \mathbb{R}^2$ defined by $\mathbf{A}$ is bijective.

(c) The linear system of equations $\mathbf{A}\mathbf{x} = \mathbf{y}$ has a unique solution $\mathbf{x} \in \mathbb{R}^2$ for arbitrary right-hand sides $\mathbf{y} \in \mathbb{R}^2$.

(d) The column vectors of $\mathbf{A}$ are linearly independent.

(e) The linear mapping $\mathbb{R}^2 \to \mathbb{R}^2$ defined by $\mathbf{A}$ is injective.

(f) *The only solution of the linear system of equations* $\mathbf{Ax} = \mathbf{0}$ *is the zero solution* $\mathbf{x} = \mathbf{0}$.

(g) $\det \mathbf{A} \neq 0$.

Proof The equivalence of the statements (a), (b) and (c) was already observed above. The equivalence of (d), (e) and (f) can easily be seen by negation. Indeed, if the column vectors are linearly dependent, then there exists $\mathbf{x} = [x_1 \ x_2]^\mathsf{T} \neq \mathbf{0}$ with $x_1\mathbf{a}_1 + x_2\mathbf{a}_2 = \mathbf{0}$. On the one hand, this means that the vector $\mathbf{x}$ is mapped to $\mathbf{0}$ by $\mathbf{A}$; thus this mapping is not injective. On the other hand, $\mathbf{x}$ is a nontrivial solution of the linear system of equations $\mathbf{Ax} = \mathbf{0}$. The converse implications are shown in the same way. Thus (d), (e) and (f) are equivalent. The equivalence of (g) and (d) is obvious from the geometric meaning of the determinant. If the determinant does not vanish then

$$\mathbf{A}^{-1} = \frac{1}{a_{11}a_{22} - a_{21}a_{12}} \begin{bmatrix} a_{22} & -a_{12} \\ -a_{21} & a_{11} \end{bmatrix}$$

is an inverse to $\mathbf{A}$, as can be verified at once. Thus (g) implies (a). Finally, (e) obviously follows from (b). Hence all statements (a)–(g) are equivalent. $\qquad\square$

Proposition B.3 holds for matrices of arbitrary dimension $n \times n$. For $n = 3$ one can still use geometrical arguments. The cross product, however, has to be replaced by the triple product $\langle \mathbf{a}_1 \times \mathbf{a}_2, \mathbf{a}_3 \rangle$ of the three column vectors, which then also defines the determinant of the (3×3)-matrix $\mathbf{A}$. In higher dimensions the proof requires tools from combinatorics, for which we refer to the literature.

B.2 Canonical Form of Matrices

In this subsection we will show that every (2×2)-matrix $\mathbf{A}$ is similar to a matrix of standard type, which means that it can be put into standard form by a basis transformation. We need this fact in Sect. 20.1 for the classification and solution of systems of differential equations. The transformation explained below is a special case of the *Jordan canonical form*[1] for $(n \times n)$-matrices.

If $\mathbf{T}$ is an invertible (2×2)-matrix, then the columns $\mathbf{t}_1, \mathbf{t}_2$ form a basis of $\mathbb{R}^2$. This means that every element $\mathbf{x} \in \mathbb{R}^2$ can be written in a unique way as a *linear combination* $c_1\mathbf{t}_1 + c_2\mathbf{t}_2$; the coefficients $c_1, c_2 \in \mathbb{R}$ are the coordinates of $\mathbf{x}$ with respect to $\mathbf{t}_1$ and $\mathbf{t}_2$. One can regard $\mathbf{T}$ as a linear transformation of $\mathbb{R}^2$ which maps the standard basis $\{[1 \ 0]^\mathsf{T}, [0 \ 1]^\mathsf{T}\}$ to the basis $\{\mathbf{t}_1, \mathbf{t}_2\}$.

Definition B.4 Two matrices $\mathbf{A}, \mathbf{B}$ are called *similar*, if there exists an invertible matrix $\mathbf{T}$ such that $\mathbf{T}^{-1}\mathbf{AT} = \mathbf{B}$.

---

[1]C. Jordan, 1838–1922.

The three standard types which will define the similarity classes of (2×2)-matrices are of the following form:

type I	type II	type III
$\begin{bmatrix} \lambda_1 & 0 \\ 0 & \lambda_2 \end{bmatrix}$	$\begin{bmatrix} \lambda & 1 \\ 0 & \lambda \end{bmatrix}$	$\begin{bmatrix} \mu & -\nu \\ \nu & \mu \end{bmatrix}$

Here the coefficients $\lambda_1, \lambda_2, \lambda, \mu, \nu$ are real numbers.

In what follows, we need the notion of eigenvalues and eigenvectors. If the equation

$$\mathbf{A}\mathbf{v} = \lambda\mathbf{v}$$

has a solution $\mathbf{v} \neq \mathbf{0} \in \mathbb{R}^2$ for some $\lambda \in \mathbb{R}$, then λ is called *eigenvalue* and $\mathbf{v}$ *eigenvector* of $\mathbf{A}$. In other words, $\mathbf{v}$ is the solution of the equation

$$(\mathbf{A} - \lambda\mathbf{I})\mathbf{v} = \mathbf{0},$$

where $\mathbf{I}$ denotes again the unit matrix. For the existence of a nonzero solution $\mathbf{v}$ it is necessary and sufficient that the matrix $\mathbf{A} - \lambda\mathbf{I}$ is not invertible, i.e.

$$\det(\mathbf{A} - \lambda\mathbf{I}) = 0.$$

By writing

$$\mathbf{A} = \begin{bmatrix} a & b \\ c & d \end{bmatrix}$$

we see that λ has to be a solution of the *characteristic equation*

$$\det \begin{bmatrix} a - \lambda & b \\ c & d - \lambda \end{bmatrix} = \lambda^2 - (a + d)\lambda + ad - bc = 0.$$

If this equation has a real solution λ, then the system of equations $(\mathbf{A} - \lambda\mathbf{I})\mathbf{v} = \mathbf{0}$ is underdetermined and thus has a nonzero solution $\mathbf{v} = [v_1 \ v_2]^\mathsf{T}$. Hence one obtains the eigenvectors to the eigenvalue λ by solving the linear system

$$(a - \lambda)\, v_1 + b\, v_2 = 0,$$
$$c\, v_1 + (d - \lambda)\, v_2 = 0.$$

Depending on whether the characteristic equation has two real, a double real or two complex conjugate solutions, we obtain one of the three similarity classes of $\mathbf{A}$.

Proposition B.5 *Every (2×2)-matrix $\mathbf{A}$ is similar to a matrix of type I, II or III.*

Proof (1) The case of two distinct real eigenvalues $\lambda_1 \neq \lambda_2$. With

$$\mathbf{v}_1 = \begin{bmatrix} v_{11} \\ v_{21} \end{bmatrix}, \quad \mathbf{v}_2 = \begin{bmatrix} v_{12} \\ v_{22} \end{bmatrix}$$

we denote the corresponding eigenvectors. They are linearly independent and thus form a basis of the $\mathbb{R}^2$. Otherwise they would be multiples of each other and so $c\mathbf{v}_1 = \mathbf{v}_2$ for some nonzero $c \in \mathbb{R}$. Applying $\mathbf{A}$ would result in $c\lambda_1\mathbf{v}_1 = \lambda_2\mathbf{v}_2 = \lambda_2 c\mathbf{v}_1$ and thus $\lambda_1 = \lambda_2$ in contradiction to the hypothesis. According to Proposition B.3 the matrix

$$\mathbf{T} = [\mathbf{v}_1 \vdots \mathbf{v}_2] = \begin{bmatrix} v_{11} & v_{12} \\ v_{21} & v_{22} \end{bmatrix}$$

is invertible. Using

$$\mathbf{A}\mathbf{v}_1 = \lambda_1\mathbf{v}_1, \quad \mathbf{A}\mathbf{v}_2 = \lambda_2\mathbf{v}_2,$$

we obtain the identities

$$\mathbf{T}^{-1}\mathbf{A}\mathbf{T} = \mathbf{T}^{-1}\mathbf{A}[\mathbf{v}_1 \vdots \mathbf{v}_2] = \mathbf{T}^{-1}[\lambda_1\mathbf{v}_1 \vdots \lambda_2\mathbf{v}_2]$$

$$= \frac{1}{v_{11}v_{22} - v_{21}v_{12}} \begin{bmatrix} v_{22} & -v_{12} \\ -v_{21} & v_{11} \end{bmatrix} \begin{bmatrix} \lambda_1 v_{11} & \lambda_2 v_{12} \\ \lambda_1 v_{21} & \lambda_2 v_{22} \end{bmatrix} = \begin{bmatrix} \lambda_1 & 0 \\ 0 & \lambda_2 \end{bmatrix}.$$

The matrix $\mathbf{A}$ is similar to a diagonal matrix and thus of type I.
(2) The case of a double real eigenvalue $\lambda = \lambda_1 = \lambda_2$. Since

$$\lambda = \frac{1}{2}\left(a + d \pm \sqrt{(a-d)^2 + 4bc}\right)$$

is the solution of the characteristic equation, this case occurs if

$$(a-d)^2 = -4bc, \quad \lambda = \frac{1}{2}(a+d).$$

If $b = 0$ and $c = 0$, then $a = d$ and $\mathbf{A}$ is already a diagonal matrix of the form

$$\mathbf{A} = \begin{bmatrix} a & 0 \\ 0 & a \end{bmatrix},$$

thus of type I. If $b \neq 0$, we compute c from $(a-d)^2 = -4bc$ and find

$$\mathbf{A} - \lambda\mathbf{I} = \begin{bmatrix} a - \lambda & b \\ c & d - \lambda \end{bmatrix} = \begin{bmatrix} \frac{1}{2}(a-d) & b \\ -\frac{1}{4b}(a-d)^2 & -\frac{1}{2}(a-d) \end{bmatrix}.$$

Note that

$$\begin{bmatrix} \frac{1}{2}(a-d) & b \\ -\frac{1}{4b}(a-d)^2 & -\frac{1}{2}(a-d) \end{bmatrix} \begin{bmatrix} \frac{1}{2}(a-d) & b \\ -\frac{1}{4b}(a-d)^2 & -\frac{1}{2}(a-d) \end{bmatrix} = \begin{bmatrix} 0 & 0 \\ 0 & 0 \end{bmatrix},$$

or $(\mathbf{A} - \lambda \mathbf{I})^2 = \mathbf{O}$. In this case, $\mathbf{A} - \lambda \mathbf{I}$ is called a *nilpotent matrix*. A similar calculation shows that $(\mathbf{A} - \lambda \mathbf{I})^2 = \mathbf{O}$ if $c \neq 0$. We now choose a vector $\mathbf{v}_2 \in \mathbb{R}^2$ for which $(\mathbf{A} - \lambda \mathbf{I})\mathbf{v}_2 \neq \mathbf{0}$. Due to the above consideration this vector satisfies

$$(\mathbf{A} - \lambda \mathbf{I})^2 \mathbf{v}_2 = \mathbf{0}.$$

If we set

$$\mathbf{v}_1 = (\mathbf{A} - \lambda \mathbf{I})\mathbf{v}_2,$$

then obviously

$$\mathbf{A}\mathbf{v}_1 = \lambda \mathbf{v}_1, \quad \mathbf{A}\mathbf{v}_2 = \mathbf{v}_1 + \lambda \mathbf{v}_2.$$

Further $\mathbf{v}_1$ and $\mathbf{v}_2$ are linearly independent (because if $\mathbf{v}_1$ were a multiple of $\mathbf{v}_2$, then $\mathbf{A}\mathbf{v}_2 = \lambda \mathbf{v}_2$ in contradiction to the construction of $\mathbf{v}_2$). We set

$$\mathbf{T} = [\mathbf{v}_1 \mathbin{\vdots} \mathbf{v}_2].$$

The computation

$$\mathbf{T}^{-1}\mathbf{A}\mathbf{T} = \mathbf{T}^{-1}[\lambda \mathbf{v}_1 \mathbin{\vdots} \mathbf{v}_1 + \lambda \mathbf{v}_2]$$

$$= \frac{1}{v_{11}v_{22} - v_{21}v_{12}} \begin{bmatrix} v_{22} & -v_{12} \\ -v_{21} & v_{11} \end{bmatrix} \begin{bmatrix} \lambda v_{11} & v_{11} + \lambda v_{12} \\ \lambda v_{21} & v_{21} + \lambda v_{22} \end{bmatrix} = \begin{bmatrix} \lambda & 1 \\ 0 & \lambda \end{bmatrix}.$$

shows that $\mathbf{A}$ is similar to a matrix of type II.

(3) The case of complex conjugate solutions $\lambda_1 = \mu + i\nu$, $\lambda_2 = \mu - i\nu$. This case arises if the discriminant $(a - d)^2 + 4bc$ is negative. The most elegant way to deal with this case is to switch to complex variables and to perform the computations in the complex vector space $\mathbb{C}^2$. We first determine complex vectors $\mathbf{v}_1, \mathbf{v}_2 \in \mathbb{C}^2$ such that

$$\mathbf{A}\mathbf{v}_1 = \lambda_1 \mathbf{v}_1, \quad \mathbf{A}\mathbf{v}_2 = \lambda_2 \mathbf{v}_2$$

and then decompose $\mathbf{v}_1 = \mathbf{f} + i\mathbf{g}$ into real and imaginary parts with vectors $\mathbf{f}, \mathbf{g}$ in $\mathbb{R}^2$. Since $\lambda_1 = \mu + i\nu$, $\lambda_2 = \mu - i\nu$, it follows that

$$\mathbf{v}_2 = \mathbf{f} - i\mathbf{g}.$$

Note that $\{\mathbf{v}_1, \mathbf{v}_2\}$ forms a basis of $\mathbb{C}^2$. Thus $\{\mathbf{g}, \mathbf{f}\}$ is a basis of $\mathbb{R}^2$ and

$$\mathbf{A}(\mathbf{f} + i\mathbf{g}) = (\mu + i\nu)(\mathbf{f} + i\mathbf{g}) = \mu\mathbf{f} - \nu\mathbf{g} + i(\nu\mathbf{f} + \mu\mathbf{g}),$$

consequently

$$\mathbf{A}\mathbf{g} = \nu\mathbf{f} + \mu\mathbf{g}, \quad \mathbf{A}\mathbf{f} = \mu\mathbf{f} - \nu\mathbf{g}.$$

Again we set

$$\mathbf{T} = [\mathbf{g} \mathbin{\vdots} \mathbf{f}] = \begin{bmatrix} g_1 & f_1 \\ g_2 & f_2 \end{bmatrix}$$

from which we deduce

$$\mathbf{T}^{-1}\mathbf{A}\mathbf{T} = \mathbf{T}^{-1}[\nu\mathbf{f} + \mu\mathbf{g} \vdots \mu\mathbf{f} - \nu\mathbf{g}]$$

$$= \frac{1}{g_1 f_2 - g_2 f_1} \begin{bmatrix} f_2 & -f_1 \\ -g_2 & g_1 \end{bmatrix} \begin{bmatrix} \nu f_1 + \mu g_1 & \mu f_1 - \nu g_1 \\ \nu f_2 + \mu g_2 & \mu f_2 - \nu g_2 \end{bmatrix} = \begin{bmatrix} \mu & -\nu \\ \nu & \mu \end{bmatrix}.$$

Thus $\mathbf{A}$ is similar to a matrix of type III.

Further Results on Continuity

C

This appendix covers further material on continuity which is not central for this book but on the other hand is required in various proofs (like in the chapters on curves and differential equations). It includes assertions about the continuity of the inverse function, the concept of uniform convergence of sequences of functions, the power series expansion of the exponential function and the notions of uniform and Lipschitz continuity.

C.1 Continuity of the Inverse Function

We consider a real-valued function f defined on an interval $I \subset \mathbb{R}$. The interval I can be open, half-open or closed. By $J = f(I)$ we denote the image of f. First, we show that a continuous function $f : I \to J$ is bijective, if and only if it is strictly monotonically increasing or decreasing. Monotonicity was introduced in Definition 8.5. Subsequently, we show that the inverse function is continuous if f is continuous, and we describe the respective ranges.

Proposition C.1 *A real-valued, continuous function $f : I \to J = f(I)$ is bijective if and only if it is strictly monotonically increasing or decreasing.*

Proof We already know that the function $f : I \to f(I)$ is surjective. It is injective if and only if

$$x_1 \neq x_2 \quad \Rightarrow \quad f(x_1) \neq f(x_2).$$

Strict monotonicity thus implies injectivity. To prove the converse implication we start by choosing two points $x_1 < x_2 \in I$. Let $f(x_1) < f(x_2)$, for example. We will show that f is strictly monotonically increasing on the entire interval I. First we observe that for every $x_3 \in (x_1, x_2)$ we must have $f(x_1) < f(x_3) < f(x_2)$. This

M. Oberguggenberger and A. Ostermann, *Analysis for Computer Scientists*,
Undergraduate Topics in Computer Science,
https://doi.org/10.1007/978-3-319-91155-7

is shown by contradiction. Assuming $f(x_3) > f(x_2)$, Proposition 6.14 implies that every intermediate point $f(x_2) < \eta < f(x_3)$ would be the image of a point $\xi_1 \in (x_1, x_3)$ and also the image of a point $\xi_2 \in (x_3, x_2)$, contradicting injectivity.

If we now choose $x_4 \in I$ such that $x_2 < x_4$, then once again $f(x_2) < f(x_4)$. Otherwise we would have $x_1 < x_2 < x_4$ with $f(x_2) > f(x_4)$; this possibility is excluded as in the previous case. Finally, the points to the left of x_1 are inspected in a similar way. It follows that f is strictly monotonically increasing on the entire interval I. In the case $f(x_1) > f(x_2)$, one can deduce similarly that f is monotonically decreasing.   □

The function $y = x \cdot 1_{(-1,0]}(x) + (1 - x) \cdot 1_{(0,1)}(x)$, where 1_I denotes the indicator function of the interval I (see Sect. 2.2), shows that a discontinuous function can be bijective on an interval without being strictly monotonically increasing or decreasing.

Remark C.2 If I is an open interval and $f : I \rightarrow J$ a continuous and bijective function, then J is an open interval as well. Indeed, if J were of the form $[a, b)$, then a would arise as function value of a point $x_1 \in I$, i.e. $a = f(x_1)$. However, since I is open, there are points $x_2 \in I$, $x_2 < x_1$ and $x_3 \in I$ with $x_3 > x_1$. If f is strictly monotonically increasing then we would have $f(x_2) < f(x_1) = a$. If f is strictly monotonically decreasing then $f(x_3) < f(x_1) = a$. Both cases contradict the fact that a was assumed to be the lower boundary of the image $J = f(I)$. In the same way, one excludes the possibilities that $J = (a, b]$ or $J = [a, b]$.

Proposition C.3 *Let $I \subset \mathbb{R}$ be an open interval and $f : I \rightarrow J$ continuous and bijective. Then the inverse function $f^{-1} : J \rightarrow I$ is continuous as well.*

Proof We take $x \in I$, $y \in J$ with $y = f(x)$, $x = f^{-1}(y)$. For small $\varepsilon > 0$ the ε-neighbourhood $U_\varepsilon(x)$ of x is contained in I. According to Remark C.2 $f(U_\varepsilon(x))$ is an open interval and therefore contains a δ-neighbourhood $U_\delta(y)$ of y for a certain $\delta > 0$. Consider a sequence of values $y_n \in J$ which converges to y as $n \rightarrow \infty$. Then there is an index $n(\delta) \in \mathbb{N}$ such that all elements of the sequence y_n with $n \geq n(\delta)$ lie in the δ-neighbourhood $U_\delta(y)$. That, however, means that the values of the function $f^{-1}(y_n)$ from $n(\delta)$ onwards lie in the ε-neighbourhood $U_\varepsilon(x)$ of $x = f^{-1}(y)$. Thus $\lim_{n \rightarrow \infty} f^{-1}(y_n) = f^{-1}(y)$ which is the continuity of f^{-1} at y.   □

C.2 Limits of Sequences of Functions

We consider a sequence of functions $f_n : I \rightarrow \mathbb{R}$, defined on an interval $I \subset \mathbb{R}$. If the function values $f_n(x)$ converge for every fixed $x \in I$, then the sequence $(f_n)_{n \geq 1}$ is called *pointwise convergent*. The pointwise limits define a function $f : I \rightarrow \mathbb{R}$ by $f(x) = \lim_{n \rightarrow \infty} f_n(x)$, the so-called *limit function*.

Example C.4 Let $I = [0, 1]$ and $f_n(x) = x^n$. Then $\lim_{n \to \infty} f_n(x) = 0$ if $0 \leq x < 1$, and $\lim_{n \to \infty} f_n(1) = 1$. The limit function is thus the function

$$f(x) = \begin{cases} 0, & 0 \leq x < 1, \\ 1, & x = 1. \end{cases}$$

This example shows that the limit function of a pointwise convergent sequence of continuous functions is not necessarily continuous.

Definition C.5 (*Uniform convergence of sequences of functions*) A sequence of functions $(f_n)_{n \geq 1}$ defined on an interval I is called *uniformly convergent* with *limit function* f, if

$$\forall \varepsilon > 0 \ \exists n(\varepsilon) \in \mathbb{N} \ \forall n \geq n(\varepsilon) \ \forall x \in I : \ |f(x) - f_n(x)| < \varepsilon.$$

Uniform convergence means that the index $n(\varepsilon)$ after which the sequence of function values $(f_n(x))_{n \geq 1}$ settles in the ε-neighbourhood $U_\varepsilon(f(x))$ can be chosen independently of $x \in I$.

Proposition C.6 *The limit function f of a uniformly convergent sequence of functions $(f_n)_{n \geq 1}$ is continuous.*

Proof We take $x \in I$ and a sequence of points x_k converging to x as $k \to \infty$. We have to show that $f(x) = \lim_{k \to \infty} f(x_k)$. For this we write

$$f(x) - f(x_k) = \big(f(x) - f_n(x)\big) + \big(f_n(x) - f_n(x_k)\big) + \big(f_n(x_k) - f(x_k)\big)$$

and choose $\varepsilon > 0$. Due to the uniform convergence it is possible to find an index $n \in \mathbb{N}$ such that

$$|f(x) - f_n(x)| < \frac{\varepsilon}{3} \quad \text{and} \quad |f_n(x_k) - f(x_k)| < \frac{\varepsilon}{3}$$

for all $k \in \mathbb{N}$. Since f_n is continuous, there is an index $k(\varepsilon) \in \mathbb{N}$ such that

$$|f_n(x) - f_n(x_k)| < \frac{\varepsilon}{3}$$

for all $k \geq k(\varepsilon)$. For such indices k we have

$$|f(x) - f(x_k)| < \frac{\varepsilon}{3} + \frac{\varepsilon}{3} + \frac{\varepsilon}{3} = \varepsilon.$$

Thus $f(x_k) \to f(x)$ as $k \to \infty$, which implies the continuity of f. $\qquad \square$

Application C.7 The exponential function $f(x) = a^x$ is continuous on $\mathbb{R}$. In Application 5.14 it was shown that the exponential function with base $a > 0$ can be defined for every $x \in \mathbb{R}$ as a limit. Let $r_n(x)$ denote the decimal representation of x, truncated at the nth decimal place. Then

$$r_n(x) \leq x < r_n(x) + 10^{-n}.$$

The value of $r_n(x)$ is the same for all real numbers x, which coincide up to the nth decimal place. Thus the mapping $x \mapsto r_n(x)$ is a step function with jumps at a distance of 10^{-n}. We define the function $f_n(x)$ by linear interpolation between the points

$$\left(r_n(x), a^{r_n(x)} \right) \quad \text{and} \quad \left(r_n(x) + 10^{-n}, a^{r_n(x)+10^{-n}} \right),$$

which means

$$f_n(x) = a^{r_n(x)} + \frac{x - r_n(x)}{10^{-n}} \left(a^{r_n(x)+10^{-n}} - a^{r_n(x)} \right).$$

The graph of the function $f_n(x)$ is a polygonal chain (with kinks at the distance of 10^{-n}), and thus f_n is continuous. We show that the sequence of functions $(f_n)_{n \geq 1}$ converges uniformly to f on every interval $[-T, T], 0 < T \in \mathbb{Q}$. Since $x - r_n(x) \leq 10^{-n}$, it follows that

$$|f(x) - f_n(x)| \leq \left| a^x - a^{r_n(x)} \right| + \left| a^{r_n(x)+10^{-n}} - a^{r_n(x)} \right|.$$

For $x \in [-T, T]$ we have

$$a^x - a^{r_n(x)} = a^{r_n(x)} \left(a^{x-r_n(x)} - 1 \right) \leq a^T \left(a^{10^{-n}} - 1 \right)$$

and likewise

$$a^{r_n(x)+10^{-n}} - a^{r_n(x)} \leq a^T \left(a^{10^{-n}} - 1 \right).$$

Consequently

$$|f(x) - f_n(x)| \leq 2a^T \left(\sqrt[10^n]{a} - 1 \right),$$

and the term on the right-hand side converges to zero independently of x, as was proven in Application 5.15.

The rules of calculation for real exponents can now also be derived by taking limits. Take, for example, $r, s \in \mathbb{R}$ with decimal approximations $(r_n)_{n \geq 1}, (s_n)_{n \geq 1}$. Then Proposition 5.7 and the continuity of the exponential function imply

$$a^r a^s = \lim_{n \to \infty} \left(a^{r_n} a^{s_n} \right) = \lim_{n \to \infty} \left(a^{r_n+s_n} \right) = a^{r+s}.$$

With the help of Proposition C.3 the continuity of the logarithm follows as well.

C.3 The Exponential Series

The aim of this section is to derive the series representation of the exponential function

$$e^x = \sum_{m=0}^{\infty} \frac{x^m}{m!}$$

by using exclusively the theory of convergent series without resorting to differential calculus. This is important for our exposition because the differentiability of the exponential function is proven with the help of the series representation in Sect. 7.2.

As a tool we need two supplements to the theory of series: Absolute convergence and Cauchy's[2] formula for the product of two series.

Definition C.8 A series $\sum_{k=0}^{\infty} a_k$ is called *absolutely convergent*, if the series $\sum_{k=0}^{\infty} |a_k|$ of the absolute values of its coefficients converges.

Proposition C.9 *Every absolutely convergent series is convergent.*

Proof We define the positive and the negative parts of the coefficient a_k by

$$a_k^+ = \begin{cases} a_k, & a_k \geq 0, \\ 0, & a_k < 0, \end{cases} \qquad a_k^- = \begin{cases} 0, & a_k \geq 0, \\ |a_k|, & a_k < 0. \end{cases}$$

Obviously, we have $0 \leq a_k^+ \leq |a_k|$ and $0 \leq a_k^- \leq |a_k|$. Thus the two series $\sum_{k=0}^{\infty} a_k^+$ and $\sum_{k=0}^{\infty} a_k^-$ converge due to the comparison criterion (Proposition 5.21) and the limit

$$\lim_{n \to \infty} \sum_{k=0}^{n} a_k = \lim_{n \to \infty} \sum_{k=0}^{n} a_k^+ - \lim_{n \to \infty} \sum_{k=0}^{n} a_k^-$$

exists. Consequently, the series $\sum_{k=0}^{\infty} a_k$ converges. $\qquad \square$

We consider two absolutely convergent series $\sum_{i=0}^{\infty} a_i$ and $\sum_{j=0}^{\infty} b_j$ and ask how their product can be computed. Term-by-term multiplication of the nth partial sums suggests to consider the following scheme:

$$
\begin{array}{ccccc}
a_0 b_0 & a_0 b_1 & \cdots & a_0 b_{n-1} & a_0 b_n \\
a_1 b_0 & a_1 b_1 & \cdots & a_1 b_{n-1} & a_1 b_n \\
\vdots & & \ddots & & \vdots \\
a_{n-1} b_0 & a_{n-1} b_1 & \cdots & a_{n-1} b_{n-1} & a_{n-1} b_n \\
a_n b_0 & a_n b_1 & \cdots & a_n b_{n-1} & a_n b_n
\end{array}
$$

---

[2] A.L. Cauchy, 1789–1857.

Adding all entries of the quadratic scheme one obtains the product of the partial sums

$$P_n = \sum_{i=0}^{n} a_i \sum_{j=0}^{n} b_j.$$

In contrast, adding only the upper triangle containing the bold entries (diagonal by diagonal), one obtains the so-called *Cauchy product formula*

$$S_n = \sum_{m=0}^{n} \left(\sum_{k=0}^{m} a_k b_{m-k} \right).$$

We want to show that, for absolutely convergent series, the limits are equal:

$$\lim_{n \to \infty} P_n = \lim_{n \to \infty} S_n.$$

Proposition C.10 (Cauchy product) *If the series $\sum_{i=0}^{\infty} a_i$ and $\sum_{j=0}^{\infty} b_j$ converge absolutely then*

$$\sum_{i=0}^{\infty} a_i \sum_{j=0}^{\infty} b_j = \sum_{m=0}^{\infty} \left(\sum_{k=0}^{m} a_k b_{m-k} \right).$$

The series defined by the Cauchy product formula also converges absolutely.

Proof We set

$$c_m = \sum_{k=0}^{m} a_k b_{m-k}$$

and obtain that the partial sums

$$T_n = \sum_{m=0}^{n} |c_m| \le \sum_{i=0}^{n} |a_i| \sum_{j=0}^{n} |b_j| \le \sum_{i=0}^{\infty} |a_i| \sum_{j=0}^{\infty} |b_j|$$

remain bounded. This follows from the facts that the triangle in the scheme above has fewer entries than the square and the original series converge absolutely. Obviously the sequence T_n is also monotonically increasing; according to Proposition 5.10 it thus has a limit. This means that the series $\sum_{m=0}^{\infty} c_m$ converges absolutely, so the Cauchy product exists. It remains to be shown that it coincides with the product of the series. For the partial sums, we have

$$\left| P_n - S_n \right| = \left| \sum_{i=0}^{n} a_i \sum_{j=0}^{n} b_j - \sum_{m=0}^{n} c_m \right| \le \left| \sum_{m=n+1}^{\infty} c_m \right|,$$

since the difference can obviously be approximated by the sum of the terms below the nth diagonal. The latter sum, however, is just the difference of the partial sum

S_n and the value of the series $\sum_{m=0}^{\infty} c_m$. It thus converges to zero and the desired assertion is proven. $\qquad\qquad\qquad\qquad\qquad\qquad\qquad\qquad\qquad\qquad\qquad$ $\square$

Let

$$E(x) = \sum_{m=0}^{\infty} \frac{x^m}{m!}, \qquad E_n(x) = \sum_{m=0}^{n} \frac{x^m}{m!}.$$

The convergence of the series for $x = 1$ was shown in Example 5.24 and for $x = 2$ in Exercise 14 of Chap. 5. The absolute convergence for arbitrary $x \in \mathbb{R}$ can either be shown analogously or by using the ratio test (Exercise 15 in Chap. 5). If x varies in a bounded interval $I = [-R, R]$, then the sequence of the partial sums $E_n(x)$ converges uniformly to $E(x)$, due to the uniform estimate

$$\left| E(x) - E_n(x) \right| = \left| \sum_{m=n+1}^{\infty} \frac{x^m}{m!} \right| \leq \sum_{m=n+1}^{\infty} \frac{R^m}{m!} \to 0$$

on the interval $[-R, R]$. Proposition C.6 implies that the function $x \mapsto E(x)$ is continuous.

For the derivation of the product formula $E(x)E(y) = E(x + y)$ we recall the *binomial formula*:

$$(x + y)^m = \sum_{k=0}^{m} \binom{m}{k} x^k y^{m-k} \qquad \text{with} \qquad \binom{m}{k} = \frac{m!}{k!(m-k)!},$$

valid for arbitrary $x, y \in \mathbb{R}$ and $n \in \mathbb{N}$, see, for instance, [17, Chap. XIII, Theorem 7.2].

Proposition C.11 *For arbitrary $x, y \in \mathbb{R}$ it holds that*

$$\sum_{i=0}^{\infty} \frac{x^i}{i!} \sum_{j=0}^{\infty} \frac{y^j}{j!} = \sum_{m=0}^{\infty} \frac{(x+y)^m}{m!}.$$

Proof Due to the absolute convergence of the above series, Proposition C.10 yields

$$\sum_{i=0}^{\infty} \frac{x^i}{i!} \sum_{j=0}^{\infty} \frac{y^j}{j!} = \sum_{m=0}^{\infty} \sum_{k=0}^{m} \frac{x^k}{k!} \frac{y^{m-k}}{(m-k)!}.$$

An application of the binomial formula

$$\sum_{k=0}^{m} \frac{x^k}{k!} \frac{y^{m-k}}{(m-k)!} = \frac{1}{m!} \sum_{k=0}^{m} \binom{m}{k} x^k y^{m-k} = \frac{1}{m!} (x+y)^m$$

shows the desired assertion. $\qquad\qquad\qquad\qquad\qquad\qquad\qquad\qquad\qquad\qquad\qquad$ $\square$

Proposition C.12 (Series representation of the exponential function) *The exponential function possesses the series representation*

$$e^x = \sum_{m=0}^{\infty} \frac{x^m}{m!},$$

valid for arbitrary $x \in \mathbb{R}$.

Proof By definition of the number e (see Example 5.24) we obviously have

$$e^0 = 1 = E(0), \quad e^1 = e = E(1).$$

From Proposition C.11 we get in particular

$$e^2 = e^{1+1} = e^1 e^1 = E(1)E(1) = E(1+1) = E(2)$$

and recursively

$$e^m = E(m) \quad \text{for} \quad m \in \mathbb{N}.$$

The relation $E(m)E(-m) = E(m - m) = E(0) = 1$ shows that

$$e^{-m} = \frac{1}{e^m} = \frac{1}{E(m)} = E(-m).$$

Likewise, one concludes from $\bigl(E(1/n)\bigr)^n = E(1)$ that

$$e^{1/n} = \sqrt[n]{e} = \sqrt[n]{E(1)} = E(1/n).$$

So far this shows that $e^x = E(x)$ holds for all rational $x = m/n$. From Application C.7 we know that the exponential function $x \mapsto e^x$ is continuous. The continuity of the function $x \mapsto E(x)$ was shown above. But two continuous functions which coincide for all rational numbers are equal. More precisely, if $x \in \mathbb{R}$ and x_j is the decimal expansion of x truncated at the jth place, then

$$e^x = \lim_{j \to \infty} e^{x_j} = \lim_{j \to \infty} E(x_j) = E(x),$$

which is the desired result. $\square$

Remark C.13 The rigorous introduction of the exponential function is surprisingly involved and is handled differently by different authors. The total effort, however, is approximately the same in all approaches. We took the following route: Introduction of Euler's number e as the value of a convergent series (Example 5.24); definition of the exponential function $x \mapsto e^x$ for $x \in \mathbb{R}$ by using the completeness of the real numbers (Application 5.14); continuity of the exponential function based on uniform convergence (Application C.7); series representation (Proposition C.12); differentiability and calculation of the derivative (Sect. 7.2). Finally, in the course of the computation of the derivative we also obtained the well-known formula $e = \lim_{n \to \infty} (1 + 1/n)^n$, which Euler himself used to define the number e.

C.4 Lipschitz Continuity and Uniform Continuity

Some results on curves and differential equations require more refined continuity properties. More precisely, methods for quantifying how the function values change in dependence on the arguments are needed.

Definition C.14 A function $f : D \subset \mathbb{R} \to \mathbb{R}$ is called *Lipschitz continuous*, if there exists a constant $L > 0$ such that the inequality

$$|f(x_1) - f(x_2)| \le L|x_1 - x_2|$$

holds for all $x_1, x_2 \in D$. In this case L is called a *Lipschitz constant* of the function f.

If $x \in D$ and $(x_n)_{n \ge 1}$ is a sequence of points in D which converges to x, the inequality $|f(x) - f(x_n)| \le L|x - x_n|$ implies that $f(x_n) \to f(x)$ as $n \to \infty$. Every Lipschitz continuous function is thus continuous. For Lipschitz continuous functions one can quantify how much change in the x-values can be allowed to obtain a change in the function values of $\varepsilon > 0$ at the most:

$$|x_1 - x_2| < \varepsilon/L \quad \Rightarrow \quad |f(x_1) - f(x_2)| < \varepsilon.$$

Occasionally the following weaker quantification is required.

Definition C.15 A function $f : D \subset \mathbb{R} \to \mathbb{R}$ is called *uniformly continuous*, if there exists a mapping $\omega : (0, 1] \to (0, 1] : \varepsilon \mapsto \omega(\varepsilon)$ such that

$$|x_1 - x_2| < \omega(\varepsilon) \quad \Rightarrow \quad |f(x_1) - f(x_2)| < \varepsilon$$

for all $x_1, x_2 \in D$. In this case the mapping ω is called a *modulus of continuity* of the function f.

Every Lipschitz continuous function is uniformly continuous (with $\omega(\varepsilon) = \varepsilon/L$), and every uniformly continuous function is continuous.

Example C.16 (a) The quadratic function $f(x) = x^2$ is Lipschitz continuous on every bounded interval $[a, b]$. For $x_1 \in [a, b]$ we have $|x_1| \leq M = \max(|a|, |b|)$ and likewise for x_2. Thus

$$|f(x_1) - f(x_2)| = |x_1^2 - x_2^2| = |x_1 + x_2||x_1 - x_2| \leq 2M|x_1 - x_2|$$

holds for all $x_1, x_2 \in [a, b]$.

(b) The absolute value function $f(x) = |x|$ is Lipschitz continuous on $D = \mathbb{R}$ (with Lipschitz constant $L = 1$). This follows from the inequality

$$\big||x_1| - |x_2|\big| \leq |x_1 - x_2|,$$

which is valid for all $x_1, x_2 \in \mathbb{R}$.

(c) The square root function $f(x) = \sqrt{x}$ is uniformly continuous on the interval $[0, 1]$, but not Lipschitz continuous. This follows from the inequality

$$\left|\sqrt{x_1} - \sqrt{x_2}\right| \leq \sqrt{|x_1 - x_2|},$$

which is proved immediately by squaring. Thus $\omega(\varepsilon) = \varepsilon^2$ is a modulus of continuity of the square root function on the interval $[0, 1]$. The square root function is not Lipschitz continuous on $[0, 1]$, since otherwise the choice $x_2 = 0$ would imply the relations

$$\sqrt{x_1} \leq L|x_1|, \qquad \frac{1}{\sqrt{x_1}} \leq L$$

which cannot hold for fixed $L > 0$ and all $x_1 \in (0, 1]$.

(d) The function $f(x) = \frac{1}{x}$ is continuous on the interval $(0, 1)$, but not uniformly continuous. Assume that we could find a modulus of continuity $\varepsilon \mapsto \omega(\varepsilon)$ on $(0, 1)$. Then for $x_1 = 2\varepsilon\omega(\varepsilon)$, $x_2 = \varepsilon\omega(\varepsilon)$ and $\varepsilon < 1$ we would get $|x_1 - x_2| < \omega(\varepsilon)$, but

$$\left|\frac{1}{x_1} - \frac{1}{x_2}\right| = \left|\frac{x_2 - x_1}{x_1 x_2}\right| = \frac{\varepsilon\omega(\varepsilon)}{2\varepsilon^2\omega(\varepsilon)^2} = \frac{1}{2\varepsilon\omega(\varepsilon)}$$

which becomes arbitrarily large as $\varepsilon \to 0$. In particular, it cannot be bounded from above by ε.

From the mean value theorem (Proposition 8.4) it follows that differentiable functions with bounded derivative are Lipschitz continuous. Further it can be shown that every function which is continuous on a closed, bounded interval $[a, b]$ is uniformly continuous there. The proof requires further tools from analysis for which we refer to [4, Theorem 3.13].

Apart from the intermediate value theorem, the *fixed point theorem* is an important tool for proving the existence of solutions of equations. Moreover one obtains an iterative algorithm for approximating the fixed point.

Definition C.17 A Lipschitz continuous mapping f of an interval I to $\mathbb{R}$ is called a *contraction*, if $f(I) \subset I$ and f has a Lipschitz constant $L < 1$. A point $x^* \in I$ with $x^* = f(x^*)$ is called *fixed point* of the function f.

Proposition C.18 (Fixed point theorem) *A contraction f on a closed interval $[a, b]$ has a unique fixed point. The sequence, recursively defined by the iteration*

$$x_{n+1} = f(x_n)$$

converges to the fixed point x^* for arbitrary initial values $x_1 \in [a, b]$.

Proof Since $f([a, b]) \subset [a, b]$ we must have

$$a \leq f(a) \quad \text{and} \quad f(b) \leq b.$$

If $a = f(a)$ or $b = f(b)$, we are done. Otherwise the intermediate value theorem applied to the function $g(x) = x - f(x)$ yields the existence of a point $x^* \in (a, b)$ with $g(x^*) = 0$. This x^* is a fixed point of f. Due to the contraction property the existence of a further fixed point y^* would result in

$$|x^* - y^*| = |f(x^*) - f(y^*)| \leq L|x^* - y^*| < |x^* - y^*|$$

which is impossible for $x^* \neq y^*$. Thus the fixed point is unique.

The convergence of the iteration follows from the inequalities

$$|x^* - x_{n+1}| = |f(x^*) - f(x_n)| \leq L|x^* - x_n| \leq \ldots \leq L^n|x^* - x_1|,$$

since $|x^* - x_1| \leq b - a$ and $\lim_{n \to \infty} L^n = 0$. $\qquad \square$

Description of the Supplementary Software

D

In our view *using and writing* software forms an essential component of an analysis course for computer scientists. The software that has been developed for this book is available on the website

https://www.springer.com/book/9783319911540

This site contains the Java applets referred to in the text as well as some source files in maple, Python and MATLAB.

For the execution of the maple and MATLAB programs additional licences are needed.

Java applets. The available applets are listed in Table D.1. The applets are executable and only require a current version of Java installed.

Source codes in MATLAB **and** maple. In addition to the Java applets, you can find maple and MATLAB programs on this website. These programs are numbered according to the individual chapters and are mainly used in experiments and exercises. To run the programs the corresponding software licence is required.

Source codes in Python. For each MATLAB program, an equivalent Python program is provided. To run these programs, a current version of Python has to be installed. We do not specifically refer these programs in the text; the numbering is the same as for the M-files.

© Springer Nature Switzerland AG 2018
M. Oberguggenberger and A. Ostermann, *Analysis for Computer Scientists*,
Undergraduate Topics in Computer Science,
https://doi.org/10.1007/978-3-319-91155-7

Table D.1 List of available Java applets

Sequences
2D-visualisation of complex functions
3D-visualisation of complex functions
Bisection method
Animation of the intermediate value theorem
Newton's method
Riemann sums
Integration
Parametric curves in the plane
Parametric curves in space
Surfaces in space
Dynamical systems in the plane
Dynamical systems in space
Linear regression

References

Textbooks

1. E. Hairer, G. Wanner, *Analysis by Its History* (Springer, New York, 1996)
2. S. Lang, *Introduction to Linear Algebra*, 2nd edn. (Springer, New York, 1986)
3. S. Lang, *Undergraduate Analysis* (Springer, New York, 1983)
4. M.H. Protter, C.B. Morrey, *A First Course in Real Analysis*, 2nd edn. (Springer, New York, 1991)

Further Reading

5. M. Barnsley, *Fractals Everywhere* (Academic Press, Boston, 1988)
6. M. Braun, C.C. Coleman, D.A. Drew (eds.), *Differential Equation Models* (Springer, Berlin, 1983)
7. M. Bronstein, *Symbolic Integration I: Transcendental Functions* (Springer, Berlin, 1997)
8. A. Chevan, M. Sutherland, Hierarchical partitioning. Am. Stat. **45**, 90–96 (1991)
9. J.P. Eckmann, Savez-vous résoudre $z^3 = 1$? La Recherche **14**, 260–262 (1983)
10. N. Fickel, Partition of the coefficient of determination in multiple regression, in *Operations Research Proceedings 1999*, ed. by K. Inderfurth (Springer, Berlin, 2000), pp. 154–159
11. E. Hairer, S.P. Nørsett, G. Wanner, *Solving Ordinary Differential Equations I. Nonstiff Problems*, 2nd edn. (Springer, Berlin, 1993)
12. E. Hairer, G. Wanner, *Solving Ordinary Differential Equations II. Stiff and Differential-Algebraic Problems*, 2nd edn. (Springer, Berlin, 1996)
13. M.W. Hirsch, S. Smale, *Differential Equations, Dynamical Systems, and Linear Algebra* (Academic Press, New York, 1974)
14. R.F. Keeling, S.C. Piper, A.F. Bollenbacher, J.S. Walker, Atmospheric CO_2 records from sites in the SIO air sampling network, in *Trends: A Compendium of Data on Global Change. Carbon Dioxide Information Analysis Center, Oak Ridge National Laboratory, U.S. Department of Energy* (Oak Ridge, Tennessy, USA, 2009). https://doi.org/10.3334/CDIAC/atg.035

© Springer Nature Switzerland AG 2018
M. Oberguggenberger and A. Ostermann, *Analysis for Computer Scientists*,
Undergraduate Topics in Computer Science,
https://doi.org/10.1007/978-3-319-91155-7

15. E. Kreyszig, *Statistische Methoden und ihre Anwendungen*, 3rd edn. (Vandenhoeck & Ruprecht, Göttingen, 1968)
16. W. Kruskal, Relative importance by averaging over orderings. Am. Stat. **41**, 6–10 (1987)
17. S. Lang, *A First Course in Calculus*, 5th edn. (Springer, New York, 1986)
18. M. Lefebvre, *Basic Probability Theory with Applications* (Springer, New-York, 2009)
19. D.C. Montgomery, E.A. Peck, G.G. Vining, *Introduction to Linear Regression Analysis*, 3rd edn. (Wiley, New York, 2001)
20. M.L. Overton, *Numerical Computing with IEEE Floating Point Arithmetic* (SIAM, Philadelphia, 2001)
21. H.-O. Peitgen, H. Jürgens, D. Saupe, *Fractals for the Classroom. Part One: Introduction to Fractals and Chaos* (Springer, New York, 1992)
22. H.-O. Peitgen, H. Jürgens, D. Saupe, *Fractals for the Classroom. Part Two: Complex Systems and Mandelbrot Set* (Springer, New York, 1992)
23. A. Quarteroni, R. Sacco, F. Saleri, *Numerical Mathematics* (Springer, New York, 2000)
24. H. Rommelfanger, *Differenzen- und Differentialgleichungen* (Bibliographisches Institut, Mannheim, 1977)
25. B. Schuppener, Die Festlegung charakteristischer Bodenkennwerte – Empfehlungen des Eurocodes 7 Teil 1 und die Ergebnisse einer Umfrage. Geotechnik Sonderheft (1999), pp. 32–35
26. STATISTIK AUSTRIA, Statistisches Jahrbuch Österreich. Verlag Österreich GmbH, Wien 2018. http://www.statistik.at
27. M.A. Väth, *Nonstandard Analysis* (Birkhäuser, Basel, 2007)

Index

© Springer Nature Switzerland AG 2018   369
M. Oberguggenberger and A. Ostermann, *Analysis for Computer Scientists*,
Undergraduate Topics in Computer Science,
https://doi.org/10.1007/978-3-319-91155-7

Printed in the United States
By Bookmasters